British Columbia Place Names

G.P.V. Akrigg and
Helen B. Akrigg

British
Columbia
Place
Names

Third Edition

UBCPress / Vancouver

First edition 1986
Second edition 1988

09 08 07 06 05 04   5 4 3 2

Printed in Canada on acid-free paper

National Library of Canada Cataloguing in Publication

Akrigg, G.P.V., 1913-
  British Columbia place names

  ISBN 0-7748-0636-2 (bound)
  ISBN 0-7748-0637-0 (pbk.)

  1. Names, Geographical – British Columbia. 2. British Columbia – History, Local.
I. Akrigg, Helen B., 1921- II. Title.

FC3806.A47 1997   917.11'003   C97-910842-x
F1086.4.A47 1997

Canadä

UBC Press gratefully acknowledges the financial support for our publishing
program of the Government of Canada through the Book Publishing Industry
Development Program (BPIDP), and of the Canada Council for the Arts, and
the British Columbia Arts Council.

UBC Press
The University of British Columbia
2029 West Mall
Vancouver, BC V6T 1Z2
604-822-5959 / Fax: 604-822-6083
www.ubcpress.ca

To Cecil Akrigg and Rodger Manning

Acknowledgements

Various of those whose assistance we acknowledged when this book was first published in 1986 have continued to help us. It is a pleasure in this our third edition to record once more our gratitude to them.

R.C. (Bob) Harris's detailed notes on the earlier text of *British Columbia Place Names* have been most valuable to us. Randy Bouchard and Dorothy Kennedy, archival retrievers *par excellence,* even when deep in their own research have always found time to answer our enquiries about Indian place names. A further acknowledgement is due to the assistance we have received over the years from Anne Yandle and Frances Woodward. R.W. (Sandy) Sandilands has continued to help us with BC coast names.

Three other persons must be added to our acknowledgements: Dr. W. Kaye Lamb, our friend of many years, has drawn on the research that went into his definitive edition of Captain Vancouver's journals and supplied us with notes and suggestions. So too has that prince of proofreaders, Martin Lynch. Finally, we owe much to Janet Mason, who presides over the provincial government's British Columbia Geographical Names office. She has not only answered our queries but, when we have been in Victoria, has opened her files for us and found us working space in her office.

Introduction

Let us commence with the pleasant duty of paying tribute to our predecessors. Standing at the head of the succession of those who have studied British Columbia's place names is the doughty figure of Captain John T. Walbran (1848-1913). During the thirteen years that he commanded the Canadian Government Steamship *Quadra,* putting into almost every bay and inlet on the BC coast while attending to an amazing variety of duties, Captain Walbran became more and more interested in its place names, discussing their origins with lighthouse-keepers, fishermen, settlers and Indians, hunting for them in the libraries of Victoria, or sending off letters of enquiry to retired naval officers in England. The outcome was the publication in 1909 of that landmark in British Columbia studies, Walbran's *British Columbia Coast Names.* The first edition of the book is now something of a rarity, but fortunately the Vancouver Public Library reprinted it in facsimile in 1971.

Even while Captain Walbran was working on his book, R.E. Gosnell was including in the editions of his *British Columbia Year Book* a few pages on the province's place names. More substantial were the labours of Frederick W. Laing (1869-1948). A pioneer school teacher in Revelstoke, Laing later became private secretary to several provincial Ministers of Agriculture. Working in Victoria, he became very interested in the government's early preemption records, and drawing upon them he compiled his *Geographical Naming Record.* Dated 1938, this survives as seventeen pages of single-spaced typing in the Provincial Archives of British Columbia (reference number I/A/L14g).

The man who, had he lived, would have given us our first book dealing with BC toponyms on a province-wide scale was A.G. Harvey (1884-1950). Born in England, Harvey was a Vancouver lawyer who served at one time as the reeve of Point Grey and later became a Vancouver alderman. At the time of his death, Harvey was well along with his *Place Names in British Columbia,* his materials for the book, with the rest of his papers, ending up in the Provincial Archives of British Columbia (Add. MSS. 1925). Using the Harvey place-name notes as a basis, PABC has added to them to produce the present place-name file kept in its map room.

We trust that indulgent readers will permit us to say something about the history of our own interest in toponyms. Philip dates his fascination with place names back to 1934, when he was spending a year of unutterable tedium at Calgary Normal School. One day, going into the school's modest library, he opened for the first time the great *Times Atlas of the World.* It was a revelation – not least because of the wonderful names that he found as he scanned its broad pages. At the time Philip was somehow subsisting on $30 a month (of which $27.50 went for room, board, and laundry). He needed money and he knew that the *Calgary Herald* paid $1 for any contribution that it printed in its 'Poets' Corner.' Presto! Out of the magical names in the atlas came verses:

> I want to sail past the Arrogant Reef,
> See Nil Desperandum, and travel in brief
> From the Gulf of Siam to Blue Mud Bay,
> Finding new marvels all of the way,
> While sailing the Malay seas.

In short order the lines were in print and Philip was the richer by one precious dollar. Thereafter he could never be insensitive to Sir Francis Bacon's dictum: 'Name, though it seem but a superficial and outward matter, yet it carrieth much impression and enchantment.'

Meanwhile, Helen was growing up in Victoria, in a family which was well aware that British Columbia consists of much more than Victoria and Vancouver. Her father, Ernest C. Manning, Chief Forester of British Columbia (see *Manning Park*), knew the province well, and had friends in many parts of it. Often these friends were guests at the family dinner table – in later years Helen was to regret that she had not asked more questions of these old-timers, such as that prince of raconteurs, Louis LeBourdais, and had not taken notes. During summers spent at her family's log cabin on Canim Lake, miles from any road, she became well acquainted with the life of pioneer homesteaders.

With a background such as this, Helen found her interest in BC history kindling rapidly when she went to Victoria College (then in Craigdarroch Castle). From here Helen proceeded to the University of British Columbia where, looking for a subject for her honours graduating essay, she went to see Dr. Kaye Lamb, the University Librarian, already winning distinction as a historian. 'Your family has a summer place in the Cariboo,' remarked Dr. Lamb, 'why don't you do your essay on Cariboo place names?' And she did.

In September 1944 Philip and Helen were married, the former with a brand-new PhD from the University of California. Ahead for him lay a career as a Shakespeare scholar teaching in UBC's Department of English. Ahead for her lay a busy life as a wife and the mother of three children. A significant event happened in 1950. Leafing through an English bookseller's catalogue for Renaissance items, he chanced on Walbran's *British Columbia Coast Names*. Remembering his wife's interest in BC toponymy, he sent away for the book and that year it was his birthday present to her. That acquisition inaugurated many years of collecting books on BC history – and not only acquiring them but using them for bedtime reading, with an increasing number of notes being taken on whatever related to place naming. By the mid-1960s, with Helen possessing a new MA in British Columbia history, reading had widened into planned research with more travel in BC, enquiries, interviews, tapings, and map studies added to our activities. We began to see within our grasp a book, the first dealing with the toponymy of the province as a whole. Thus it was that in 1969 we set up Discovery Press and published our *1001 British Columbia Place Names*. This little book proved a great success, going through three editions before we replaced it in 1986 with the first edition of this more ambitious work.

The number of officially recognized BC place names has increased over the years. The second edition of the BC volume of the *Gazetteer of Canada*, published in 1966, listed some 35,000 place names. The third edition, published in 1985, listed over 40,000. In this present book we have had room for about 2,400 names. We have chosen them using two principal criteria: (1) the geographical prominence of the feature described, and (2) the cultural or historical importance of the name it bears. To these must be added two other considerations – when a name is redolent of BC life, or carries with it an anecdote that we would not want our readers to miss, we have popped it in also.

The most perfunctory glance at the map of British Columbia reveals that a very considerable number (perhaps 20 percent) of our place names have been derived from Indian names or words: Cassiar, Chilcotin, Cowichan, Kamloops, Kelowna, Kootenay, Lillooet, Nootka, Okanagan, Omineca, Saanich, Sechelt, Shuswap ... one could go on and on.

The great advances of recent years in the careful study of Indian languages have done much for the study of Native toponymy. The glib, hackneyed 'translations' given by Chambers of Commerce and real estate developers who want almost any Indian name to mean 'shimmering waters' or 'beautiful valley' have again and again been disproved, and fascinating insights have been given into the Indians' perception of their world.

In recent years aboriginal people have shown renewed interest in their languages. Often in response to representations from the First Nations, an increasing number of Indian names have become official for both settlements and geographic features. If anything, a danger has arisen that zeal for Indian names may create problems for the vast majority of British Columbians who are ignorant of the pronunciation of complicated aboriginal toponyms. For example, consider the plight of the sole survivor of an avalanche who can only gasp into his cell phone that the rest of his party are buried near the foot of Zexwzaxw Glacier (to use an actual name, which means 'soft snow' glacier).

Complicating the study of aboriginal place names is the number of Native languages involved. Because of the way that our province's rugged terrain separated the areas of Indian settlement, we have today about thirty different Indian languages, encompassing numerous dialects (see following page). Almost an additional language is the Chinook jargon, once used throughout the Pacific Northwest as a lingua franca. This was a fascinating mixture of

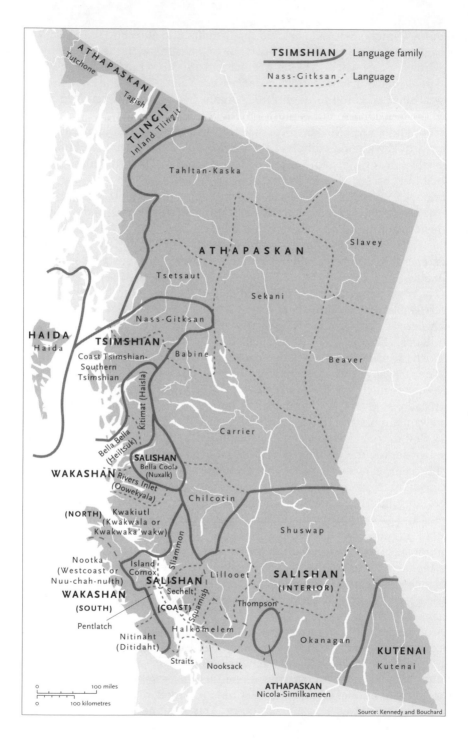

TSIMSHIAN ⌒ Language family
Nass-Gitksan ⌒ Language

ATHAPASKAN
Tutchone
Tagish
TLINGIT
Inland Tlingit
Tahltan-Kaska
Slavey
ATHAPASKAN
Tsetsaut
Sekani
Nass-Gitksan
HAIDA
Haida
TSIMSHIAN
Coast Tsimshian-
Southern
Tsimshian
Babine
Beaver
Kitimat (Haisla)
Bella Bella
(Heiltsuk)
Carrier
SALISHAN
Bella Coola
(Nuxalk)
WAKASHAN
Rivers Inlet
(Oowekyala)
Chilcotin
(NORTH)
Kwakiutl
(Kwakwala or
Kwakwaka'wakw)
Shuswap
Nootka
(Westcoast or
Nuu-chah-nulth)
Island
Comox
SALISHAN
Sliammon
Lillooet
SALISHAN
(INTERIOR)
WAKASHAN
(SOUTH)
Sechelt
(COAST)
Squamish
Thompson
Pentlatch
Nitinaht
(Ditidaht)
Halkomelem
Okanagan
KUTENAI
Kutenai
Straits
Nooksack
ATHAPASKAN
Nicola-Similkameen

0   100 miles
0   100 kilometres

Source: Kennedy and Bouchard

Table 1

Linguistic subdivisions

Ethnic division	Language	Dialect
Haida	Haida	Masset/Skidegate
Tsimshian	Tsimshian	Coast/Southern Nass (Nisgha)/Gitksan
Kwakiutl	North Wakashan	Haisla or Kitimat Heiltsuk or Bella Bella Kwakwala
Nootka	South Wakashan	Westcoast/Nitinat
Bella Coola	Bella Coola	
Coast Salish	Comox/Pentlatch (extinct) Sechelt/Squamish/Halkomelem Straits Salish	
Interior Salish	Thompson Lillooet Shuswap Okanagan	Okanagan Lakes (extinct in Canada)
Kootenay	Kutenai	
Athapaskan	Chilcotin/Carrier/Sekani Tahltan/Kaska/Slave/Beaver Tsetsaut (extinct)/Nicola-Similkameen (extinct)	
Tlingit	Inland Tlingit	

Indian pronunciations of English and French words, plus a large admixture of words borrowed from the languages of the Chinook Indians (living around the mouth of the Columbia River) and the Nootka or Nuu-chah-nulth Indians.

Often when we speak of Indian place names on the map of British Columbia we are hardly speaking of Indian names at all – rather we are dealing with Indian names or words that have been so mangled by whites that even those Indians who retain their language have difficulty in recognizing them. A non-aboriginal writing down an Indian name can generally manage only a rough approximation of the original sounds.

Overwhelmingly Indian place names are descriptive either of natural features or of the food and other resources of the area. Occasionally names relate to events in Indian legends. One way or another, Indian place names are an inextricable part of Indian culture, and here we can do no better than repeat the words of the late Wilson Duff: 'A place name is a reminder of history, indelibly stamped on the land. To enquire about it is to reawaken memories of the history that produced it. To write about it is to retell some of that history. To work with Indian place names is to learn something about the Indian versions of what happened in history.'[1]

Let us turn now from British Columbia's 'Indian names' to her place names in general. From the outset of our studies we have sought to guard against uncritical acceptance of ill-founded reports, though nothing had prepared us for the sheer volume of misinformation afloat about BC's place name origins. Gosnell told us that Westham at the mouth of the Fraser is named for Westham in Essex – there is no Westham in Essex though there is in Sussex. The *Cominco Magazine* told us Wynndel in the Kootenays is named after Wynndel in England, but there is no Wynndel anywhere in England and the place is named after a Mr. Wynndel. Lukin Johnston reported that Williams Lake was named after William Pinchbeck, but the preemption records make it plain that there was a Williams Lake before Mr. Pinchbeck took out his land there.

Especially we have come to distrust that fascinating but treacherous lady, Oral History. Monte Creek provided us with a learning experience here. Speaking to an old gentleman whose great-uncle had acquired the Monte Creek Ranch in 1863, we obtained his story of the origin of the name. That great-uncle had purchased the ranch from a Mexican named Monte who was wanted for horse-stealing in the United States, had been hiding in British Columbia, but believed that it was time to move on. We were given, in all good faith, a circumstantial account of the sale of the ranch by Monte, of how the great-uncle had poured out of a miner's 'poke' and into Monte's hand the amount of gold dust needed to buy it. Unsuspecting, we printed the story but soon received our first intimation that things were badly wrong when Mary Balf, then curator of the Kamloops Museum, wrote to us that she had the Hudson's Bay Company's Kamloops ledgers for the period but could find no mention in them of 'Monte' – a very curious fact since any rancher living as close as Monte Creek would almost certainly have traded with the HBC store. In the end Monte proved a ghost, though it was not until several years later that we discovered the true origin of 'Monte Creek' (see *Monte Creek* in the directory portion of this book).

When we first commenced our study of BC names we were powerfully struck by the element of historical 'stratigraphy' in the toponymy of the province, how each era chronologically has brought its own distinctive layer of names before being succeeded by the next. We may illustrate with the following table.

---

1 'The Fort Victoria Treaties,' *BC Studies* 3 (1969):3.

Table 2

Major sources for BC place names

Source	Names
Aboriginal Indians	Ahousat, Botanie Mountain, Chilcotin
Spanish explorers	Estevan Point, Galiano Island, Texada Island
Royal Navy (1)	
Captain Vancouver	Point Grey, Howe Sound, Rivers Inlet
Maritime fur traders	Cape Scott, Queen Charlotte Islands
Overland explorers and fur traders	
Mackenzie	West Road River
Fraser	Quesnel River, Thompson River
Thompson	Fraser River, Mount Nelson
Others	Lac la Hache, Vaseux Lake, Yale, Tête Jaune Cache, Fort Langley
Royal Navy (2)	
Captain Richards and others	Alert Bay, Mayne Island, Thormanby Islands
Royal Engineers	Queensborough, Sapperton
Gold rush miners	
Fraser	Boston Bar, Hills Bar, Jackass Mountain
Cariboo	Barkerville, Jack of Clubs Lake
Elsewhere	Wild Horse Creek, McDame Creek
Canadian Pacific Railway	Mount Stephen, Revelstoke
Grand Trunk Pacific Railway (CNR)	Vanderhoof, Prince George
Canadian Northern Railway (CNR)	Port Mann, Irvine
Pacific Great Eastern Railway (BCR)	Chetwynd, D'Arcy
Boer War	Ladysmith, Natal
World War I	Mount Haig, Mont des Poilus
World War II	Battle of Britain Range, Dieppe Mountain

Our early interest in the historical 'stratigraphy of nomenclature' has been matched by a growing interest in the 'psychology of nomenclature' – in the human element that enters into the choice of names. Categories can be set up in terms of the motives of name givers.

(1) Descriptive Names. These constitute the simplest and most obvious category. Names such as Stinking Lake, Straggling Islands, Marble Canyon, and Clearwater River come to mind. Such names, needing no explanation, have generally been excluded from this book. However, not all descriptive names are obvious, an example being provided by Italy Lake, whose name describes its shape, that of the Italian peninsula.

(2) Metaphorical Names. These are the names that originated in simile or metaphor. The two can hardly be separated – somebody observes that a particular mountain looks like a black tusk and almost at once it is 'The Black Tusk.' Many of the names that Don Munday gave in the Mount Waddington area are splendidly metaphoric, names such as those of Silverthrone Mountain and Dais Glacier.

(3) Possessive Names. Many features get named after their owners. Here we have Gibson's Landing, Woodward's Slough, and Webster's Corners, to give these places their names in the original forms.

(4) Nostalgic Names. Settlers in a new land being apt to think back on their homelands, it is not surprising that many a place in British Columbia takes its name from some cherished spot in 'the Old Country.' Ashcroft Manor, and hence nearby Ashcroft, commemorates the Cornwall family's ancestral home in England. Creston is named after Creston, Iowa, and Paldi after a village in the Punjab province of India.

In one sense the oldest place names in British Columbia may not have been bequeathed to us by Indians but imported by whites. Consider Lavington in the Okanagan. Many centuries ago there was an Anglo-Saxon chief named 'Lav.' A follower of his would have been called a 'Laving,' and the 'tun' ('town' or settlement) in which he and the other Lavings lived would be 'Lavington.'

(5) Episodic Names. These follow the familiar 'This is the place where' formula, by which a place is identified through linkage with something that happened there. Lac la Hache is where the axes were lost. Maiden Creek is where the betrayed Indian maiden died of grief. Broken Leg Lake speaks for itself. Because these names so often tie in with extended stories, one is tempted to describe them as 'anecdotal names.'

(6) Memorial Names. Here we have names given to keep alive memories of departed persons. These range from Ernest C. Manning, once British

Columbia's Chief Forester, to the young wife Sheila Leonard (see *Manning Park* and *Sheila Lake*).

Important memorial names are those commemorating the British Columbians killed in the two world wars. A book of names of the war dead is maintained in Victoria, and every Remembrance Day it provides names for a few more of the province's not yet officially named features. Some years ago a dedicated Chilliwack alpinist set himself the task to ascend each local mountain so named and to plant a cross on the summit of each. (Unfortunately storms have already swept away some of the crosses.)

Much might be said about the hundreds of persons whose uninhibited egos have led them to memorialize themselves by having geographical features in British Columbia named after them. Such persons were detested by W. Fleet Robertson, the pungent gentleman who for many years was British Columbia's representative on the old Geographic Board of Canada. Writing to the Board in 1912 about an American who wished to give his name to what became Mount Toby, Robertson observed:

> Mr. Gleason is *scarcely* a discoverer – he found good and well-worn trails all the way. I was over most of the ground twelve years ago, but had something better to do than climb a barren peak – to be able to say I had. It is slightly irritating to have B.C. *re*-discovered and renamed by parties of ladies and gentlemen out on a summer holiday, while the pioneers are overlooked.

The memorializing of undeserving persons has long been found objectionable. Walter Wicks in his *Memories of the Skeena* (p. 88) has some telling words on this subject:

> Engineers and surveyors came to the surrounding land areas and changed the [Indian] names of our bays, lakes, rivers and mountains in honour of white men who were credited for their discoveries. Some of these men worked in the area and did deserve honour for their efforts in opening up the country but many more men in plush offices in the southern cities had their feet on the desk and not on the locations that bore their names. They never became rain-soaked or sweated on the mountain slopes or sailed a boat on the waters named after them ...

This petty scandal ended with the adoption of a rule, one that has seldom been broken, that features on the map must not be named after living persons. Of course, by the time that this salutary principle was established, rather too many BC toponyms immortalized petty politicians, mediocre civil servants, or nonentities with friends in high places (see *Goodwin Falls*).

(7) Honorific Names. Closely resembling memorial names, but often distinguishable from them, are those bestowed to honour recipients. One thinks here of the royal family, even its minor members having had features named

after them. Thus in southeastern BC we have the Royal Group of mountains which includes not only mountains named after King George V and his consort, Queen Mary, but others named after their children: Prince Edward, Prince George, Prince Henry, Prince John, and Princess Mary.

Among the names that must be called 'honorific' is that of Queens Bay on Kootenay Lake. Encountering this name, one asks which queens may have had this bay named after them. The answer is that, despite the plural form, Queens Bay was named after just one queen, the great Victoria – originally it was Queen's Bay.

This brings us to a sad development in the history of Canadian toponymy. The trouble can be traced back to 1890, when the United States set up in Washington its Board of Geographic Names charged with regularizing, recording, and making official the country's geographic names. One of the arbitrary rules that the Americans set for themselves was that geographical names must not contain apostrophes.[2] In 1897 the Geographic Board of Canada came into being and proceeded to make official for Canadian names the no-apostrophe rule introduced by the Americans. By 1961, when the provinces took control of their own geographical nomenclature, the forms without apostrophes had become so established on maps, in legal documents, and in everyday use, that it was hardly feasible to return to the original forms.[3]

These official spellings without apostrophes set a very nice problem for the authors of a book such as this, for not only are they ungrammatical but they frequently obscure the meaning of a name, and in a way falsify history. Adam's Lake, where Chief Adam of the local Indians once lived, is a much more vivid name than the pallid 'Adams Lake,' with its suggestion of an eponymous though nonexistent Mr. Adams. Similarly 'Blackeye's Creek' properly identifies the stream with the Indian 'Blackeye,' whereas the official 'Blackeyes Creek' merely leaves one wondering what brawl occurred there. Our solution in such cases has been to give first the governmental form, lacking the apostrophe, and then, in equally bold print but within square brackets, the name with its apostrophe.

Too often in the past civil servants have not been content to collect and record names actually in use but have set themselves up as arbiters of taste to decide what names the public should use. A case in point is 'God's Pocket,'

---

2 In one case at least, the Americans have ignored their own rule – Martha's Vineyard is still Martha's Vineyard. British gazetteers and maps continue to use the apostrophe from Land's End in Cornwall to St. Mary's in the Orkneys.

3 Since the Canadian Permanent Committee on Geographical Names, successor to the Geographic Board of Canada, respects the names given by other jurisdictions, it has officially accepted a few apostrophes. Thus, in the most recent BC volume of the *Gazetteer of Canada*, we have not only Deadmans Lake but also Deadman's Creek Indian Reserve and Dead Man's Island Park. We applaud the municipality that officially named itself Hudson's Hope and so preserved the old form of the name even though Ottawa made its post office Hudson Hope.

a wonderful name used by our West Coast fishermen for a snug anchorage where, in a storm, one would be as safe as if one were in God's pocket. W. Fleet Robertson, BC's representative on the old Geographic Board of Canada, found God's Pocket a 'bizarre' name, and so one of God's Pockets became Haven Cove and another Pocket Inlet. With this sort of mentality deciding things for us, there was absolutely no chance that 'Thank God Point' would become the official name for a promontory on Cortes Island, so named by the locals since the Rev. Mr. Boas of the Columbia Coast Mission, rounding it in rough weather, would always utter, 'Thank you, God, for a safe trip home.'

Another form of censorship indulged in by government officials dealt with matters sexual, no matter how innocent. Thanks to their vigilance, 'Sheba's Breasts' is no longer the name of a very prominent two-pointed peak near the forks of Gun Creek. Instead it is now Mount Sheba.

More recently there has been a drive against names that are considered racist. To take one example, 'nigger' has been expunged from maps and the gazetteer. Nigger Bar Creek is now Goodeve Creek. One Nigger Creek has become Negro Creek, and another has become just one more of half a dozen Pine Creeks. Nigger Mountain has become Mount Jeldness. As for Niggertoe Mountain north of Penticton, this has become Mount Nkwala, though many locals keep using their own name of Jerry Mountain for it.

Failure to keep official nomenclature mirroring the names actually used has given rise to a whole unofficial nomenclature, one whose names are often more vivid, spirited, and imaginative than the official ones. For examples of unofficial names in use today, one may consider the following, all on the BC coast:

China Hat, officially Cone Island
Disappointment Inlet, officially Lemmens Inlet
The Gap, officially Marcus Passage
Hookass Bay, officially Patrician Bay
The Glory Hole, on the Skeena (no official name)
The Boneyard, also officially unnamed, a bank on the Skeena River where the bleached remains of trees washed up at high water look like a jumble of whale bones.

One may regret that the preceding names lack official approval, but fortunately there is no lack of colourful BC names that have won governmental acceptance. Leafing through the gazetteer one finds the very essence of pioneer British Columbia in names such as:

Badshot Mountain	Brinkman's Terror Rapids
Beaverhut Lake	(name now rescinded)
Belly Up Canyon	China Nose Mountain
Blue Joe Creek	Claimjumper Creek
Boomchain Bay	Cougar Smith Creek

Cutfoot Creek	Moosehorn Bay
Fur Thief Creek	Mooseskin Johnny Lake
Gunboat Passage	One Ace Mountain
Hairtrigger Lake	Packer Tom Creek
Hardscrabble Creek	Powerline Pass
Hooch Lake	Rawhide Lake
Hungry Moose Creek	Ripsaw Glacier
Jack of Clubs Lake	Sick Wife Creek
Lean-to Creek	Speculator Creek
Looncry Lake	Top of the World (pass)
Mink Trap Bay	Weary Ridge

At times a welcome spirit of fun illuminates BC names. Thus we have the inimitably named Block Head, the happy aura surrounding Comedy Creek, and the genesis of Damfino Creek when a packer, asked the name of the creek, replied 'Damned if I know.' And what are we to make of Fizzle Mountain, Guess-again Lake, and Tightfit Lake? British Columbia continues to come up with ingenious, surprising, and amusing names. Among recent official adoptions are Old Ecks Lake and Zippermouth Lake. The former is named after a veteran trapper who always signed his name with an *x*. The latter (once Walkin Lake because one had to walk in to get there) got its new name from locals who have noticed how rarely the fish there bite.

British Columbians are stronger on humour than poetry. Hardly anything in BC is named after a poet. The Shakespeare Banks are named after a Victoria alderman, and the community of Shelley after a railway contractor. Mount Milton is named after an itinerant English nobleman. Wordsworth, Tennyson, Spenser, Browning, Donne, and Chaucer have no memorial. Scotland's bard, however, has Bobbie Burns Mountain, and a very minor Scottish versifier has Nevay Island.

Regrettably no records survive for the origins of a great many of our British Columbia place names. As one veteran surveyor observed to us years ago, 'Victoria wanted us to give names, but couldn't have cared less where we got our names from.' Our quest for the origins of particular names has too often ended in failure. Mr. W.C. McDougall could clearly recall naming Elgin, BC, but could not for the life of him remember why he had called it Elgin.

Finding names for the tens of thousands of places and features in British Columbia has not always been an easy task. Occasionally unimaginative souls have settled for appellations such as No Name Creek, or No. 1 Lake, No. 2 Lake, or One Fifty Creek, One Fifty-One Creek. And we have Another Lake with nearby And Another Lake, names that must have been awarded by a surveyor who was totally fed up. One solution has been to adopt a distinctive naming motif for a whole series of places in one particular area. Thus we have the Arthurian names – Galahad Point, Lancelot Point, Grail Point – off

Theodosia Inlet; and the musical names – Melody Creek, Harmony Creek, Singing Pass – of Garibaldi Park. Features in the Chilko Lake area were named after the British ships involved in the Battle of Coronel in World War I – Canopus Mountain, Good Hope Mountain, Glasgow Mountain, and Otranto Mountain, along with Cradock Mountain for their admiral and, of course, Coronel Mountain for the battle itself. Years later other peaks in the area were named for German warships in the same battle – Dresden Mountain, Leipzig Mountain, and Scharnhorst Mountain.

In 1964 a party of mountaineers, members of the Alpine Club of Canada, were rained in for two or three days in the same general Chilko area. To their horror they found that they had only one piece of reading material among the lot of them – a pocketbook biography of Alfred the Great. Their solution was to have the owner of the book tear out each page after he had read it, pass it to the next person, who read it and passed it to the next person, and so on right down the line. The interesting consequence of all this was that, when the mountaineers came to name the peaks that they had climbed, they used names from that biography of King Alfred, and so we got Mount Athelstan, Mount Ethelweard, Mount Guthrum, not to mention Athelney Pass.

About 1889 G.M. Dawson named the Valhalla Mountains (now the Valhalla Ranges). Following Dawson's lead, R.W. Brock, later Dean of Applied Science at the University of British Columbia, carrying a little handbook of Norse mythology with him on his survey, created a Scandinavian pantheon as he named peaks in the Valhallas after the old gods. In the Tantalus Range we have, similarly, names taken out of Greek mythology, while the party that explored and named the Pantheon Range in 1964 did so with the announced intention 'to name the peaks after deities from various mythologies' and so gave us Siva Glacier, Mount Astarte, Hermes Peak, Manitou Peak, Osiris Peak, Mount Thor, Mount Vishnu, and Mount Zeus. Finally we may note the activities of a party of officers and NCOs from the Royal Fusiliers (City of London Regiment) who, coming out to northern BC in 1960, named the Tower of London Range after their regimental headquarters, and individual peaks after lesser towers in the Tower of London. In a remarkably sprightly report to their colonel, the leader of the expedition, Captain M.F.R. Jones, noted that they had not named anything after the Bloody Tower though several peaks undoubtedly qualified.

Name association may add interest to particular names. Close to the Lawyer Islands is Bribery Islet; and the Doctor Islets, Lady Island, and Clapp Passage are in interesting proximity.

Many features in British Columbia share their names with adjacent features. Thus we have not only Cowichan Lake but Cowichan Bay, Cowichan Head, Cowichan River, Cowichan Station, and Cowichan Valley. Amiskwi River has associated with it Amiskwi Falls, Amiskwi Lake, Amiskwi Pass, and Amiskwi Peak. In cases such as these we have limited our entries to just one of these places, supplying a note for what seems to be the principal feature, and leaving our readers to make the connection with nearby features of the same name.

Over the years we have been sustained in our labours by our love for British Columbia, its mountains, lakes, and streams. Pursuing information, we have come to know a great variety of British Columbians, almost all of whom were kindly and cooperative. And we have been repeatedly entertained by stories that go with the giving of names. For those who share our taste for such tales we suggest, as they begin to browse in this book, a glance at:

Billy Whiskers Glacier	Money Makers Rock
Character Cove	Phyllis's Engine
Clemretta	Sin Lake
Elephant Crossing	Sob Lake
Flourmill Creek	Sylvester Peak
Houdini Needles	Ta Ta Creek
Lecture Cutters, The	10 Downing Street
Liberated Group	Tickletoeteaser Tower
Likely	Why Not Mountain
Miniskirt	

Map References

The bracketed letters and numbers that follow each place name are grid references to the map of British Columbia in this book (pp. xxvi-xxvii). The grid references for the rivers normally locate either the mouth of the river or the point at which it flows out of the province.

Each square in the grid corresponds to a map sheet in the National Topographical System, scale 1:250,000 (approximately four miles to the inch). The NTS uses its own idiosyncratic system of numbering, but the following conversion table will enable readers to identify the sheets that correspond to our grid references. For a few minor features, it will be necessary to consult the NTS maps using the scale 1:50,000.

Table 3

Conversion table

Our grid reference	National Topographical System map sheet		Our grid reference	National Topographical System map sheet	
A-7	92B-C	Victoria	E-7	93C	Anahim Lake
A-8			E-8	93B	Quesnel
B-6	92E	Nootka Sound	E-9	93A	Quesnel Lake
B-7	92F	Alberni	E-10	83D	Canoe River
B-8	92G	Vancouver	E-11	83C	Brazeau Lake
B-9	92H	Hope	F-3	103F	Graham Island
B-10	82E	Penticton	F-4	103G	Hecate Strait
B-11	82F	Nelson	F-5	103H	Douglas Channel
B-12	82G	Fernie	F-6	93E	Whitesail Lake
C-5	92L	Alert Bay	F-7	93F	Nechako River
C-6			F-8	93G	Prince George
C-7	92K	Bute Inlet	F-9	93H	McBride
C-8	92J	Pemberton	F-10	83E	Mount Robson
C-9	92I	Ashcroft	G-3	103K	Dixon Entrance
C-10	82L	Vernon	G-4	103J	Prince Rupert
C-11	82K	Lardeau	G-5	103I	Terrace
C-12	82J	Kananaskis Lakes	G-6	93L	Smithers
D-5	92M	Rivers Inlet	G-7	93K	Fort Fraser
D-6			G-8	93J	McLeod Lake
D-7	92N	Mount Waddington	G-9	93I	Monkman Pass
D-8	92O	Taseko Lakes	H-4	103P	Nass River
D-9	92P	Bonaparte Lake	H-5		
D-10	82M	Seymour Arm	H-6	93M	Hazelton
D-11	82N	Golden	H-7	93N	Manson River
E-3	103B-C	Moresby Island	H-8	93O	Pine Pass
E-4			H-9	93P	Dawson Creek
E-5	103A	Laredo Sound	I-4	104B	Iskut River
E-6	93D	Bella Coola	I-5	104A	Bowser Lake

Our grid reference	National Topographical System map sheet		Our grid reference	National Topographical System map sheet	
I-6	94D	McConnell Creek	K-5	104I	Cry Lake
I-7	94C	Mesilinka River	K-6	94L	Kechika
I-8	94B	Halfway River	K-7	94K	Tuchodi Lakes
I-9	94A	Charlie Lake	K-8	94J	Fort Nelson
J-3	104F	Sumdum	K-9	94I	Fontas River
J-4	104G	Telegraph Creek	L-1	114P	Tatshenshini River
J-5	104H	Spatsizi	L-2	104M	Skagway
J-6	94E	Toodoggone River	L-3	104N	Atlin
J-7	94F	Ware	L-4	104O	Jennings River
J-8	94G	Trutch	L-5	104P	McDame
J-9	94H	Beatton River	L-6	94M	Rabbit River
K-2	104L	Juneau	L-7	94N	Toad River
K-3	104K	Tulsequah	L-8	94O	Maxhamish Lake
K-4	104J	Dease Lake	L-9	94P	Petitot River

Abbreviations

BC	British Columbia
BCLS	British Columbia Land Surveyor
BCR	British Columbia Railway
CE	Civil Engineer
CEF	Canadian Expeditionary Force
CNR	Canadian National Railway
CPR	Canadian Pacific Railway
Cr.	Creek
DLS	Dominion Land Surveyor
DSO	Distinguished Service Order
E & N	Esquimalt and Nanaimo Railway
FRGS	Fellow of the Royal Geographical Society
FRS	Fellow of the Royal Society
FSA	Fellow of the Society of Antiquaries
GTPR	Grand Trunk Pacific Railway
HBC	Hudson's Bay Company
HM	Her Majesty (or, if a king is reigning, His Majesty)
HMS	Her Majesty's Ship (or His Majesty's Ship)
I.	Island
KCMG	Knight Commander of the Order of St. Michael and St. George
L.	Lake
MC	Military Cross
MLA	Member of the Legislative Assembly
MP	Member of Parliament
Mt.	Mount
Mtn.	Mountain
NCO	Non-commissioned officer
NP	National Park
NWMP	North West Mounted Police
OMI	Oblates of Mary Immaculate
PABC	Provincial Archives of British Columbia
Pen.	Peninsula
PGE	Pacific Great Eastern Railway
PLS	Provincial Land Surveyor
QCI	Queen Charlotte Islands
R.	River
RCAF	Royal Canadian Air Force
RCMP	Royal Canadian Mounted Police
RE	Royal Engineers
RN	Royal Navy
RNWMP	Royal North West Mounted Police
UBC	University of British Columbia
VC	Victoria Cross
VI	Vancouver Island

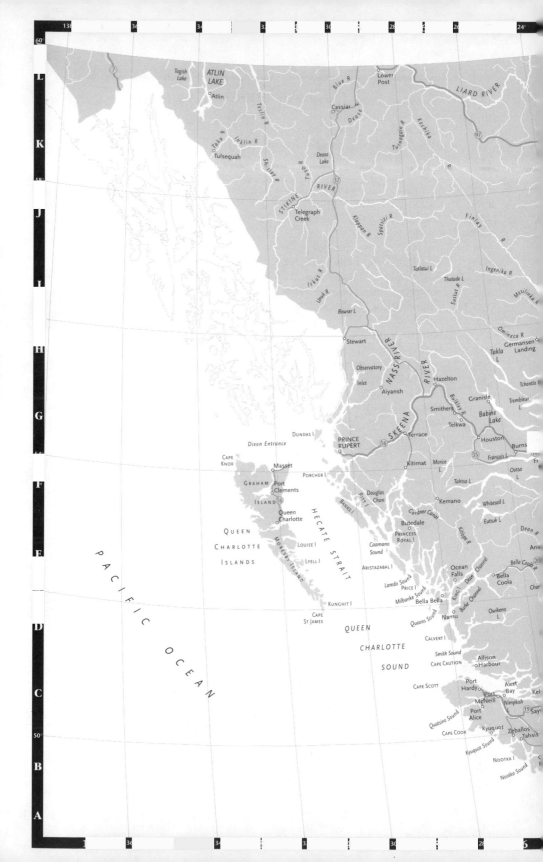

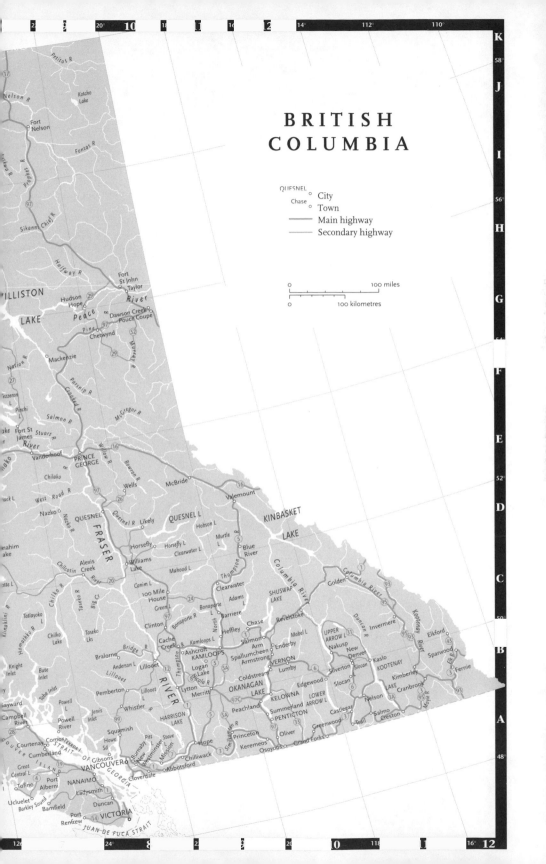

A

ABBOT PASS, Yoho NP (D-11). After Philip Stanley Abbot of the Appalachian Mountain Club, killed by a fall on adjacent Mount Lefroy, 1896.

ABBOTSFORD, Fraser Valley (B-8). J.C. Maclure, who chose the original townsite, is sometimes said to have named the town after Abbotsford, the baronial mansion that Sir Walter Scott built with the profits from his novels. In a letter of 1924, however, Maclure declared that, when the town was laid out in 1889, he named it after Harry Abbott, general superintendent of the Pacific Division of the Canadian Pacific Railway. Abbott, in his day a great man in the province, was a brother of Sir John Abbott, Prime Minister of Canada in 1891-2.

ABBOTT, MOUNT, Glacier NP (D-11). After Harry Abbott. (See *Abbotsford*.)

ABERDEEN HILLS, Kamloops (C-9). These were named after John Campbell Gordon, seventh Earl of Aberdeen, Governor-General of Canada from 1893 to 1898, who visited Kamloops in November 1894. The name has also been applied to a subdivision and shopping centre in the southwestern section of the city.

Also named after Lord Aberdeen were ABERDEEN LAKE and ABERDEEN MOUNTAIN (now Silver Star Mtn.) in the Vernon area, where Lord Aberdeen bought the Coldstream Ranch from F.G. Vernon in 1891.

ACTIVE PASS, between Galiano I. and Mayne I. (A-8). After the US survey ship *Active,* a paddle steamer of 750 tons with two guns, commanded by Lieutenant-Commander James Alden.

In 1858 Captain G.H. Richards named this stretch of water Plumper Pass after his ship HMS *Plumper,* an action to which he refers in a letter of 11 October of that year to the Hydrographer of the Royal Navy:

> I have had some correspondence with Mr. Campbell [the American boundary commissioner] on the subject of naming the passage ... It appears that the 'Active' sent her boats into this Channel a month before I arrived, and named it 'Active passage.' I have informed Mr. C. that had I known the circumstance I should have had much pleasure in retaining the name – & that I would request you to substitute 'Active' for 'Plumper' on the Chart sent home, but I have at the same time fully explained to him that the right of naming all places on our side rests solely with us ... I shall be much obliged if you will alter the name to Active.

In a later letter to the Hydrographer, Richards confided that he suspected Campbell had territorial claims in mind when he pressed for the American name. The locals, despite the Royal Navy's change to Active Pass, continued to speak of Plumper Pass. And the official name of the post office on Mayne Island was Plumper Pass until 1 April 1900.

ADAM RIVER, flows N. into Johnstone Strait (C-6). This river probably received its name from Adam Cove, a name in use as early as 1853. Note the name of the main tributary: Eve River. The Kwakwala Indian name for Adam River is 'He-la-de,' meaning 'land of plenty' with lots of berries, birds, animals, and salmon.

ADAMS RIVER [ADAM'S RIVER], flows S. into Shuswap L. (D-10). Also ADAMS LAKE. Appears as 'choo choo ach' (see *Chu Chua*) on Archibald McDonald's 1827 map of 'Thompson's River District.' Subsequently renamed after a Shuswap Indian chief, Sel-howt-ken, who was baptized in 1849 by Father Nobili and given the name of Adam. Walter Moberly reports that, when he visited the lake in 1865, he 'made the acquaintance of Adam and Eve, an Indian and his wife.'

AENEAS LAKE, W. of Skaha L. (B-10). After 'Little Aeneas,' a wizened little Indian who lived at the south end of the lake. He died c. 1946, aged over 100.

AGAMEMNON CHANNEL, NW of Sechelt Pen. (B-7). After HMS *Agamemnon*, sixty-four guns, commanded in 1793 by Captain Horatio Nelson.

AGASSIZ, S. of Harrison L. (B-9). After Captain Lewis N. Agassiz. Having sold his commission in the Royal Welch Fusiliers, he emigrated first to Prince Edward Island and then to Ontario. In 1858 he arrived in Victoria among the thousands of gold-seekers pouring in from California. After some time in the Cariboo goldfields, he turned to farming in the Fraser Valley. Preempting land in the Agassiz Valley in 1862, Captain Agassiz and his family finally took up residence here in 1867, giving his place the ultra-English name of Ferny Coombe. When the CPR was built, the station here was named Agassiz.

AHBAU LAKE, NE of Quesnel (F-8). After an old Chinese prospector and trapper who lived on this lake and had workings on Ahbau Creek. He is said to have had a weakness for poker and whisky.

AHNUHATI RIVER, flows SE into Knight Inlet (C-7). A Kwakwala Indian word meaning 'where the humpback [pink] salmon go up.'

AHOUSAT, NW of Tofino (B-6). From a Nootka Indian word probably meaning 'facing opposite from the ocean.' Captain Walbran (in his fine *British Columbia Coast Names*) notes that the name can be translated as 'people living with their backs to the land and mountains' and explains that the original home of the Ahousat band was on the exposed western shore of Vargas Island.

John Jewitt, who survived the massacre of the crew of the *Boston* in 1803, mentions that among the bands that came to Nootka to congratulate Maquinna on his capture of the ship were the 'Ah-owz-arts.' In 1864 the Ahousats themselves murdered the crew of the trading schooner *Kingfisher*.

AINSWORTH, W. side of Kootenay L. (B-11). After George J. Ainsworth, an American capitalist from San Francisco. Through the claim-jumping of his agent, Thomas Hammil (who got murdered by his victim – see *Sproule Creek*), Ainsworth and his associates obtained possession of the fabulous Bluebell mine. Ainsworth is also remembered as the promoter of the Kootenay Railway and Navigation Company. Early settlers called this place on Kootenay Lake simply Hot Springs Camp until Ainsworth acquired the site in 1883. It then became Ainsworth Hot Springs.

AIYANSH, E. of Nass R. (H-5). From the Nisgha Indian word meaning 'early leaves,' or 'leafing early.'

AKAMINA PASS, Alberta-BC boundary (B-12). From the Kootenay Indian word meaning 'mountain pass.'

AKIE RIVER, flows SW into Finlay R. (J-7). Noted surveyor Frank Swannell recorded in his journal (2 October 1914) that he was 'camping above Akie (Ah-ki-ce, Cut-bank river).'

ALAVA, MOUNT, N. of Nootka Sound (B-6). After Brigadier-General Don José Manuel de Alava, the Spanish officer who handed over to the British in 1795 his country's base on Nootka Sound.

ALBAS, Seymour Arm, Shuswap L. (D-10). After Al Bass, an early trapper in the area northwest of Shuswap Lake.

ALBERNI INLET, W. coast VI (B-7). Also PORT ALBERNI. After Don Pedro de Alberni, who commanded the tiny military force that Lieutenant Francisco Eliza carried with him when he sailed from Mexico in 1790 to establish a Spanish base at Nootka. Alberni was the first agriculturalist in the area, successfully growing vegetables in a garden at Nootka. In 1793 the little fort here was garrisoned by two corporals and eighteen soldiers detached from Alberni's company, but Alberni himself had returned to Mexico.

ALBERT CANYON, W. of Glacier NP (D-11). After Albert L. Rogers, nephew of Major A.B. Rogers (see *Rogers Pass*), who assisted his uncle in his explorations.

ALBERT HEAD, SW of Victoria (A-8). After Prince Albert, consort of Queen Victoria.

ALBION, SE of Haney (B-8). When a name was needed for the post office to be opened here, a Mr. Paine suggested Albion since it was a name for 'the old country,' Britain.

ALBREDA RIVER, flows S. into N. Thompson R. (E-10). In 1863, when Viscount Milton and Dr. Cheadle travelled from the Yellowhead Pass to the North Thompson River on their overland journey to the Pacific, they named the lake (now filled in) at the head of this river after Milton's Aunt Albreda (Lady Lyveden).

ALDERGROVE, W. of Abbotsford (B-8). From the abundant second growth of alder. Until 1919 the name was written 'Alder Grove.'

ALDRIDGE, S. end of Moyie L. (B-12). Named Moyelle when founded in 1898 during the building of the CPR's Crowsnest Pass line. The railway had been unable to obtain all the land it wanted at Moyie, two miles away, and so established this rival settlement. The village was subsequently renamed after W.H. Aldridge, general manager of the Consolidated Mining and Smelting Company. When the railway finally came to terms with the main landowner in Moyie, it moved its station and offices here.

ALERT BAY, Cormorant I. (C-6). Named by Captain G.H. Richards, RN, in 1860 after HMS *Alert,* a screw corvette with seventeen guns, at that time serving on the Pacific Station.

ALEXANDRA BRIDGE, Fraser Canyon (B-9). The original bridge, opened in 1863, was named after Alexandra, Princess of Wales.

ALEXANDRIA, Fraser R., S. of Quesnel (E-8). After Sir Alexander Mackenzie. The settlement is at the most southerly point on the Fraser River reached by Mackenzie in 1793, before he partly retraced his steps and struck out west to reach the Pacific. Fort Alexandria was the last fort to be established by the NWC before it merged in 1821 with the Hudson's Bay Company.

ALEXIS CREEK, W. of Williams L. (E-8). After Chief Alexis, a leader of the Chilcotin Indians. After the Chilcotin War of 1864, Governor Seymour led a force into the chief's territory. He described his meeting with the chief thus: 'Alexis and his men came on at the best pace of their horses, holding their muskets over their heads to show they came in peace. Having ascertained which was the Governor, he threw himself from his horse and at once approached me. He was dressed in a French uniform, such as one sees in pictures of Montcalm.' Ten years later the CPR surveyor Marcus Smith met him: 'A ride of fourteen miles ... brought us to the Alexis lakes, near one of which the chief has a rough log-house, his headquarters ... The chief Alexis looks fully fifty years of age, rather under the middle height, has small black restless eyes, expressive of distrust ...'

ALEZA LAKE, NE of Prince George (G-8). After an old Indian woman who lived in the area.

ALICE ARM, Observatory Inlet (H-5). After Alice, wife of the Reverend Robert Tomlinson. Immediately after their wedding day in Victoria in 1868, Alice and her new husband journeyed for twenty-four days in a large Haida canoe manned by eight Indians (and an Indian woman to serve the bride) before arriving at their mission station at Kincolith.

ALICE LAKE, N. of Squamish (B-8). After Alice, wife of Charles Rose, who settled in the district about 1888.

ALKALI LAKE, S. of Williams L. (D-8). Got its name from a landmark patch of alkali here.

ALLIFORD BAY, SE of Skidegate (F-4). After Able Seaman William Alliford, RN, coxswain of the ship's boat from HM hired surveying vessel *Beaver,* which sounded the bay in 1866.

ALLISON PASS, Manning Park (B-9). After John Fall Allison (1825-97). English by birth, Allison moved to the United States with his family while still a boy. In 1849 he travelled to the California goldfields, where he prospered, but in 1858 he joined the rush to the Fraser River gold area. At Governor Douglas's suggestion, he set out in 1860 to prospect along the Similkameen River and discovered Allison Pass through the Cascade Mountains. In September of that year, Allison and others preempted land, which later became the townsite of Princeton, but they failed to complete their titles. Thereafter Allison began his career as a farmer and rancher. By 1894 he owned 5,000 acres but, dogged by debts, still sought wealth through mining ventures. Mrs. Allison's memoirs have been published by UBC Press.
    ALLISON LAKE, north of Princeton, is also named after J.F. Allison.

ALOUETTE LAKE, NE of Haney (B-8). This was Lillooet Lake until 1917, when the name was changed to avoid confusion with the much larger Lillooet Lake north of Garibaldi Park. Presumably the new name was chosen because it harmonizes with the old. *Alouette* is the French for 'lark.'

ALTA LAKE, N. of Squamish (C-8). Originally Summit Lake, its name was changed to avoid confusion with other Summit Lakes. *Alta* (Latin for 'high') indicates, like 'Summit,' that this is the highest of the lakes along this stretch of the British Columbia Railway. Alta Lake Station was originally Mons.

AMBLESIDE, West Vancouver (B-8). Named by Morris Williams, who settled here in 1912. He had earlier lived in Ambleside in the English Lake District.

AMISKWI RIVER, flows into Kicking Horse R. (D-11). From the Cree word meaning 'beaver tail.'

AMOR DE COSMOS CREEK, flows N. into Johnstone Strait (C-7). Also AMOR LAKE and MOUNT DE COSMOS. After Amor De Cosmos, second Premier of British Columbia (1872-4). Born William Alexander Smith in Nova Scotia in 1825, he worked as a photographer in the California goldfields. In 1854, by act of the California legislature, he changed his name to Amor De Cosmos, a bastard mixture of Latin, French, and Greek that he mistranslated as 'Lover of the World.'

Arriving in Victoria in 1858, he established the *British Colonist,* now the *Times-Colonist.* In his first issue, he began his attack on Governor Douglas, charging that under him 'the offices of the Colony are filled with toadyism, consanguinity, and incompetency, compounded with whitewashed Englishmen and renegade Yankees.' Recalling the period of De Cosmos's own participation in politics, Dr. Helmcken observed: 'He always "took a little" before appearing on a public platform. At this time he was a radical and a demagogue, a good speaker, who knew all the captivating sentences for the multitude, well-read, a free-thinker in religion, a sort of socialist, and uncommonly egotistical.'

De Cosmos worked hard to persuade British Columbia to join Canada. After the defeat of his government in Victoria, he remained active as an MP in Ottawa until 1882. He died in 1897, and with him died his resplendent name.

ANAHIM LAKE, Western Chilcotin (E-7). After a Chilcotin Indian chief. In 1861 Ranald Macdonald and John G. Barnston, describing a recent trip from Alexandria to Bella Coola, mentioned 'Lake Anawhim.' The Indian name was 'Na-coont-loon,' meaning 'a fence built across' or a 'fish trap.' ANAHIM PEAK was a well-known source of obsidian, for which the Indians came from far and wide.

ANARCHIST MOUNTAIN, E. of Osoyoos (B-10). After Richard G. Sidley, an Irishman who arrived in the Osoyoos district around 1889. His extreme political views ultimately resulted in the cancellation of his appointments as justice of the peace and customs officer at Sidley. Before the anarchist came to the Okanagan, this mountain was known as Larch Tree Hill.

ANDERSON LAKE, W. of Lillooet (C-8). After Alexander Caulfield Anderson, born in Calcutta in 1814 and educated in England. He entered the service of the HBC and arrived at Fort Vancouver on the Columbia River in 1832. During the 1840s he made important explorations of the country between Fort Kamloops and Fort Langley. He left the service of the HBC in 1854. In 1858 Governor Douglas commissioned him to open up a route from the lower Fraser River, via Harrison Lake and Lillooet, to the upper Fraser. Douglas was so pleased with this survey that he expressed a wish that Anderson would bestow his own name upon this lake, which was an important part of the route.

ANDERSON RIVER, which flows westward into the Fraser near Boston Bar, is also named after A.C. Anderson.

ANDY GOOD CREEK, flows into Michel Cr. NW of Corbin (B-12). After Andrew Good, Crowsnest hotel proprietor in the 1900s.

ANGLEMONT, Shuswap L. (C-10). The name was coined in 1914 by W.A. Hudson, first postmaster here. Hudson derived Anglemont from nearby Angle Mountain, so named about 1877 by G.M. Dawson.

ANGLESEY, E. of Cache Creek (C-9). After the sixth Marquess of Anglesey. (See *Wulhachin.*)

ANGUS HORNE LAKE, Wells Gray Park (E-10). Named after a prospector in the adjacent Blue River country. He was a big, raw-boned Scot with tremendous physical strength.

ANGUSMAC, on the BCR north of Prince George (G-8). After a man of Indian-Scottish descent named Angus or Angus Mac.

ANMORE, N. of Ioco (B-8). Named by F.J. Lancaster, a local 'part-time homesteader.' He had a wife, Annie, and a daughter, Leonore, and from their names he coined 'Annore,' which the locals modified to Anmore.

ANNACIS ISLAND, SW of New Westminster (B-8). A slurred form of 'Annance's Island,' named after François Noel Annance. Annance was an HBC clerk who accompanied James McMillan when the latter sailed up the Fraser in the *Cadboro* in 1827 to found Fort Langley. He was also on the earlier reconnaissance of the Fraser Valley in 1824.

ANNIEVILLE, S. of New Westminster (B-8). In 1871 James Symes and his wife, Annie, were with a party in a boat looking for a suitable cannery site. They found a likely looking place, but the shallowness of the water prevented them from landing the boat. Annie then waded ashore while somebody shouted, 'Annie will make it.' The first cannery on the Fraser River was subsequently built here, and the settlement was named Annieville.

ANSTEY ARM, Shuswap L. (D-10). The son of a master at Rugby, the famous English public school, Francis Senior Anstey became, during the building of the CPR, the first large-scale lumberman on Shuswap Lake. He had a logging camp on Anstey Arm as early as 1882.

ANTHONY ISLAND, S. of Moresby I. (E-4). After Anthony Denny, Archdeacon of Ardfert, Ireland, father of a midshipman on HMS *Virago,* which was here in 1853. The Indian name 'Skang'wai' means 'Red Cod Island.' The Indians abandoned the island after the 1862 smallpox epidemic.

ANTLE ISLANDS, W. of Banks I. (F-4). After Captain the Reverend John Antle, founder of the Columbia Coast Mission.

In 1903 Vancouver was shocked when the steamer *Cassiar* arrived with the bodies of four dead loggers aboard, victims of the lack of medical services upcoast. The next year, determined to investigate, Antle (the rector of the Anglican parish of Holy Trinity, Vancouver) sailed in his little boat *Laverock* to survey the needs of the Indians, loggers, fishermen, and settlers living to the east of Vancouver Island. The next year his mission ship, the *Columbia,* was built, and Antle began his years as a missionary, founding four hospitals and ministering to an area of 10,000 square miles

Antle had learned his seamanship off Newfoundland. His language at times could be decidedly unclerical. At the age of seventy-five, he sailed his yacht *Reverie* from Falmouth, England, to Victoria.

ANTLER CREEK, E. of Barkerville (F-9). Reporting on the new discoveries of gold in the Cariboo, the *British Colonist* noted on 18 April 1861: 'Antler Creek, on which new and rich diggings have recently been struck, received its name from the fact that a pair of very large antlers were found on its banks by the first miners who ascended that stream.'

ANVIL ISLAND, Howe Sound (B-8). Captain Vancouver gave this island its name because of 'the shape of the mountain that composes it.'

ANYOX CREEK, flows S. into Observatory Inlet (H-5). From the Nisgha word meaning 'place of hiding.' Close to where this stream flows into the inlet, at Granby Bay, was an important copper mine and smelter operated by the Granby Mining Company.

ANZAC, BCR junction, N. of Prince George (G-8). After the nearby Anzac River, named for the Australian and New Zealand Army Corps of World War I renown.

APODACA COVE, Bowen I., Howe Sound (B-8). Apodaca was the name given to Bowen Island by José Maria Narvaez in 1791. (See *Bowen Island.*)

ARAWANA, N. of Naramata (B-10). Earlier Naramata Siding, but the name was changed to avoid confusion with Naramata when railway tickets were written. The new name was possibly taken from a popular song, 'Arrah Wannah,' written by Theodore F. Morse around 1906.

ARGENTA, near the head of Kootenay L. (C-11). A simplified or mistaken form of *argentea,* the Latin adjective for 'silver.' The community originated during the silver-mining boom in the 1890s.

ARISTAZABAL ISLAND, Central Coast (E-5). Named by Caamaño after Gabriel de Aristazabal (1743-1805), a Spanish naval officer under whom he had served.

ARMSTRONG, N. of Vernon (C-10). The first settlers chose the name of Aberdeen; however, when the Shuswap and Okanagan Railway (a subsidiary of the CPR) was built, the station here was named after W.C. Heaton-Armstrong (1853-1917), head of the London banking house that had floated the bonds to build the railway. Around 1892 Armstrong visited the little settlement that had been given his name.

ARROW LAKES, E. of Okanagan L. and W. of Kootenay L. (B-10 and 11, C-10 and 11). From Arrow Rock, which, in the days before the Keenleyside Dam flooded the area, was a high cliff on the east side of Lower Arrow Lake. Three large crevices, approximately forty feet above the water, once held embedded hundreds of Indian arrows. Aemilius Simpson, passing that way in 1826, referred to the rock by name and noted that the Indians used it for target practice. A subsequent account, however, says that the arrows were shot to determine an Indian's luck: an arrow that lodged in the rock signified good fortune, while one that fell into the water signified bad fortune. HBC men amused themselves by firing volleys to bring down arrows.

ARROWSMITH, MOUNT, E. of Port Alberni (B-7). After the famous English cartographers Aaron Arrowsmith (1750-1823) and his nephew John Arrowsmith. As the excellence of their maps was thoroughly appreciated, it is not surprising that in 1858 Captain G.H. Richards, RN, recommended adoption of the name to the Hydrographer of the Royal Navy.

ARROWSTONE HILLS, N. of Cache Cr. (C-9). The Indians made arrowheads out of the fine black basalt found here.

ARTABAN, MOUNT, Gambier I., Howe Sound (B-8). Takes its name from the nearby Camp Artaban (Anglican Church in Canada). This camp, in turn, was named after a character in Henry Van Dyke's *The Story of the Other Wise Man.*

ASH RIVER, N. of Great Central L. (B-7). After Dr. John Ash, the Victoria physician who was one of the sponsors of the Vancouver Island Exploration Expedition of 1864. Later Dr. Ash became Provincial Secretary and the first Minister of Mines of British Columbia.

ASHCROFT, S. of Cache Cr. (C-9). In 1862 Clement Francis Cornwall, a Cambridge BA and a barrister of the Inner Temple, arrived in British Columbia accompanied by his brother, Henry Pennant Cornwall, also a graduate of Cambridge. That year they established a ranch, which they named Ashcroft after their family home in England. They grew wheat, installed a mill, and sold flour to the packers and miners passing by on the Cariboo Road. Ashcroft became a major stopping-place for travellers between Kamloops and Spence's Bridge, and the Cornwalls grew famous for their hospitality. Hundreds of people arrived for the annual races held on Cornwall Flats (the brothers had imported an Arabian stud). The Ashcroft Hunt pursued the coyote instead of the fox, but its hounds had been brought around the Horn. When the CPR named its nearby station Ashcroft, the Cornwalls added 'Manor' to the name of their home to distinguish between the two Ashcrofts.

During the time of the Crown colony, C.F. Cornwall was a member of the Legislative Council. When British Columbia entered the Canadian federation, he became one of the senators representing the province in Ottawa. In 1881 he resigned his senatorship and began a six-year period as Lieutenant-Governor of British Columbia.

ASHNOLA RIVER, SW of Keremeos (B-9). This name is found, spelled Ashtnoulou, as early as 1861 and is from the Okanagan Indian language. The meaning of the suffix is 'land, place, etc.' The meaning of the root is less certain, but it could be '(ex)change.' Thus, Ashnola might be translated as 'place of trading.'

ASHTON CREEK, E. of Enderby (C-10). After Charles Ashton, who came to British Columbia from England in 1865 and settled here in 1887.

ASKOM MOUNTAIN, between Texas Cr. and Fraser R. (C-9). From the Lillooet Indian word for 'mountain.'

ASP CREEK, W. of Princeton (B-9). After Charles Asp, miner and rancher. The Reverend John Goodfellow remembered him as 'a tough hombre who looked like Father Time without his scythe.'

ASPEN GROVE, SE of Merritt (B-9). After a grove of trembling aspen, *Populus tremuloides*. According to the newsletter of the Nicola Valley Archives Association, 'Aspen Grove has moved about over the years depending on where someone could be found to operate the Post Office.'

ASSINIBOINE, MOUNT, Alberta-BC boundary (C-12). After the Assiniboine Indians, who used to hunt in the Rockies. In turn their name is simply the Cree Indian word for 'stone,' which is said to have been applied to them because they cooked using hot stones.

ATCHELITZ CREEK, flows N. into Chilliwack Cr. (B-9). This Halkomelem word means 'bottom.' The creek goes around the bottom of Chilliwack Mountain, which may be the reason for its name.

ATHABASCA PASS, Alberta-BC boundary (E-10). This pass through the Rockies takes its name from the Athabasca River. *Athabasca*, from a Cree word meaning 'a place where there are reeds,' refers to the delta where the river enters Lake Athabasca in northern Alberta.

ATHALMER, N. end of Windermere L. (C-11). The Honourable F.W. (Fred) Aylmer was a civil engineer with an aristocratic English background. When he came to lay out a townsite at the locality hitherto known as Salmon Beds, he remembered that his surname derived from two Anglo-Saxon words, *athel*, meaning 'noble,' and *mere*, meaning 'lake,' and named the settlement Athalmer in honour of his ancestors and himself.

ATHLONE ISLAND, Milbanke Sound (E-5). Originally Smyth Island, it was renamed to honour the Earl of Athlone, who was appointed Governor-General of Canada in 1940.

ATKINSON, POINT, Howe Sound (B-8). Captain Vancouver recorded that he had named this point after a 'particular friend.' Unfortunately he neglected to identify the friend. The most likely candidate appears to be Thomas Atkinson, RN, who was later master of HMS *Victory* during the Battle of Trafalgar.

ATLIN LAKE, Yukon-BC boundary (L-3). From the Inland Tlingit Indian word meaning 'big lake.'

ATNA RANGE, N. of Babine R. (H-6). Derived from the Carrier Indian word meaning 'strangers,' or 'other people.' G.M. Dawson, more than a century ago, wrote that nearby 'Atna Pass is here so called as being that used by the Coast Indians as an avenue to the interior country.'

ATNARKO RIVER, flows W. into Bella Coola R. (E-6). From a Chilcotin Indian word meaning 'river of strangers' (i.e., the Indians from the Coast).

ATTACHIE, junction of Halfway R. and Peace R. (I-9). After a Beaver Indian named Attachie, who signed Treaty No. 8 and is buried near Attachie. His descendants still live on the Doig River Reserve.

AUGIER LAKE, S. of Babine L. (G-7). Named by Father Morice after Father Cassien Augier (1846-1927), Superior-General of the Oblates of Mary Immaculate.

AUNTEATER GLACIER, SW of Windy Craggy Mtn. (L-1). Noting this glacier's long, thin tongue, a prospector was reminded of an anteater's tongue. A play on words probably led to this, the official name of the glacier.

AUSTRALIAN, S. of Quesnel (E-8). Takes its name from the Australian Ranch, preempted by Stephen Downes, William H. Downes, and Andrew Anderson in 1863. Stephen and William Downes had earlier been in the Australian goldfields.

AVOLA, N. Thompson R. (D-10). A post office was opened here in 1913 during construction of the Canadian Northern Railway. Since there was already a Stillwater post office elsewhere in British Columbia, the existing name of Stillwater Flats was discarded and that of Avola in Sicily borrowed.

B

BABINE LAKE, Central BC (G-6 and 7, H-6). *Babine* (the French word for a large or pendulous lip) was applied by the early voyageurs to the Indians living around the lake. Once female Indians reached puberty, they wore a labret (a plug of bone or wood) in the lower lip, gradually distending it far beyond its normal shape.

Old Fort Babine (or Fort Kilmaurs) was founded in 1822, partly as a fur-trading post but more importantly as a source for dried salmon should the Fraser River salmon run fail. Babine Lake was known to the Carrier Indians as 'Na-taw-bun-kut,' or 'Long Lake.'

BAEZAEKO RIVER, flows N. into West Road R. (F-8). Derived from the Carrier Indian word meaning 'basalt river,' with special reference to the black basalt from which arrowheads were made.

BAJO POINT, Nootka I. (B-6). *Bajo* is Spanish for 'below.' Malaspina gave the point this name in 1791, probably with reference to the very dangerous reef here.

BAKER CREEK, flows E. into Fraser R. (E-8). After Auguste Boulanger (August Baker), known locally as Augie Baker. He settled in the Quesnel area in 1903, and in 1910-11 he operated the government ferry across the Fraser here.

BAKER, MOUNT, SE of Cranbrook (B-12). After Colonel James Baker. (See *Cranbrook*.)

BALDONNEL, E. of Fort St. John (I-9). J. Abbot, a local justice of the peace who began homesteading here in 1923, is said to have named this settlement after his home in Ireland. Gazetteers show no Baldonnels but several Ballydonnels in Ireland.

BALDY HUGHES, MOUNT, S. of Prince George (F-8). Named around 1928 by Forin Campbell when he set up a triangulation point on the mountain, at the base of which 'Baldy' Hughes, an old-time stage-driver turned trapper, had lived in a cabin since about 1915.

BALFOUR, W. side Kootenay L. (B-11). In 1895 Lady Aberdeen noted in her diary, 'Opposite Pilot Bay a gentleman of the name of Balfour has built a private residence just because of the beauty of the place. And he is fully justified.' Balfour was superintendent of bridges during construction of the CPR's Crowsnest line.

BALL, CAPE, E. coast of Graham I. (F-4). Possibly from the Haida Indian name for the point, 'Gaalhíns kwún,' meaning 'point with a flat place higher up.'

BALLENAS ISLANDS, NW of Nanaimo (B-7). Named Islas de las Ballenas ('Islands of the Whales') by Narvaez in 1791.

BAMBERTON, W. side of Saanich Inlet (A-8). After H.K. Bamber, managing director of British Portland Cement Manufacturers Ltd., who came from England to help with construction of a cement plant here.

BAMFIELD, S. shore of Barkley Sound (A-7). A slightly corrupted form of the surname of William Eddy Banfield, who came to the Pacific coast on HMS *Constance* in 1846 and later became a trader and an Indian agent on the west coast of Vancouver Island. In 1862 he drowned in a canoe accident, which some suspected was murder.

BANKS ISLAND, North Coast (F-4). Named by Captain Duncan of the *Princess Royal* in 1788 after Sir Joseph Banks, president of the Royal Society.

BARKERVILLE, E. of Quesnel (F-9). After Billy Barker, who struck it rich here. Barker, a Cornish potter turned sailor, jumped ship in 1858 to join the gold rush in the Fraser Canyon. On 21 August 1862, he was digging in the Cariboo – fifty feet down and still no gold. He was ready to give up, but two feet deeper he hit the pay dirt. The find started a spree that left everybody drunk except one 'well-brought-up' Englishman. In the period that followed, Billy and his partners took out $600,000 worth of gold (very much more in modern values).

Barker has been described as 'a man of less than average height, stout with heavy body, short, slightly bowed legs ... His face was partially hidden beneath a bushy black beard, plentifully streaked with grey.' When Barker entered a saloon, particularly if he was already primed with a few drinks, he would do a little dance while singing:

I'm English Bill,
Never worked, an' never will.
Get away girls,
Or I'll tousle your curls.

He soon lost his fortune (marriage to a gold digger of another kind accelerated the process), and he ended his life in dire poverty, dying in the Old Men's Home in Victoria in 1894.

BARKLEY SOUND, SW Coast of VI (A-7). Named after himself by Captain Charles William Barkley of the *Imperial Eagle,* a British trading vessel sailing under Austrian colours. During this voyage of 1787, Barkley was accompanied by his seventeen-year-old bride, Frances, the first white woman to see British Columbia.

BARNARD CREEK, S. of Ashcroft (C-9). Apparently after Francis Jones

Barnard (1829-89), proprietor of the famous Barnard's Express (BX), whose stagecoaches were the chief means of transportation on the Cariboo Wagon Road in the gold rush days and afterwards. Barnard started his business in 1861 by walking 760 miles from Yale to Cariboo and back, delivering letters and other papers.

BARNES CREEK, flows into Thompson River, S. of Ashcroft (C-9). After John Christopher Barnes, who preempted land above Ashcroft in 1868.

BARNET, Burrard Inlet (B-8). In 1890 the North Pacific Lumber Company, owned by Eastern Canadian timber magnate James Maclaren, opened a sawmill here. The settlement was named for Maclaren's mother, Elizabeth, whose maiden name had been Barnet.

BARNHARTVALE, E. of Kamloops (C-9). After Peter Barnhart, conductor on the first CPR train to enter Kamloops from Montreal (3 July 1886). In 1909 Barnhart became the first postmaster here and annoyed the locals by having the post office named after himself.

BARNSTON ISLAND, Fraser R., E. of confluence with Pitt R. (B-8). After George Barnston, an HBC clerk, one of the party that founded Fort Langley in 1827. He became a Chief Factor in 1847 and retired in 1863. Governor Simpson described him as 'A well educated man, very active and high spirited to a romantic degree, who will on no account do what he considers an improper thing.'

BARRETT LAKE, NW of Houston (G-6). After Charlie Barrett, a packer between Ashcroft and Hazelton around 1900. Later he established one of the largest ranches in the Smithers area.

BARRIÈRE, N. Thompson R. (D-9). Takes its name from the Barrière River, named as early as 1828. The 'barrière' was almost certainly a fish trap such as G.M. Dawson saw here and described as 'two weirs or fences, each of which stretched completely across the stream.'

BASQUE, S. of Ashcroft (C-9). After the Basque Ranch, established and named by a Basque, Antoine Minaberriet, in the 1860s. He sold out in 1883 and returned to France.

BATCHELOR HILLS, N. of Kamloops (C-9). After Owen Salisbury Batchelor, reputedly 'the biggest man in the British Army.' He took part in the Klondike gold rush, became a jailer in Kamloops, went with his sons to World War I, and owned various mines that failed to make him rich.

BATNUNI LAKE, expansion of Euchiniko R. (F-7). From the Carrier Indian word meaning 'abundance of char.'

BATTLE BLUFF, Kamloops L. (C-9). Writing of Dr. A.R.C. Selwyn's survey party, which passed this bluff in 1871, Benjamin Baltzly, the photographer who accompanied them, recorded:

> Here, at the foot of this rock, a naval battle was fought about a hundred years ago between two Indian tribes – at least so the Indians say. The victorious tribe stained or painted a large projecting rock, which is about 15 feet above water, with some kind of red material to commemorate the place. Many of the pre sent Indians have superstitious notions in relation to this place. The Bluff had no name, although it is the most prominent point on the lake, so we named it Battle Bluff.

BATTLE MOUNTAIN, Wells Gray Park (D-10). After a battle fought about 1875 between Chilcotin and Shuswap Indians. At stake were the caribou hunting grounds in the area.

BATTLE MOUNTAIN, W. of Alexis Creek (E-8). G.M. Dawson noted the name in the 1870s. According to J.A. Teit, below this mountain 'may be seen a number of boulders which, according to tradition, are the transformed bodies of Alexandria warriors who strayed over the cliff in the dark while on the way to attack a camp of Chilcotin who lived in the vicinity' *(The Shuswap, p. 784).*

BATTLE OF BRITAIN RANGE, S. of Muncho L. Park (K-7). Whereas the name of the range commemorates the valour of the Royal Air Force in repelling the German aerial onslaught on Britain in 1940, individual mountains in it commemorate not only the allied leaders (Mount Churchill and Mount Roosevelt) and the places of their meetings (Teheran Mountain, Yalta Peak), but also the battles in which Canadian troops served (Dieppe Mountain, Falaise Mountain, and Ortona Mountain).

BAUZA COVE, S. side of Johnstone Strait (C-6). After Felipe Bauzá, map-maker on the Malaspina expedition. Named by Galiano and Valdes in 1792.

BAYNES LAKE, SW of Elko (B-12). After Andrew Bain, who settled here in 1896.

BAYNES PEAK, Saltspring I. (A-8). After Rear-Admiral Robert Lambert Baynes, who commanded the Royal Navy's Pacific Station from 1857 to 1860.

BAZAN BAY, E. side of Saanich Pen. (A-8). After Cayetano Valdes y Bazan, who commanded the *Mexicana* on Galiano's expedition of 1792. (See *Valdes Island.*)

BEACON HILL, Victoria (A-8). The 'beacon' was a navigational aid to help ships find Victoria harbour. Captain Walbran says that back in the 1840s the beacon consisted of two tall masts, one surmounted by a triangle and the other by a square or barrel. He notes that when an observer saw the latter through the former he was on Brotchie Ledge – a rather curious way of proceeding, which could have contributed to a number of shipwrecks! Edgar Fawcett recalls in *Some Reminiscences of Old Victoria* that it was one of the pastimes on Christmas Day to go and take potshots at the barrel on top of the mast.

The Saanich Indians called the Beacon Hill bluffs 'big belly,' since from the sea they look like a big bellied person lying on his back

BEALE, CAPE, S. entrance of Barkley Sound (A-7). Named by Captain Barkley in 1787 after John Beale, his purser on the *Imperial Eagle*. Later in the same year, Beale was murdered by Indians.

BEATON, NE arm of Upper Arrow L. (C-11). Originally Thomson's Landing, after James William Thomson. When Thomson left in 1907, his partner, Malcolm Beaton, took over.

BEATTON RIVER, flows SE into Peace R. (I-9). After Frank Wark Beatton (1865-1945), the Orkneyman in charge of the HBC post at Fort St. John for many years.

BEAUFORT RANGE, SE of Comox L. (B-7). After Sir Francis Beaufort (1774-1857), Hydrographer of the Royal Navy.

BEAVER HARBOUR, E. of Port Hardy (C-6). After the HBC's little ship *Beaver*, launched in London in 1835, wrecked off Stanley Park in 1888, the first steamship on this coast. The story of how the ship gave its name to the bay was told thus by S.F. Tolmie:

> When my father was stationed at Fort McLoughlin, a number of Indians from the north end of Vancouver Island came to trade. The blacksmith was at work at his forge, and when he went to put more coal on the fire the Indians were very curious. They asked him where the coal came from, and he explained that it took six months to bring it by ship from Wales. He noticed that they were greatly amused and asked what was so funny about it. The Indians replied that it seemed funny that the white men should carry this soft black stone so far when it could be had without expense close at hand. The blacksmith called Dr. Tolmie, and the Indians told him there were places on Vancouver Island where he could get all the soft black stone he wanted as it cropped out of the ground. My father then notified Dr. McLoughlin, at Fort Vancouver, who ordered the steamer *Beaver* to stop on her next voyage to see if the Indians were telling the truth. The result was the discovery of the coalfield at Beaver Harbour. (*Journals of William Fraser Tolmie*, pp. 394-5)

BEAVERDELL, West Kettle R. (B-10). Beaverton and Rendell were originally two settlements only about a mile apart. When they united, they adopted this composite name.

BEAVERMOUTH, W. of Donald (D-11). Close to the mouth of the Beaver River, where it joins the Columbia River.

BECHER BAY, SW of Victoria (A-8). After Commander Alexander Becher, RN (1796-1876), a surveying officer.

BECKER LAKE, W. of Canim L. (D-9). When Emile Becker applied for water rights on an as yet officially unnamed lake, he referred to it as Loon Lake (translating its Shuswap Indian name, 'iswelh'). Due to a misreading of his handwriting, this was entered as Love Lake on his permit. Becker did not like the name and, when a survey party arrived, he asked them to change it. The surveyors obligingly made Love Lake into Becker Lake.

BEDAUX PASS, Kwadacha Wilderness Park (J-7). This and Mount Bedaux commemorate the Bedaux Sub-Arctic Expedition of 1934. That year Charles E. Bedaux, a Franco-American entrepreneur, arrived at Dawson Creek with limousines, loads of baggage, champagne, asbestos tents, rubber pontoons, river boats, horses, axemen and packers, Mrs. Bedaux, her maid Josephine, her friend Madame Chiesa, cameramen, and five Citroen half-track trucks. One of the purposes of the expedition was to demonstrate the versatility of the Citroens by having them travel across the wilderness between Fort St. John and Telegraph Creek. Trouble began almost at once with the half-tracks taking up to four hours to travel a quarter of a mile through the gumbo. Ninety miles from Fort St. John, the Citroens were abandoned, two of them being driven over a cliff to provide sensational motion picture footage. Continuing by horse, the expedition got as far as Citroen Peak in the Cassiar before turning back.

BEDNESTI LAKE, E. of Cluculz L. (F-8). Father Morice, the well-known Oblate priest who spent so many years among the Carrier Indians, noted that this name means 'the bull trout got glutted.'

BEDWELL HARBOUR, S. Pender I. (A-8). After Edward Parker Bedwell, second master on HMS *Plumper* (1857-60). He was promoted to master in 1860 and subsequently served on HMS *Hecate*. (Also BEDWELL BAY, Indian Arm.)

BEECE CREEK, flows NW into Taseko R. (D-8). *Beece* is the Chilcotin Indian word for obsidian, a sharp black stone much prized for making arrowheads.

BEECHEY HEAD, SE of Sooke Inlet (A-8). After Captain F.W. Beechey, RN (1796-1856), Arctic navigator.

BEEF TRAIL CREEK, Tweedsmuir Park (E-7). So named since the trail that parallels this creek had been used to drive cattle from the Chilcotin to Bella Coola.

BEGBIE, MOUNT, SW of Revelstoke (C-10). After Sir Matthew Baillie Begbie

(1819-94), Chief Justice of British Columbia, the famous 'hanging judge' who by firmness, impartiality, and sheer power of personality maintained British law and order when the mining camps of the Cariboo and other gold-mining areas were flooded with American riffraff, fresh from the lynch-law camps to the south.

The son of a colonel in the Royal Engineers, Begbie was a highly civilized man who spoke both French and Italian (he had visited Italy) and had a taste for music. He received his MA from Cambridge in 1844 and in the same year was called to the bar at Lincoln's Inn. After some years as an impoverished young lawyer and a man-about-town in London, he decided to emigrate, possibly as a consequence of a disappointment in love. In September 1858 he was appointed 'Judge in Our Colony of British Columbia.' Tall (six foot four), his long black cloak swirling behind him, his eyes gleaming between the stylish rake of a gaucho hat and the carefully trimmed lines of his Van Dyke beard, Begbie was a commanding figure. He never held court unless attired in the robes of an English judge. Inevitably he became the centre of numerous stories and legends.

Vacationing in Salt Lake City, he met an American who had served on a jury in the Cariboo and who remarked, 'You certainly did some hanging, judge.' To which Begbie pungently replied, 'Excuse me, my friend, I never hanged any man. I simply swore in good American citizens, like yourself, as jurymen, and it was you that hanged your own fellow-countrymen.' Both Begbie and J.C. Haynes have been credited with warning the gold miners at Wild Horse Creek: 'Boys, if there is any shooting at Kootenay, there will be hanging at Kootenay.' Begbie believed in flogging too. 'My idea is that if a man insists upon behaving like a brute, after a fair warning, & won't quit the Colony, treat him like a brute & flog him.' Angered by the acquittal of a holdup man, he said to the prisoner, 'The jurymen say you are not guilty, but with that I do not agree. It is now my duty to set you free and I warn you not to pursue your evil ways, but if you ever again should be so inclined, I hope you select your victim from the men who have acquitted you.'

Begbie became Chief Justice of mainland British Columbia in 1869 and Chief Justice of all British Columbia in 1870. He was knighted in 1875. Sir Matthew died in Victoria in 1894.

BELCARRA, Indian Arm, Burrard Inlet (B-8). From two Gaelic words, *bal*, meaning 'the sun,' and *carra*, meaning 'cliff' or 'rock.' Judge Norman Bole, who gave the name in the 1870s, declared that it meant 'The Fair Land of the Sun.'

BELGO CREEK, E. of Kelowna (B-10). The Belgo-Canadian Fruit Lands Co., incorporated in 1909, built an irrigation dam here.

BELLA BELLA, NW side of Denny I. (E-5). Its name is that of the local Indians, the Bella Bella (Bil-Billa or Bel-Bella) band. That name may derive from a Heiltsuk word meaning 'flat point(ed),' describing the village's original location on McLoughlin Bay near the HBC's Fort McLoughlin (1833-43). NEW BELLA BELLA (see *Waglisla*) is on nearby Campbell Island.

BELLA COOLA, head of N. Bentinck Arm (E-6). From the Heiltsuk (Bella Bella) Indian word that means 'person from Bella Coola.' The word refers to the entire Bella Coola ethnic region, not to any village in particular.

Early versions of the name show a wild divergence in spellings, as white men tried to cope with the guttural sounds of the Indian tongue – Billa Whulha, Bell-houla, Billichoola, Bill-Whoalla, etc. Sir Alexander Mackenzie called the settlement of Bella Coola 'Rascal's Village.'

BELL-IRVING RIVER, flows SE into Nass R. (I-5). After D.P. Bell-Irving, BCLS. A month after the outbreak of World War I, he sent his father a telegram from a cabin on the Yukon Telegraph Line: 'Hear there is war. Who is fighting who?' He returned to Vancouver, joined up as befitted a graduate of the Royal Military College, Kingston, and became the first Canadian officer to be killed in action.

BELLY UP CANYON, between Peters L. and Spectrum L. (C-10). A guide, Eugene Foisy, lost a horse when it fell over a precipice and into the canyon. When Foisy descended into the canyon, he found the horse alive but 'belly up.'

BENDOR RANGE, S. of Carpenter L. (C-8). After the Ben d'Or Mines Co. claims. *Ben d'or*, a Scots-French hybrid, means 'mountain of gold.'

BENNETT BAY, Mayne I. (A-8). After Thomas and Alice Bennett, who settled here about 1879. Weary after a long journey from Scotland, Mrs. Bennett opined that they had arrived at 'the arse end of the world.'

BENNETT LAKE, Yukon-BC boundary (L-2). This was the Boat Lake of the white gold-seekers until Lieutenant Frederick Schwatka of the US Army passed this way on his Yukon expedition in 1883 and renamed it after James Gordon Bennett (1841-1918), proprietor of the *New York Herald*.

BENSON, MOUNT, W. of Nanaimo (B-7). After Dr. A.R. Benson, a physician in the service of the HBC between 1857 and 1862. Lieutenant Charles Wilson, RE, remembered Dr. Benson as 'a great character, never seen without a pipe in his mouth & his rooms ... crowded with Indian curiosities, bird skins, geological specimens, books & tobacco in the most inextricable confusion.'

The Nanaimo Indian name for Mount Benson, 'Wakesiah,' means 'mysterious, sinister, or forbidding.'

BENTINCK ARMS, Central Coast (E-6). North Bentinck Arm and South Bentinck Arm, leading off Burke Channel, were named by Captain

Vancouver in 1793 after William Henry Cavendish Bentinck, Duke of Portland, Prime Minister of Great Britain in 1783 and again from 1807 to 1809. Observed one of his contemporaries: 'Without any apparent brilliancy, his understanding is sound and direct, his principles most honourable, and his intentions excellent.'

BENVOULIN, E. of Kelowna (B-10). When George G. McKay laid out the townsite here around 1891 in the hope that it would become the southern terminus of the Shuswap and Okanagan Railway, he named it after his home in Scotland.

BERESFORD, S. of Kamloops (C-9). After Admiral Lord Charles Beresford, who visited Kamloops in 1911 and made property investments here. He was an absentee landlord whose holdings were administered by a local agent.

BERG LAKE, Mt. Robson Park (F-10). So named for the many small icebergs that topple into it from the foot of Mount Robson's Berg Glacier.

BERLAND, MOUNT, N. of Radium Hot Springs (C-11). After Edward Berland, the HBC guide who saw Sir George Simpson through this part of the country in 1841. A pious Catholic, he gave religious instruction to the Indians and taught them how to keep track of the months and days by cutting notches in wooden sticks.

BERNARD CREEK, flows into Peace Arm of Williston L. (I-8). In September 1828 the HBC *engagé* carrying Sir George Simpson on his back from the canoe to the shore slipped, and the two fell into the water. Simpson named the stream after the man.

BESA RIVER, flows NE into Prophet R. (J-8). From the Beaver Indian word for 'knife.'

BESSETTE CREEK, flows NE into Shuswap R. (C-10). After Pierre (Peter) Bessette, a member of the early French-Canadian settlement in the Lumby area. He preempted land in 1877.

BILLY WHISKERS GLACIER, Battle Range (C-11). After N.H. Brewster and his party had spent a stormy night on a nearby mountain, they awoke to find the footprints of Billy Whiskers (mountain goat) where he had sheltered in the lee of their tent.

BIRKENHEAD RIVER, NW of Lillooet L. (C-8). Also BIRKENHEAD LAKE and BIRKEN. Famous in the annals of the British Army is the story of the *Birkenhead*. In February 1852 this troopship struck a rock near the Cape of Good Hope. Since a number of families were travelling with the men, the places in the few lifeboats were given to the women and children. While the boats rowed away, the soldiers stood lined up in their ranks on the deck.

When the ship began to sink, the captain cried out, 'Every man for himself.' Trained in the iron discipline of the British Army, not a man moved from his place. Instead, they awaited an order of dismissal from their officers. That order was not given, two of the officers explaining that, should the soldiers be dismissed from parade and swim out to the boats, they would certainly capsize them in their endeavours to be saved. Firm until the end, the men stood in their ranks until the ship sank under them. Of the 638 men aboard the *Birkenhead*, 454 drowned, but the boats got ashore safely, and every woman and child who had been aboard the ship was saved. The King of Prussia was so impressed by the event that he ordered an account of it to be read to every regiment in his army.

The officer commanding the soldiers on the *Birkenhead*, and one of those who perished, was Lieutenant-Colonel Alexander Seton, a relative of A.C. Anderson, the HBC officer and explorer who named the lake and river after him. (See *Anderson Lake* and *Seton Lake*.)

BISHOP, MOUNT, W. of Indian Arm, Burrard Inlet (B-8). After Joseph C. Bishop, first president of the BC Mountaineering Club (1907-10). He was killed in July 1913, at the age of sixty-two, when he fell into a crevasse on Mount Baker.

BLACK MOUNTAIN, N. of West Vancouver (B-8). At one time a forest fire left this mountain a mass of charred trees, giving it a black appearance. The name first appears on an admiralty chart of 1865.

BLACK TUSK, Garibaldi Park (B-8). Descriptive of this peak's lava top. The Squamish Indian name for it signified 'landing place of the thunderbird.' The Squamish people believed that the magic thunderbird lived on top of Black Tusk, flapping its wings to cause thunder and shooting lightning bolts from its eyes at anyone who came too close.

BLACKCOMB PEAK, Whistler (C-8). Takes its name from a serrated edge of black rock.

BLACKEYES CREEK [BLACKEYE'S CREEK], W. of Princeton (B-9). After an Indian who helped A.C. Anderson during his search for a route through the Cascade Mountains in 1846. Blackeye's Trail ran by the headwaters of this stream.

BLACKIE SPIT, NW of White Rock (B-8). After a Scot, Walter Blackie, the first blacksmith in New Westminster, who in 1875 bought the spit (six and a half acres) for a dollar an acre.

BLAEBERRY RIVER, N. of Golden (D-11). David Thompson's 'Portage Creek.' One of the earliest routes for explorers entering British Columbia from the east was through Howse Pass and down the Blaeberry to the Columbia River. In 1859 James Hector named this river Blaeberry, or Blueberry, because of the abundance of berries in its valley. His 'blaeberries' were probably our huckleberries.

BLAKEBURN, W. of Princeton (B-9). Compounded out of the names of W.J. Blake Wilson and Pat Burns, who acquired the Coalmont colliery here in 1917, finding a major buyer in the Kettle Valley Railway. The last coal was mined in 1940.

BLANSHARD, MOUNT, N. of Haney (B-8). In 1859 the admiralty surveyor Captain Richards named this mountain behind Haney after Richard Blanshard, first Governor of Vancouver Island (1850-1). Blanshard's tenure was brief and unhappy. He found Victoria entirely ruled by the HBC, which had provided no proper quarters for him. He had no staff to enable him to set up his own government, and after a few months he returned to England in disgust. A visitor to Fort Victoria in 1850 records: 'We found Governor Blanshard smoking a very thick pipe with a very long stem. He was a comparatively young man, of medium height, with aquiline, aristocratic features, set off by a large military moustache.' (See *Golden Ears*.)

BLIGH ISLAND, Nootka Sound (B-6). After Captain Bligh of *Mutiny on the Bounty* fame. William Bligh was master of HMS *Resolution* on Captain Cook's third expedition and thus visited Nootka with him in 1778.

BLIND BAY, Shuswap L. (C-10). Among the meanings of the word *blind* are 'out of sight' and 'concealed from sight' (*OED* I:920). Because of the angle at which this bay joins Shuswap Lake, it is easy to travel down the lake without noticing the bay at all.

BLUBBER BAY, Texada Island (B-7). Once a rendezvous for whaling ships. Back around 1890 Elijah Fader used to cut up whales in this bay.

BLUE RIVER, flows E. into the N. Thompson R. (E-10). The Reverend George M. Grant, describing his journey down the valley of the North Thompson in 1872, observed: 'Blue River gets its name from the deep soft blue of the distant hills, which are seen from its mouth well up into the gap through which it runs.'

BLUNDEN HARBOUR, N. of Queen Charlotte Strait (C-6). After Edward R. Blunden, second master of HM hired surveying vessel *Beaver* when the ship was making a survey here in 1863.

BOBBIE BURNS CREEK, flows E. into Spillimacheen River, S. of Golden (C-11). After the Bobbie Burns mine, presumably named for Scotland's national poet.

BOB QUINN LAKE, Highway 37, S. of Kinaskan L. (I-4). After an operator at the relay cabin here on the old government telegraph line to the Yukon.

BODEGA HILL, Galiano Island (A-8). After Juan Francisco de la Bodega y Quadra. (See *Quadra Island*.)

BOER MOUNTAIN, NE of Burns L. (G-7). After George ('The Boer') Wallace, who filed for land near Burns Lake in 1908, his wife being the first white woman in the area. Both came from South Africa and spoke Zulu fluently.

BOITANO LAKE, S. of Williams L. (D-8). After Augustine Boitano, who owned the adjacent Springhouse Ranch.

BOLEAN CREEK, flows SE into Salmon R. (C-10). A misreading of Rivière Boleau (Birch River) produced this name.

BONANZA LAKE, E. of Nimpkish L. (C-6). So named by Eustace Smith, a timber cruiser, because of the high value of the fine stands of trees around this lake.

BONAPARTE RIVER, flows S. past Cache Cr. (C-9). Since this is mentioned by name as early as 1826, it would seem that the river was named directly after the great Napoleon Bonaparte (1769-1821) rather than the Kanaka labourer who is known to have been nicknamed Napoleon.

BONILLA POINT, W. entrance to Strait of Juan de Fuca (A-7). Named by Quimper in 1790, almost certainly after Antonio Bonilla, secretary to the Spanish royal government in Mexico.

BONNEVIER CREEK, E. boundary, Manning Park (B-9). After the Swede (Charles Bonnevier) who was the first homesteader at the eastern end of today's Manning Park. He did a lot of prospecting and in the early 1940s at the age of eighty was still to be seen with shovel and goldpan, and a pot of mush or mulligan, prospecting in the hills.

BONNEY, MOUNT, Glacier NP (D-11). After the geologist Professor T.G. Bonney, FRS, when he was president of the Alpine Club of England.

BONNINGTON FALLS, W. of Nelson (B-11). After Bonnington Linn, a waterfall on the Scottish estate of Sir Charles Ross, baronet (1872-1942), first president of the West Kootenay Power & Light Co. Ross designed the dam and power plant here. He also invented the Ross rifle, so unpopular with Canadian troops in World War I.

BOO MOUNTAIN, W. of Decker L. (G-7). Named after Donald Boo. Mrs. Turkki reports in her history of Burns Lake that 'here resided Lucy Boo, her son Donald, his wife, Sarah, and son Leno. The other inhabitants were One-eyed Emily and Square-Ass Jenny Boo ... The word "boo" meaning "wolf" in the native language.'

BOOKWORMS, THE (SPIRES), Garibaldi Park (B-8). In 1970 Dr. Neal M. Carter wrote to the authors:

In 1922, while returning from a climb of Castle Towers Mountain, my companion C.T. Townsend and I were trudging along the Sphinx Glacier névé past three rocky pinnacles that stuck up all by themselves out of the névé. As we looked up at them, one of them struck us as having a silhouette of a cowled monk, and a thin blade of rock projected at an angle where he would be holding up a book as though reading. So we dubbed this pinnacle 'The Bookworm' and tarried long enough to climb it (it's only about 150 feet high).

The name has been extended to include large rock formations in the vicinity and 'has received unwarranted prominence on maps' – a 1980 Ministry of the Environment map of Vancouver-Kamloops region gives names of only ten mountains in Garibaldi Park, and one of them is The Bookworms.

BOOTHROYD, N. of Boston Bar (B-9). After George Boothroyd (1829-1902), who, with his brother, kept a roadhouse here during the days of the old Cariboo Road.

BORDEN GLACIER, NE of Terrace (G-5). Also MOUNT SIR ROBERT. After Sir Robert Borden, Prime Minister of Canada 1911-20.

BORLAND CREEK, E. of Williams L. (E-9). After Robert Borland, founder of the Borland Ranch, who preempted here in 1865.

BOSTON BAR, Fraser Canyon (B-9). So named because of the numerous Americans washing for gold in the bar in the Fraser River here. Since the first American ships off our coast were almost invariably from Boston, the Indians took to calling the Americans 'Boston men.' Similarly the Indians called the British 'King George men.' The Americans, carrying into British Columbia their tradition that the only good Indian is a dead Indian, treated the Natives abominably. In 1859 Arthur Bushby noted in his journal that it 'is quite strange to see how soon the Indians detect the Boston men & how they dislike them and how much they like King Georgie man.' Provoked by the Americans, the Indians in the Fraser Canyon began to murder isolated parties of whites washing the bars of the Fraser. The trouble culminated in the so-called 'Battle of Boston Bar' on 14 August 1858. A correspondent of the *San Francisco Bulletin* who was present reported that the fight 'lasted three hours, and resulted in the complete rout of the savages. Seven of the Indians are known to have been killed, and a number wounded. Only one white man was wounded, and that slightly in the arm. About 150 white men were in the fight.' According to some authorities, the Battle of Boston Bar actually took place at Spuzzum.

The original Indian village at Boston Bar was named Koia'um, often spelled Quayome, meaning 'to pick berries.'

BOSWELL, E. side of Kootenay L. (B-11). When Earl Grey, Governor-General of Canada, visited Kootenay Lake in 1906, he was so taken with the area that he purchased some land here for his son, Lord Howick. The land was purchased from a James Johnstone, and the surveyor employed by Earl Grey was named Boswell. Apparently there was a viceregal jest about Boswell having followed Dr. Johnson. In any event Grey named the newly acquired property the Boswell Ranch.

BOTANICAL BEACH, SW of Port Renfrew (A-7). As early as 1900 the abundant intertidal marine life led the University of Minnesota's Dr. Josephine Tildon to establish a marine research station here. Access was difficult (steamship from Victoria to Port Renfrew, then a long, very muddy, and narrow trail), and the station closed in 1907 when promised trail improvements failed to materialize.

BOTANIE MOUNTAIN, N. of Lytton (C-9). The meaning of the name of this important food-gathering area of the Thompson Indians has frequently been given as 'perpetual root place.' However, the true meaning, according to Annie York, who spoke the Thompson language, is 'covered' (by plants with edible roots). An alternative translation given by an elderly Thompson woman is that it means 'walled, enclosed all around.' Certainly this is a good description of the deep valleys that surround most of Botanie Mountain.

James Teit, in his fine work *The Thompson Indians of B.C.*, wrote that 'Botani Valley ... has been from time immemorial a gathering-place for the upper divisions of the [Thompson] tribe, chiefly for root-digging during the months of May and June. Sometimes over a thousand Indians, representing all the divisions of the tribe, would gather there.' G.M. Dawson remarked in 1891 that the root chiefly sought was that of the tiger lily (*L. Columbianum*).

Note the spelling of Bootahnie Indian Reserve, which gives a clue to the Indian pronunciation of the name.

BOUCHIE LAKE, NW of Quesnel (F-8). After 'Billy' Boucher (Bouchier), a homesteader who at one time operated a ferry across the Fraser River. Boucher was a son of Jean Baptiste Boucher, one of Fraser's party on his famous journey to the Pacific in 1808, later known as 'Waccan,' the HBC's feared enforcer of discipline.

BOUNDARY BAY, W. of White Rock (B-8). So named since it is intersected by the international boundary. This is Galiano's Ensenada del Engaño ('Mistake Bay'), the mistake being thinking that there was a passage at the end of the bay.

BOUNDARY CREEK, flows S. past Greenwood (B-10). So named because it enters the Kettle River at the international boundary.

BOWEN ISLAND, Howe Sound (B-8). After Rear-Admiral James Bowen (1751-1835). He was master of HMS *Queen Charlotte,* flagship of Lord Howe during the British naval victory known as 'The Glorious First of June,' 1794. Named by Captain Richards in 1860. Back in 1791 the Spanish explorer Narvaez had named this island and the one to the west 'the isles of Apodaca,' after Sebastian Ruiz de Apodaca, a Spanish naval official.

BOWRON LAKE, NE of Barkerville (F-9). Named after John Bowron (1837-1906). A native of Huntingdon, Quebec, he was one of the Overlanders of 1862 who travelled to the Cariboo goldfields by way of the Yellowhead Pass. When the first library in the Cariboo was established, on a subscription basis, at Camerontown in 1864, Bowron was the librarian. Subsequently he and the library moved to Barkerville, where he became a leading member of the Cariboo Amateur Dramatic Association, which put on its plays in the old Theatre Royal.

In 1866 Bowron became postmaster at Barkerville, in 1872 mining recorder, in 1875 government agent, and in 1883 gold commissioner. He retired in 1905.

BOWSER, NW of Qualicum Bay (B-7). After William John Bowser, Premier of British Columbia 1915-16.

BOWYER ISLAND, Howe Sound (B-8). After Admiral Sir George Bowyer (1740-1800). Serving as rear-admiral under Lord Howe during the British naval victory known as 'The Glorious First of June,' he lost a leg but won a pension of £1,000 and a baronetcy.

BOXER POINT, Nigei Island (C-6). After Alexander F. Boxer, master of HMS *Alert*. (See *Alert Bay*.)

BOYA LAKE PARK, NE of Dease L. (L-5). After Louis Boya, a local Indian.

BOYLE POINT, Denman I. (B-7). After David Boyle, first lieutenant on HMS *Tribune* when she was on this coast in 1859-60. He later became Governor of New Zealand and president of the Institute of Naval Architects.

BRABANT CHANNEL, Clayoquot Sound (B-6). After Father Augustus J. Brabant, who arrived on the west coast of Vancouver Island from Belgium in 1869. He spent the next forty years here as a missionary priest, acquiring extensive knowledge of the Indians and mastering their language. In 1908 he became Apostolic Administrator of the Diocese of Victoria.

BRACKENDALE, N. of Squamish (B-8). After Thomas Hirst Bracken, first postmaster (1906-12). He returned to England after his hotel here burned down. The Squamish Indian word for the Brackendale area is 'See-yeah-chum,' meaning 'full.'

BRADNER, NW of Abbotsford (B-8). After Thomas Bradner, who settled here about 1895.

BRALORNE, S. of W. end of Carpenter L. (C-8). In 1897 the Lorne mineral claim was staked, and it became the Lorne mine. In 1931 the Bralco Development and Investment Company acquired the mine, and a new company, Bralorne Mines Limited, was formed to operate the mine. The new name, Bralorne, was a compound of Bra, for Bralco, and Lorne, the name of the mine. Bralco consists of the first two letters of British, Alberta, and Columbia.

BRANDYWINE FALLS, SW of Whistler (C-8). The word *brandy* is a shortened form of the word *brandywine*, which comes from the Dutch *brandewijn*, meaning burnt (i.e., distilled) wine. Readers have their choice of two explanations as to how these falls came to be associated with brandy. The first is that, when the Howe Sound and Northern Railway had a survey party here in 1910, Jack Nelson, who was in charge, made a bet with Bob Mollison, an axeman, as to who could estimate more accurately the height of these falls. Each wagered a bottle of brandy. When the height (195 feet) was measured with a chain, Mollison won. Nelson handed over his brandy and named the falls Brandywine Falls. The other story is that two old-timers, Charles Chandler and George Mitchell, passed out here around 1890 after lacing their tea too generously with brandy.

BREAKENRIDGE, MOUNT, E. of Harrison L. (B-9). After Sapper A.T. Breakenridge, RE, who was engaged in a survey at the north end of Harrison Lake in 1859.

BRENT MOUNTAIN, W. of Penticton (B-10). After the family of Frederick Brent, a former US Army cavalryman and Indian scout, who first came to the Okanagan in 1865. Locals know Brent Mountain as 'Snow Mountain.'

BRENTWOOD BAY, W. side of Saanich Pen. (A-8). John Sluggett settled here in 1876, and the post office, which opened in 1920, was Sluggett until 1925, when it was renamed Brentwood Bay. The new name was taken from Brentwood in Essex, England, the home of R.M. Horne-Payne, president of the British Columbia Electric Company. The company operated an interurban railway linking the area with Victoria and had a powerhouse here.

BREW, MOUNT, S. of Lillooet (C-9). After Chartres Brew (1815-70), Crimean War veteran and inspector of constabulary, Cork, Ireland, who was appointed in 1858 first inspector of police for the infant colony of British Columbia. A year later he was appointed Chief Gold Commissioner. He ended up a county court judge at Barkerville. The epitaph, said to have been written by Judge Begbie, on Brew's grave in the Barkerville cemetery reads: 'A man imperturbable in courage and temper, endowed with a great and varied administrative capacity, a most ready wit, and most pure integrity and a most human heart.'

BRIDESVILLE, E. of Osoyoos (B-10). After David McBride, who settled here in the 1880s. Reportedly, when the Great Northern Railway ran a line into the district, McBride said that he would grant land for a townsite but only if the settlement were given his name. It seems that the railroad met McBride halfway.

BRIDGE RIVER, flows E. into Fraser R. N. of Lillooet (C-8). Commander R.C. Mayne, RN, who was here in 1859, records in *Four Years in British Columbia and Vancouver Island*: 'This river takes its English name from the fact of the Indians having made a bridge across its mouth, which was afterwards pulled down by two enterprising citizens, who constructed another one, for crossing which they charged the miners twenty-five cents' (p. 131). The Indian bridge is noted on Archibald McDonald's map of 1827.

BRIDGEPORT, Richmond (B-8). At the end of 1889, two bridges linked Richmond with the mainland, and locals hoped that a port would develop here.

BRIGADE LAKE, S. of Kamloops (C-9). On the old HBC trail linking Kamloops with Tulameen and Hope. Probably the fur brigades camped here, because of the excellent pasturage for the horses, on their first night out from Kamloops.

BRIGHOUSE, Lulu I. (B-8). After Sam Brighouse, born in Huddersfield, Yorkshire, in 1836. Brighouse and his cousin John Morton came to British Columbia in 1862 and spent a brief time in the Cariboo goldfields. Later in the same year, together with William Hailstone, they purchased 550 acres of land, that portion of present-day Vancouver lying between Burrard Street and Stanley Park. In 1864 Brighouse bought 697 acres on Lulu Island and established the Brighouse Ranch. He later became a Richmond councillor and a Vancouver alderman. In 1907 he made one of his fields into a racetrack, Minoru Park (named after a horse owned by Edward VII), which later became Brighouse Park. Brighouse died in England in 1913. (For more on him, see *Vancouver*.)

BRILLIANT, E. of Castlegar (B-11). According to A.M. Evalenko, when the Doukhobors came to the Kootenays, they gave their first settlement 'the name of the Valley of Consolation, Village of Brilliant, from a brilliant diamond of the first water, on account of the great river Columbia flowing through the land.' Actually, though near the Columbia, Brilliant is on the Kootenay River, and a little piece in the *Cominco Magazine* says that it was 'the bright racing waters' of the latter river that suggested this name to H.B. Landers, who was working with the Doukhobor settlers.

BRISCO RANGE, between Kootenay NP and Columbia R. (C-11). After Captain Arthur Brisco, 11th Hussars, one of the heroes of the Charge of the Light Brigade. He was a great friend of John Palliser and joined him during his explorations in 1858-9.

BRITANNIA BEACH, Howe Sound (B-8). Takes its name from the Britannia Range, which rises behind it. About 1859 Captain Richards of the Royal Navy's survey ship *Plumper* named these mountains after HMS *Britannia*. This ship, the third of her name, was never in these waters. She served in the Battle of St. Vincent in 1797 and in the Battle of Trafalgar in 1805.

BRITISH COLUMBIA. The British were first aware of this part of the world as a northward extension of Sir Francis Drake's 'New Albion' (California and Oregon). In 1792-4 Captain Vancouver gave diverse names to various parts of the future province of British Columbia. To Vancouver Island he gave the name of Quadra and Vancouver's Island. The coastal parts of northern Washington and the southern British Columbia mainland he named New Georgia, while he called the central and northern coastal areas of British Columbia New Hanover. These names failed to secure acceptance.

The evolution of the name of British Columbia is easily traced. In 1792 Captain Robert Gray from Boston rediscovered the river that the Spaniards had named Rio de San Roque some seventeen years earlier. Ignorant of the Spaniards' prior discovery, Gray named the river after his ship, the *Columbia*, and so the Columbia River entered history. In the following years, it was natural enough that the vast area drained by the mighty Columbia should be referred to increasingly as the Columbia country. When the Hudson's Bay Company set up two administrative areas west of the Rockies, it named the more northerly New Caledonia (q.v.) and the more southerly Columbia. After the Treaty of Washington in 1846 fixed the forty-ninth parallel of latitude as the Anglo-American boundary from the Rockies to the Pacific, most of the old HBC Department of Columbia became American. Somebody was bound to think of using 'British Columbia' as a name for what was left north of the new boundary line.

The person who took this final step was Queen Victoria. In a royal letter of 1858 to Sir Edward Bulwer-Lytton, the Colonial Secretary, we find the earliest mention of 'British Columbia.' In this letter the naming of a new Crown colony in the Pacific Northwest is discussed:

> The Queen has received Sir E. Bulwer Lytton's letter. If the name of 'New Caledonia' is objected to as being already borne by another colony or island claimed by the French, it may be better to give the new colony, west of the Rocky Mountains, another name. New Hanover, New Cornwall, New Georgia, appear from the maps to be names of sub-divisions of that country, but do not appear on all maps. The only name which is given to the whole territory in every map the Queen has consulted is 'Columbia,' but as there exists a Columbia in South America, and the citizens of the United States call their country also 'Columbia,' at least in poetry, 'British Columbia' might be, in the Queen's opinion, the best name.

The new colony of British Columbia was officially proclaimed at Fort Langley on 19 November 1858. In 1863 Stikine Territory was made part of British Columbia, and on 19 November 1866 Vancouver Island became part of the united colony of British Columbia.

BRITTAIN RIVER, flows SE into Princess Royal Reach, Jervis Inlet (C-7). After Rowland Brittain, BC's first patent attorney, who owned land at the mouth of the river around 1901.

BROCKLEHURST, NW of Kamloops (C-9). After Ed Brocklehurst, an Englishman who was a successful orchardist locally. Arriving around 1896, he returned to England in 1907.

BROCKTON POINT, Vancouver (B-8). After F. Brockton, senior engineer of HM surveying ship *Plumper.* In 1859 Captain Richards of the *Plumper* reported to Governor Douglas that Brockton, exploring the area of today's Stanley Park, had found coal in nearby Coal Harbour.

BROMLEY ROCK PARK, E. of Princeton (B-9). After John Hatton Bromley, a former gold prospector who started farming in this area in the 1890s.

BROOKMERE, NW of Princeton (B-9). After Harry Brook, a happy-go-lucky cattleman who took up land here before World War I.

BROOKS PENINSULA, between Brooks Bay and Checleset Bay (C-6). In August 1787 Captain Colnett, commanding the *Prince of Wales,* a ship owned by the King George's Sound Company, discovered Nasparti Inlet and named it Port Brooks after one of the proprietors of the company. In 1862 Captain Richards named Brooks Peninsula and Brooks Bay north of it, apparently under the wrong impression that Klaskish Inlet, an easterly arm of the bay, was Colnett's Port Brooks.

BROTCHIE LEDGE, off Victoria (A-8). After Captain William Brotchie (1799-1859), captain of various HBC ships. He was in command of the barque *Albion* in 1849 when she struck this ledge, previously known as Buoy Rock. (See also *Beacon Hill.*)

BROUGHTON ISLAND, S. of Kingcome Inlet (C-6). After Lieutenant William Robert Broughton, RN, who commanded the armed tender *Chatham,* which accompanied HMS *Discovery* during Vancouver's explorations. Early in 1793 Vancouver sent him back to England from Monterey, California, with despatches.

BROWN CREEK, N. of Grand Forks (B-10). It and nearby Volcanic Creek commemorate 'Volcanic' Brown, a colourful prospector who perished while seeking a 'lost mine.'

BROWN, MOUNT, Athabasca Pass, Alberta-BC boundary (E-10). Named in

1827 by David Douglas after Robert Brown (1775-1858), the first keeper of the botanical department of the British Museum. Douglas grossly overestimated the height of Mount Brown. His ascent of it is the earliest recorded climb in the Canadian Rockies.

BROWN PASSAGE, S. of Melville I., North Coast (G-4). Named in 1793 by Captain Vancouver after the captain of the *Butterworth,* who had sent out a whaleboat to guide Vancouver to the safe harbour where the *Butterworth* and two other English ships were anchored.

BROWNING ENTRANCE, N. of Banks I. (F-4). After George A. Browning, second master on HMS *Hecate,* 1861-2, and assistant surveying officer on HM hired surveying vessel *Beaver,* 1863-8.

BROWNS RIVER [BROWN'S RIVER], W. of Courtenay (B-7). Reporting on his explorations on Vancouver Island during 1864, Dr. Robert Brown wrote: 'The party insisted on naming the river after me ... I hope you will not accuse me of egotism, if at the earnest solicitation of the expedition, I allow the seat of this rich coal field to bear the name of Brown's River.' Brown was a Fellow of the Royal Geographical Society and the author of numerous elementary geography textbooks. His work on Vancouver Island has not been sufficiently remembered. A member of this expedition wrote of him, 'He was genial and kindly, and his conversation was remarkably copious, easy and entertaining, full of anecdotes and incidents derived from his years of travel in all latitudes, and of rare and curious and out-of-the-way knowledge.'

BRUCE CREEK, W. of Invermere (C-11). After R. Randolph Bruce, a former lieutenant-governor of British Columbia who was active in promoting the Columbia Valley Irrigated Fruit Lands Company.

BRUIN BAY, N. shore of Graham I. (G-3). In 1853, as HMS *Virago* was sailing offshore here, she opened fire with a howitzer but dismally failed to hit her target, a bear ambling along the beach.

BRUNETTE RIVER, New Westminster area (B-8). Named around 1860 by William Holmes, the first settler in Burnaby. He chose this name because the peat lands near the head of the stream had given a dark colour to its water.

BRUNSWICK, Howe Sound (B-8). Takes its name from nearby Brunswick Mountain, which was named after HMS *Brunswick,* one of the ships that fought under Earl Howe on 'The Glorious First of June,' 1794.

BUCCANEER BAY, South Thormanby I. (B-8). After the racehorse Buccaneer, which won the Royal Hunt cup at Ascot in 1861.

BUCKINGHORSE RIVER, flows E. into Sikanni Chief R. (J-8). A horse belonging to a trapping party bucked off its load here.

BUFFALO CREEK, NE of 100 Mile House (D-9). This mysterious Cariboo buffalo was actually an ox, one of a team used by an old Frenchman who freighted on the Cariboo Road.

BUGABOO CREEK, flows E. into upper Columbia R. (C-11). Also BUGABOO PASS. Probably named after the Bugaboo mining claim at the crest of the pass. Writing in 1906 to James White, the federal Chief Geographer, an informant declared, 'Bugaboo was named by a Scotchman on account of the loneliness of the place.'

BULKLEY RIVER, flows NW into Skeena R. (H-6). After Colonel Charles S. Bulkley, US Army, chief engineer of the Western Union Extension Company, 1864-6, during construction of the Collins Overland Telegraph. This project to link the Russian and American telegraph systems by way of Alaska was abandoned upon the success of the trans-Atlantic cable in 1866. The line at this time reached only into central British Columbia. Colonel Bulkley has been described as an able and experienced man, 'universally respected and trusted.'

BULKLEY HOUSE on Takla Lake is also named after Colonel Bulkley.

BULL HARBOUR, Hope I. off northern VI (C-6). The name was in use as early as 1841. It probably refers to the many large and fierce sea lion bulls that frequent this area.

BULL RIVER, flows SW into Kootenay R. (B-12). According to local report, named after J. Bull, an early prospector.

BULMAN, MOUNT, SE of Kamloops (C-9). After Joe Bulman, pioneer rancher in the Westwold area. His son is the author of *Kamloops Cattlemen*.

BUNTZEN LAKE, E. of Indian Arm (B-8). Formerly Trout Lake. After Johannes Buntzen, a native of Denmark, in 1897 appointed the first general manager of the B.C. Electric Railway Company.

BURGESS, MOUNT, N. of Field (D-11). After A.M. Burgess, at one time federal Deputy Minister of the Interior. This mountain is geologically famous for the very ancient fossils in the Burgess Shales, which have been declared a world heritage site because of the remarkable degree of preservation of the fossils.

BURGOYNE BAY, Saltspring I. (A-8). Named in 1859 by Captain Richards, RN, after Commander Hugh Talbot Burgoyne, VC, of HMS *Ganges*. In 1870, in command of HMS *Captain,* Burgoyne perished with more than 500 of his crew when the ship capsized.

BURKE CHANNEL, Central Coast (D- and E-6). Named by Captain Vancouver after Edmund Burke (1729-97), the famous parliamentary orator. He is remembered today chiefly for his speech advocating conciliation with the rebellious American colonies and for his *Reflections on the Revolution in France,* a classic statement of the philosophy of conservatism.

BURKE MOUNTAIN, N. of Coquitlam (B-8). Named after Edmund Burke, the orator, by Captain Richards about 1860. Now forms the southern part of the 38,000 ha Pinecone Burke Provincial Park established in 1995.

BURNABY (B-8). Takes its name from Burnaby Lake. The latter was named after Robert Burnaby (1828-78), the fourth son of the Reverend Thomas Burnaby of Galby, Leicestershire. After seventeen years of service in the Customs House in London, Burnaby came to British Columbia in 1858 with a letter of introduction from the Colonial Secretary, Sir Edward Bulwer-Lytton, and secured an appointment as private secretary to Colonel Moody. In this capacity he was closely associated with the survey that the Royal Engineers were making of the land around New Westminster, and the lake, whose existence had been known previously, was named after him. Burnaby left Moody's employ in August 1859 and joined Walter Moberly in a search for coal around Burrard Inlet.

Dr. Helmcken described Burnaby as a 'myrthful active honest pleasant little fellow.' After engaging in a variety of business enterprises, chiefly in Victoria, he returned to England in 1874.

BURNHAM, MOUNT, W. side of Upper Arrow L. (C-10). After Brigadier-General Frederick L. Burnham, MD (1872-1955). During World War I, he had a distinguished career as a medical officer with the army of Montenegro. In 1924 he acquired the Halcyon Hot Springs Hotel on the east side of the lake, perishing in the fire that destroyed it many years later. He was one of the really notable eccentrics in the area.

BURNS LAKE, S. of Babine L. (G-7). After Michael Byrnes, an explorer for the abortive Collins Overland Telegraph scheme. Byrnes passed this lake around 1866 while surveying a route from Fort Fraser to Hagwilget.

BURNT BRIDGE CREEK, flows SW into Bella Coola R. (E-6). According to the local Indians, the bridge was accidentally set on fire by a white man fearful of bears, who lit a fire at either end of the bridge before camping overnight at the centre. (See *Ulkatcho Stories of the Grease Trail.*)

BURQUITLAM, Fraser Valley (B-8). From the names of adjacent Burnaby and Coquitlam.

BURRARD INLET, N. of Vancouver (B-8). Named by Captain Vancouver in June 1792 after his friend Sir Harry Burrard, RN (1765-1840), a former shipmate aboard HMS *Europa* in the West Indies in 1785. Burrard changed his name in 1795 to Burrard-Neale in consequence of marriage to the heiress Grace Elizabeth Neale, lady-in-waiting to Queen Charlotte. Sir Harry was promoted to vice-admiral in 1814 and served as commander-in-chief of the Mediterranean fleet from 1823 to 1826. He became admiral in 1830.

The Spaniards, who explored the inlet about the same time as Vancouver, called it Boca de Floridablanca in honour of their Prime Minister. To the Indians it was, apparently, Sasamat. (See *Sasamat Lake*.)

BURRELL CREEK, flows S. into Granby R., N. of Grand Forks (B-10). After the Hon. Martin Burrell, a fruitgrower who became, successively, mayor of Grand Forks, an MP, federal Minister of Agriculture, and Librarian of the Parliamentary Library in Ottawa.

BURTON, Lower Arrow L. (B-11). After Reuben Burton, who preempted here in 1893 and became the village's first postmaster. The village was moved to its present site in consequence of the building of the Keenleyside Dam.

BUTE INLET, South Coast (C-7). Named by Vancouver in 1792 after John Stuart, third Earl of Bute (1713-92). An early favourite of George III, Bute was his Prime Minister in 1762-3. He was a civilized man with interests ranging from literature to botany.

BUTLER RANGE, W. of Finlay Reach, Williston L. (I-7). After Captain W.F. Butler, who ascended the Peace River in 1872. He wrote *The Wild Northland* and *The Great Lone Land*.

BUTTLE LAKE, SW of Campbell R., VI (B-7). After John Buttle, who came to British Columbia with the Royal Engineers. He served as a naturalist on Dr. Brown's Vancouver Island Exploration Expedition of 1864 and discovered Buttle Lake the following year while conducting his own exploration party into central Vancouver Island.

BUTZE RAPIDS, Prince Rupert (G-4). After A. Butze, purchasing agent for the Grand Trunk Pacific Railway when it founded Prince Rupert as its western terminus.

C

CAAMAÑO SOUND, Central Coast (E-5). After Jacinto Caamaño, commander of the Spanish corvette *Aranzazu,* which was in these waters in 1792.

CACHE CREEK, flows W. into Bonaparte R. (C-9). Commander R.C. Mayne, in his account of his inland journey in 1859, mentions camping 'by the side of Rivière de la Cache, a small stream flowing into the Bonaparte.' Cache Creek was earlier noted on David Douglas's sketch map of 1833. This latter mention of Cache Creek, many years before the discovery of gold, demolishes Gosnell's explanation that miners cached provisions here, and similarly the story, told in loving detail in Winnifred Futcher's *The Great North Road to the Cariboo,* of how a lone gunman, having murdered a miner travelling south from Barkerville and stolen his eighty pounds of gold, was seriously wounded by a pursuing settler, cached his stolen gold, and disappeared forever, leaving only a riderless horse with a bloody saddle as evidence of his fate. The story has all the marks of a fine Cariboo yarn but is nothing more. All we can say is that, at some time in or before 1833, somebody cached something in the vicinity of Cache Creek. Mary Balf, formerly of the Kamloops Museum, may be right in suggesting that there was a collection point at Cache Creek for furs en route to Thompson's River Post (Fort Kamloops).

Today the word *cache* often refers to a place where supplies have been deposited on a raised platform out of the reach of wild animals. The meaning of the word in French, however, is 'a hiding place,' and the cache of an early fur trader was exactly that. A round piece of turf about eighteen inches across was removed, leaving the mouth for a large bottle-shaped excavation. This excavation was lined with dry branches, and the cached goods were then inserted. Finally some earth and the round piece of turf were put on top and the surplus earth all carefully removed. If the job had been done expertly, possible marauders would see no evidence that they were passing a cache.

CADBORO BAY, E. of Victoria (A-8). After the HBC's brigantine *Cadboro,* the first ship to anchor here. Built at Rye in England in 1824, she took her name from nearby Cadborough, a village that no longer exists in consequence of the silting in of Cadborough Bay. The brigantine first arrived on this coast in 1827. That year she was the first ship to enter the Fraser River and in 1842 was the first ship to enter Victoria harbour. She was wrecked near Port Angeles in 1862.

CADWALLADER CREEK, near Bralorne (C-8). After a Welshman, Evan Cadwallader, who built a sawmill at Lillooet in 1862 and in 1863 guided a company of Italian miners into the valley through which the creek runs, the site later of the Pioneer and Bralorne mines.

CAESARS [CAESAR'S], W. side of Okanagan L. (C-10). After Henry Caesar, pioneer Okanagan settler and tugboat captain.

CALL INLET, near Knight Inlet (C-7). After Sir John Call (1732-1801), engineer-in-chief under Clive of India.

CALLBREATH CREEK, flows W. into Stikine R. (J-4). The Callbreaths were one of the dynastic families of northern British Columbia. When G.M. Dawson travelled through the area in 1887, he found J.C. Callbreath living at Telegraph Creek.

CALVERT ISLAND, Central Coast (D-5). Named by Captain Charles Duncan in 1788. Calvert was the family name of the Lords Baltimore, but the identity of Duncan's Calvert is unknown.

CAMBIE, E. of Sicamous (C-10). Henry J. Cambie (1836-1928) deserves more than this whistle stop on the CPR. Born in County Tipperary, Ireland, he came to Canada while still a boy and learned surveying and railway construction on the Grand Trunk and Intercolonial railways. He came to British Columbia in 1874 expecting to build the Esquimalt and Nanaimo Railway but, when this project was shelved, was placed in charge, from 1876 to 1878, of all the surveys between Burrard Inlet and Yellowhead Pass for possible routes for the CPR. In 1879 he personally made a survey of the Peace River country. During actual construction of the CPR, he was in charge of the Fraser Canyon stretch and of that between Savona and Shuswap Lake. The transcontinental line completed, Cambie became the engineer in charge of the Pacific Division and built the New Westminster and Nicola extensions. He was a fine man, and the men who worked for him always remembered the care he took to see that they were properly fed and housed. Cambie Street in Vancouver is also named after him.

CAMERON LAKE, E. of Port Alberni (B-7). After David Cameron, Chief Justice of Vancouver Island. Cameron started work as a cloth merchant in Perth, Scotland. Later he grew sugarcane in Demerara (Guyana), where he married Cecilia, sister of Sir James Douglas. In 1853 Douglas obtained for him a post in the management of the HBC coal mines at Nanaimo. Soon after Cameron and his family arrived at Nanaimo, Douglas appointed him as a temporary judge. In 1856 the Colonial Office, on Douglas's recommendation, approved Cameron's appointment as Chief Justice of Vancouver Island, an action that contributed to the charges of nepotism levied against Douglas. In fact, Cameron was a successful judge, even though he had never been trained in the law.

CAMPANIA ISLAND, Central Coast (F-5). Named by Caamaño in 1792. In classical times a large area in southern Italy was called Campania.

CAMPBELL BAY, Mayne I. (A-8). After Samuel Campbell, MD, assistant surgeon on HMS *Plumper* from 1857 to 1861. (See *Campbell River, Samuel Island.*)

CAMPBELL CREEK, E. of Kamloops (C-9). Originally the San Poel (Sans Poil?) River. Now named after Louis (or Lewis) Campbell, an American cattle drover who began building a fine ranch here in the 1860s, one that ultimately extended for about six miles along the south bank of the South Thompson River.

CAMPBELL RIVER, N. of Courtenay (C-7). Although certain proof cannot be found, it seems likely that this river, and consequently the town near its mouth, were named after Dr. Samuel Campbell, assistant surgeon on HM survey ship *Plumper*. Dr. Campbell was on the *Plumper* during the period 1857-61, when she surveyed Johnstone Strait along with other parts of our coast. (See *Campbell Bay, Samuel Island.*)

CAMP McKINNEY, S. of Baldy Mtn. (B-10). After Alfred McKinney, who discovered the Cariboo mine here. The post office, which opened in 1895, closed in 1912. Thereafter Camp McKinney became one of British Columbia's more noted ghost towns.

CANAL FLATS, S. of Columbia L. (C-12). This expanse of low ground separating the Kootenay River from nearby Columbia Lake (source of the Columbia River) was originally named McGillivray's Portage by David Thompson, who passed this way in 1808. Its present name commemorates the canal, parts of which can still be seen, completed in 1889 by William Adolph Baillie-Grohman, British sportsman and capitalist. This canal was part of a perfectly feasible scheme to divert some of the water of the upper Kootenay River into the Columbia system and thus to lower the level of Kootenay Lake sufficiently to reclaim the rich alluvial plain adjacent to Creston. Unfortunately, under pressure from the CPR, concerned about its Columbia River crossings, and from settlers around Golden who feared that their hay meadows would be flooded, the Canadian government so modified the original plans as to render the canal more costly and limited in usefulness. Baillie-Grohman finally abandoned the whole project in disgust. In 1894 the *Gwendoline* and in 1902 the *North Star* successfully passed from the Kootenay River to the Columbia using Baillie-Grohman's canal.

CANFORD, Nicola R. (C-9). Named by a local rancher, Theophilus Hardiman, after Canford Manor, near Bournemouth, England.

CANICHE PEAK, Mount Robson Park (E-10). A.O. Wheeler suggested the name Poodle Peak because the summit resembles in shape the head of a poodle. Subsequently the French word for 'poodle' was chosen to give the name more class.

CANIM LAKE, NE of 100 Mile House (D-9). In the Chinook jargon, *canim* means 'canoe.'

CANOE, NE of Salmon Arm (C-10). Reputedly early white travellers were impressed by the number of Indian dugout canoes drawn up on the fine beach here.

CANOE PASS, S. of Westham I. (B-8). Gold miners of the 1858 rush who hoped to elude naval patrols checking on mining licences are said to have used this minor entrance into the Fraser. Perhaps they tried to sneak by in canoes.

CANOE RIVER, flows E. into Kinbasket L. (E-10). Given this name in 1811 by David Thompson. At Boat Encampment, near the confluence of this river and the Wood River with the Columbia, Thompson built the canoes in which he travelled to the Pacific. Since the building of Mica Dam, the lower Canoe River has been swallowed up in Canoe Reach of Kinbasket Lake, and Boat Encampment is no more.

CAPILANO RIVER, flows S. into Burrard Inlet (B-8). This is the anglicized form of a Squamish and Musqueam Indian personal name, for which no meaning is known. The last man known as Chief Capilano, who died about 1870, was of Squamish and Musqueam ancestry, and one of his homes was beside the river that now bears his name.

CAPTAIN HARRY LAKE, E. side of Tweedsmuir Park (F-7). In remote Indian villages, the 'captain' is the religious assistant who serves in the place of the priest, who can make a visit only once or twice a year. When Chess Lyons passed through this country in 1941, he found Captain Harry the Indian with whom one dealt on the Algacho reserve.

CAREN RANGE, Sechelt Pen. (B-8). Caren is a long-standing misprint or misinterpretation of 'Carew,' the range being named after Sir Benjamin Hallowell Carew (see *Hallowell, Mount*). The Caren Range is noted for its ancient trees – one yellow cedar is over 1,800 years old.

CARIBOO, THE, east-central BC (D and E-8 and 9). A regional name first applied to the goldfield area around Quesnel and Barkerville but now generally extended to cover the country between Cache Creek and Prince George. Visitors to the province should note that it is an unforgivable solecism to spell the name 'the Caribou.' In a despatch of September 1861 to the Duke of Newcastle, Governor Douglas mentioned: '... [the] Cariboo country, in speaking of which I have adopted the popular term and more convenient orthography of the word, though properly it should be written "Cariboeuf," or "Reindeer," the country having been so called from its being the favourite haunt of that species of the deerkind.' As Douglas indicates, 'Cariboo' derives from *cariboeuf* or *cerfboeuf*, which is a French folk etymology for *xalibu*, an Algonquin Indian word meaning the 'pawer' or 'scratcher.'

CARIBOU HIDE, junction of Stikine R. and Moyez Cr. (J-6). Tommy Walker noted in his book *Spatsizi* that 'other Indians referred to the Sekanis as the Caribou Hide people, which accounts for the name of Caribou Hide ... the site of their first settlement.'

CARL BORDEN, MOUNT, SE of Tahtsa L. (F-6). After Dr. Charles E. Borden (1905-78), Professor of German at UBC, who, taking up archaeology as a hobby in middle life, made it his life work and became the father of scientific archaeology in British Columbia.

CARLIN, N. of Salmon Arm (C-10). After Michael Burns Carlin (1857-1934), manager of the Columbia River Logging Co., who established a logging operation here, complete with a logging railroad.

CARMANAH POINT, entrance to Strait of Juan de Fuca (A-7). This point was named after the nearby Nitinaht Indian village, whose name can be translated as 'thus far upstream.'

CARMI, West Kettle R. (B-10). After the Carmi Mine, which James Dale named after his birthplace, Carmi, in Illinois.

CARNARVON, MOUNT, Yoho NP (D-11). After Lord Carnarvon (1831-90), British Colonial Secretary. His 'Carnarvon Terms' of 1874 settled the disputes that had arisen between British Columbia and the dominion government during the years immediately after the province's entry into Confederation.

CARNES CREEK, flows W. into Columbia R. (D-10). After H.C. (Hank) Carnes, one of the miners here during the Big Bend gold rush of 1865.

CARO MARION, MOUNT, N. of Ocean Falls (E-6). After Caro Audres and Marion Black, two of four young people who climbed to its summit in 1911.

CARP LAKE, NE of Fort St. James (G-8). Simon Fraser makes mention in his journal, in June 1806, of Carp Lake, 'where there are immense numbers of fish of the Carp kind.'

CARPENTER LAKE, NW of Seton L. (C-8). After E. Carpenter, who came to British Columbia from the United States in 1909. He did much of the engineering design in connection with the Bridge River power project between 1927 and 1931.

CARQUILE, NW of Cache Cr. (C-9). A misspelling of Carguile. After Henry Carguile, who acquired the Hat Creek Ranch and ran a roadhouse here.

CARRIER LAKE, E. of Stuart L. (G-8). After the Carrier Indians, who received their name from a curious custom. When a man died, his widow was required to gather up the few pieces of bone remaining after his cremation, put them in a leather bag, and carry them on her back everywhere until the kinsmen of the deceased had accumulated enough food and property to hold a potlatch in his memory.

CARROLLS LANDING [CARROLL'S LANDING], E. bank of Columbia R. (C-11). After 'Mike' Carroll of Nakusp, who owned land here.

CARRUTHERS CREEK, flows SE into Omineca R. (I-6). After Dr. Carruthers, an Englishman who got lost here but was found.

CARSON, MOUNT, N. of Pavilion L. (C-9). After Robert Carson, a young Scot who worked as a packer during the Cariboo gold rush. In 1872 he took up land here and founded the Carson Ranch. One of his sons, E.C. Carson, became provincial Minister of Public Works, while another, R.H. Carson, became Speaker of the Legislative Assembly.

CARTIER, MOUNT, SE of Revelstoke(C-10). After Sir George-Étienne Cartier (1814-73), the leading French-Canadian champion of Confederation.

CASCADE RANGE, from S. of Lytton to US border (B- and C-9). This range separates the Fraser Canyon from the Okanagan. It takes its name from 'the Cascades' in the United States, where the Columbia River breaks through this range to reach the sea.

CASEY COVE, Prince Rupert (G-4). After Major W.A. Casey, topographical engineer with the Grand Trunk Pacific Railway, who camped on nearby Casey Point in 1906. He was killed in France in 1916.

CASSIAR MOUNTAINS, S. of Yukon-BC border (K- and L-5). Derived from the word *kaska*, a corruption of the Indian name of McDame Creek (a tributary of Dease River), where the Kaska Indians assembled in summer to fish and trade. One anthropologist says that *kaska* means 'old moccasins,' a term of scorn that Tahltan Indians applied to the neighbouring Kaska Indians in the Dease River area.

CASSIDY, S. of Nanaimo (B-8). After Tom Cassidy, who arrived here from Iowa in 1878 and was soon joined by his wife. A son later took over the farm and worked it until most of the land was needed for the airport.

CASTLEGAR, near confluence of Columbia R. and Kootenay R. (B-11). The available evidence favours the derivation of Castlegar from 'Castle Garden,' an immigration station in New York state. However, there is a Castlegar in Ireland.

CATALA ISLAND, Esperanza Inlet, VI (B-6). Magin Catala was an exemplary Franciscan friar who served as chaplain to the Spanish garrison at Nootka.

CATFACE RANGE, N. of Tofino (B-7). Takes its name from the cat's-face appearance of a mineral deposit showing on a principal mountain.

CATHEDRAL LAKES, SW of Keremeos (B-9). Take their name from Cathedral Peak, just south of the international boundary. The latter was named in 1901 by two brothers, Carl and George Smith, members of a US Geological Survey party. They made the first ascent of the mountain, which impressed them as looking 'something like a big church.'

CATHEDRAL MOUNTAIN, E. of Field (D-11). The appearance of this peak, one of the most sublime in the Canadian Rockies, won it this name in 1884.

CAULFEILD, West Vancouver (B-8). After Francis William Caulfeild, a most attractive English gentleman in the old tradition. In 1899, won by the beauty of the wild flowers, woods, and beaches of the area, he bought the land between Cypress Creek and Point Atkinson. Himself a lover of beautiful English villages, he sought to create one here. Rejecting the uncompromising grid universally imposed on North American settlements, he laid out winding streets that followed the natural contours of the wooded slopes. He made a public park out of the strip of land on the waterfront. Caulfeild died in 1934, aged ninety.

CAUTION, CAPE, N. of Queen Charlotte Strait (D-6). Wrote Captain Vancouver, 'This cape, from the dangerous navigation in its vicinity, I distinguished by the name of Cape Caution.'

CAWSTON, SE of Keremeos (B-10). After R.L. Cawston (1849-1923), pioneer rancher and stipendiary magistrate.

CAYCUSE RIVER, flows W. into Nitinaht L. (A-7). From the Nitinaht Indian word meaning 'place where they fix up canoes.'

CAYOOSH CREEK, flows into Seton R. (C-9). A variant of *cayuse*, a word widely used to refer to an Indian pony. The story goes that one day an Indian from Mount Currie rode to Lillooet on horseback and that his horse dropped dead at what is now Cayoosh Creek. (See *Lillooet*.)

CEDARVALE, Skeena R., N. of Terrace (H-5). This settlement began as Minskinish, a Victorian missionary village like William Duncan's Metlakatla. The benevolent despot here was the Reverend R. Tomlinson, who founded the village in 1888. The Sabbath was strictly observed in Minskinish. No work of any sort was done on Sunday, no visitor arrived or departed, and if a riverboat arrived that day, it merely left the mail on the bank of the Skeena, where it was not touched until Monday. It is not surprising that the village was nicknamed 'Holy City.' *Minskinish* means 'at the base of the jackpine.' The present name of Cedarvale comes from the cedars that grow around the landing.

CEEPEECEE, N. of Nootka I. (B-6). From the initials of the California Packing Corporation, which opened a fish-packing plant here in 1927.

CELISTA, Shuswap L. (C-10). When a post office was to be established here in 1908 and the residents could not agree on a suitable name for it, the postal authorities suggested that it be named after Celesta Creek. Although this creek is many miles away, the name was adopted, but due to an error in reading handwriting, the name came out as 'Celista.' Celesta Creek is an anglicization of a well-known Shuswap ancestral name, Selesta, which appears on the 1877 Indian reserve census by George Blenkinsop, with people of this name noted as living at Little Shuswap, Neskainlith, and Kamloops. Prominent among the bearers of this name was William Celesta (see Bouchard and Kennedy, *Shuswap Stories,* p. 126), a 'doctor' who came from Salmon Arm, had land on a Neskainlith reserve west of Chase, and died in the late 1940s.

CENTURY SAM LAKE, headwaters of Comox Cr. (B-7). Commemorating Comox district's Sid Williams, who, as 'Century Sam,' was the province's mascot during its 1958 centennial. He not only dressed up as an old-time gold prospector, but actually had a small gold claim on Forbidden Plateau.

CHABA PEAK, Alberta-BC boundary (E-11). Stoney Indian word for 'beaver.'

CHAMPION LAKES PARK, NE of Trail (B-11). These lakes are named after James W. Champion. A 1910 directory lists him as a fruit grower at Fruitvale, also in this district.

CHANCELLOR PEAK, Yoho NP (D-11). After Sir John Boyd (1837-1916), Chancellor of Ontario, who was one of the arbitrators of the dispute between the federal government and the CPR in 1886. (Until 1916 Ontario had a Chancery Division to its Supreme Court. This, like the Chancery Court in England, was presided over by a chancellor and based its judgments upon equity rather than the letter of the law.)

CHANTSLAR LAKE, NW of Tatla L. (E-7). From a Chilcotin Indian word meaning 'steelhead lake.'

CHAPMAN CAMP, S. of Kimberley (B-12). After F. Chapman, at one time the Consolidated Mining and Smelting Company's superintendent of construction at Kimberley.

CHAPMANS [CHAPMAN'S], Fraser Canyon (B-9). Named after Chapman's Bar, which in turn was named after the owner of a roadhouse on the old Cariboo Road. The roadhouse stood on the site of today's Alexandra Lodge.

CHAPPERON LAKE, NE of Douglas L. (C-9). After François Chapperon, a packer who used to keep his horses near here. In 1883 he sold the property, which eventually became part of the Douglas Lake Ranch.

CHARACTER COVE, E. of Imperial Eagle Channel, Barkley Sound (A-7). When a university course entailing diving and water collecting after midnight was given at nearby Bamfield Marine Station, the class's enthusiasm began to wane. To hearten them Professor Brian Marcotte kept chanting 'Oceanography builds character.' Hence the name given to this little bay.

CHARLIE LAKE, NW of Fort St. John (I-9). The last of the Beaver Indian prophets (see *Prophet River*) was Charlie Yahey, who died in the early 1970s. This lake was named after his father, who was known as Charlie Chok or Big Charlie.

CHARLIEBOY, MOUNT, NW of Puntzi L. (E-7). After an old chief of the Redstone Indians.

CHARLOTTE LAKE, NW of Kleena Kleene (E-7). Said to be named after the wife of a surveyor.

CHASE, W. end of Little Shuswap L. (C-10). After Whitfield Chase (1820-96). A native of New York state, Chase came west over the Oregon Trail in 1852. For some years he worked as a carpenter and unprofitably prospected for gold on the Fraser and Thompson Rivers. In 1865 he established a ranch where the South Thompson runs out from Little Shuswap Lake. He married Elizabeth, a Shuswap Indian woman, 'a wonderful person.' He prospered and raised a large family. In 1908, when the Adams River Lumber Company laid out the townsite, James A. Magee, secretary of the company, declined to have the new settlement named after himself and insisted that it should be given the name of the first settler in the district.

CHASE RIVER, S. of Nanaimo (B-8). In November 1852 Peter Brown, a young Orkneyman employed as a shepherd in the Saanich area, was murdered by two Indians. One of them, a Cowichan, was soon captured. The other, belonging to the Nanaimo band, was pursued up this stream by nine of the HBC's French-Canadian scouts, aided by sixteen sailors from the Royal Navy. After a desperate chase, the Nanaimo tried to shoot it out on the banks of the river. However, his powder was wet, and he was soon captured. With his accomplice he was hanged on Gallows Point near Nanaimo, his chase and death adding two more items to our list of British Columbia place names.

CHATFIELD ISLAND, N. of Bella Bella (E-5). After Captain Alfred J. Chatfield, commanding HMS *Amethyst* when the Earl of Dufferin, Governor-General of Canada, and his countess made their cruise to British Columbia's northern waters in the summer of 1876.

CHATHAM ISLANDS, E. of Oak Bay (A-8). After HMS *Chatham,* an armed tender of 135 tons, carrying four three-pounder cannons and six swivel guns, the consort of HMS *Discovery* during Captain Vancouver's expedition of 1791-5. Not named by Vancouver.

CHATHAM SOUND, between Portland Inlet and Porcher I. (G-4). After John Pitt, second Earl of Chatham, First Lord of the Admiralty 1788-94.

CHEADLE, MOUNT, NE of Blue R. (E-10). After W.B. Cheadle (1835-1910), the English physician who travelled this way in 1863 with Viscount Milton during their overland journey to the Pacific. Like many other travellers through British Columbia, they found little or no game down in the river valleys in summer and almost died from hunger before they reached the HBC fort at Kamloops.

CHEAKAMUS RIVER, N. of Squamish (B-8). From the Squamish Indian word meaning 'salmon weir place.'

CHEAM, E. of Chilliwack (B-9). Cheam (pronounced Chee-ám) is derived from the Halkomelem word meaning '[place to] always get strawberries.' This name refers specifically to the island across from the village and present reserve. Whites have also used the name for today's Cheam Peak, but to the Indians the latter is Thleethleq, the wife of Mount Baker, who came to Thleethleq with her sisters, dog, and daughter and was turned to stone.

CHECLESET BAY, E. of Brooks Pen. (C-6). From the Nootka Indian word meaning 'people of cut on the beach.'

CHEEKYE RIVER, flows W. into Cheakamus R. (B-8). From the Squamish Indian name for Mount Garibaldi, meaning 'dirty place' (perhaps because of the appearance of old snow on the mountain). This name was later applied to the river.

CHEEWHAT RIVER, S. of Nitinaht L. (A-7). From a very old Nitinaht Indian word meaning 'having an island nearby.'

CHEHALIS RIVER, flows S. into Harrison R. (B-9). Chehalis is also found as a place name in the Cassiar and in southern Washington state. There is a great range of meanings offered, the two most probable ones being 'the place one reaches after ascending the rapids,' and 'where the "chest" of a canoe grounds on a sandbar' – the 'chest' of a dugout being its widest part, directly behind the bow. Linguistic evidence favours the second interpretation, but two old Indian women have quite independently proffered the first.

CHELASLIE ARM, Natalkuz L. (F-7). From the Carrier Indian language. Probably means 'lake surrounded by timber.'

CHEMAINUS, SE of Ladysmith (A-8). From the Island Halkomelem word meaning 'bitten breast.' The horseshoe shape of the bay reminded Indians of the bite that a frenzied shaman would inflict upon a spectator during certain tribal ceremonies.

CHERRY CREEK, flows NW into Kamloops L. (C-9). Originally Cherry Bluff Creek, the stream took its name from the chokecherries found in abundance locally. In 1860 two retired HBC men founded the first ranch here.

CHESLATTA LAKE, S. of François L. (F-7). From the Carrier Indian word meaning either 'top of small mountain' or 'small rock mountain at east side.'

CHETWYND, W. of Dawson Cr. (H-9). Formerly Little Prairie but renamed after Ralph L.T. Chetwynd. The son of an English baronet, Chetwynd came to Canada at the age of eighteen and became manager of the Marquess of Anglesey's holdings at Walhachin (q.v.). He won the Military Cross during World War I. Entering politics, he became Minister of Railways. Speaking on a proposed extension of the PGE Railway (now the BC Railway), he bet a hat that the first train for the Peace River would leave North Vancouver on 11 June 1956 at 4:15 p.m. Since he soon had bets for $800 worth of Stetson hats, it was fortunate for him that the train did pull out on time.

CHEZACUT, N. of Chilcotin L. (E-7). This settlement preserves the earlier name of Chilcotin Lake. Chezacut is derived from a Chilcotin Indian word meaning 'birds without feathers' – geese go to moult here (during which time they cannot fly).

CHIC CHIC BAY, S. coast of Calvert I. (D-5). *Chic chic* is Chinook jargon for any wheeled vehicle, such as a wagon. It is rather unlikely that this name owes anything to the chickens that Franz Heinrich, an early settler here, raised.

CHIKAMIN RANGE, between Whitesail L. and Eutsuk L. (F-6). *Chikamin* is Chinook jargon for money or silver. Presumably the name was given because of mineral wealth sought in the area, or maybe because of the glittering appearance of the snow-capped mountains.

CHILAKO RIVER, flows NE into Nechako R. (F-8). From the Carrier Indian word meaning 'beaver hand river.'

CHILANKO RIVER, flows SE into Chilcotin R. (E-8). From a Chilcotin Indian word meaning 'many beaver river.'

CHILCOTIN RIVER, flows E. into Fraser R. (D-8). The name can be translated as 'ochre river people.' Ochre here refers not to the colour but to the mineralized substance (usually red or yellow) much prized by the Indians for use as a base for paint or dye.

CHILKAT PASS, NW of Haines, Alaska (L-l). A Tlingit Indian word meaning 'salmon storehouse.'

CHILKO RIVER, flows NE into Chilcotin R. (E-8). From a Chilcotin Indian word meaning 'ochre river.' (See *Chilcotin River.*)

CHILLIWACK, Fraser Valley (B-9). The name of the local Indian band and of various geographical features. This Halkomelem word has the sense of 'quieter water on the head,' or 'travel by way of a backwater or slough.' The name was originally pronounced ch.ihl-KWAY-uhk. Early spellings of the name include Chillwayhook, Chil-whey-uk, Chilwayook, and Silawack.

CHIMNEY CREEK, flows W. into Fraser R., S. of Williams L. (E-8). Possibly after a chimney-like rock at the mouth of the creek, or from an old chimney, all that survives of a former building.

CHINA BAR, Fraser Canyon (B-9). After the Chinese men who once washed for gold here. Many Chinese worked in the goldfields. Often they came after the whites had moved on to new areas, and by their diligence they obtained considerable gold in reworking supposedly exhausted areas.

CHINOOK COVE, N. Thompson R. (D-9). This CNR station was formerly named Genier. Its more recent name probably refers to chinook (spring) salmon, which travel hundreds of miles from the sea to spawn.

CHOATE, S. of Yale (B-9). After James Z. Choate, CPR bridge-construction foreman in early days. (See *Dogwood Valley*.)

CHOELQUOIT LAKE, N. of Chilko L. (D-7). From a Chilcotin Indian word meaning 'fishtrap lake.'

CHONAT BAY, NW end of Quadra I. (C-7). From the Kwakwala Indian word meaning 'where coho salmon are found.'

CHRISTIAN VALLEY, Kettle R. (B-10). After Joseph Christian, the first settler here.

CHRISTINA LAKE, NE of Grand Forks (B-10). Named, apparently around 1860, after Christina, the mixed-blood daughter of Angus McDonald, the HBC factor at Fort Colvile. She often accompanied her father on his field trips. On one of them, she plunged into the flooded stream running out from the lake and rescued her father's papers. In recognition of her courage, the lake was named after her.

CHU CHUA, on the CNR, N. of Barrière R. (D-9). This name is the plural form of the Shuswap Indian term for 'creek.'

CHUCKWALLA RIVER, flows S. into Rivers Inlet (D-6). From the Oowekyala Indian word meaning 'short river' – as opposed to nearby Kilbella River, meaning 'long river.'

CHURCH HOUSE, SE approach to Bute Inlet (C-7). Formerly only a small and unimportant Indian village, but after severe Bute Inlet winds destroyed a village on nearby Sonora Island, the homeless people moved to this more sheltered spot. It then became the main Homathko settlement, a new church was built, and the village acquired its new name of Church House.

CHUTE LAKE, NE of Penticton (B-10). From the French word for a waterfall or rapid.

CHUTINE RIVER, flows SE into Stikine R. (J-4). *Chutine* is an Indian word meaning 'half-people' (i.e., the population in the area was half-Tlingit and half-Tahltan).

CINEMA, N. of Quesnel (F-8). Asked why he named this settlement Cinema, Dr. Lloyd Champlain is reported to have replied, 'Cinema means action. Cinema is pictures in motion, and that is what we are, action.' On the other hand, according to a local history published by the Hixon Women's Institute, Dr. Champlain and his housekeeper hit on the name to commemorate a trip they made to Hollywood around 1920. Cinema post office opened on 2 January 1924 and closed on 12 January 1964.

CINNEMOUSUN NARROWS, Shuswap L. (C-10). In July 1865 Walter Moberly camped 'at the narrows called Cium-moust-un, which means "come and go back again."' The meaning of this anglicized Shuswap Indian word given by G.M. Dawson in the next decade was 'the bend.'

CLACHNACUDAINN RANGE, NE of Revelstoke (D-10). After the lozenge-shaped rock that for centuries was the palladium of Inverness in Scotland. Clach-na-Cudainn (Stone of the Tubs) was so named since the women of the town rested their tubs on it when bringing water from the river.

CLANWILLIAM LAKE, SW of Revelstoke (C-10). After the Earl of Clanwilliam, who married a daughter of Governor Kennedy of Vancouver Island. (See *Gilford Island*.)

CLAYBURN, NE of Abbotsford (B-8). Named by J.C. and F.S. Maclure, who established the clayworks here in 1904.

CLAYHURST, E. of Fort St. John (I-9). Named in 1932 after W. Clay, the first English-speaking settler in what had been a predominantly Ukrainian area.

CLAYOQUOT SOUND, N. of Tofino (B-6). After the Clayoquot Indians, whose name has been given a remarkable range of translations. G.M. Sproat (see *Sproat Lake*) said that it meant 'another [i.e., different] people,' while Dr. Brown (see *Brown's River*) thought that it meant 'other or strange house.' The missionary Father Brabant told Captain Walbran that Clayoquot meant 'people who are different from what they used to be,' and he noted a tradition that these Indians were originally quiet and peaceful but later became quarrelsome and treacherous. A modern linguist, John A. Thomas of the Nitinahts, says that it means 'people of the place where it becomes the same even when disturbed.'

John Jewitt, at the beginning of the nineteenth century, described the Klaa-oo-quates as 'fierce, bold and enterprizing.' Clayoquot was first known to white men as Port Cox, after John Henry Cox of Canton, who backed several fur-trade ships sent to this area.

CLEARBROOK, W. of Abbotsford (B-8). Originally named Pinecrest after the nearby Pinecrest Auto Court. The name had no sooner been made official in late 1952 than local people began to protest. Accordingly the name was quickly changed to Clearbrook.

CLEMENCEAU ICEFIELD, S. of Jasper NP (E-10). After Georges ('The Tiger') Clemenceau, Premier of France 1906-9 and 1917-19.

CLEMINA, S. of Valemount (E-10). After Clemina Buckle, wife of Wilfred Buckle, a surveyor employed during construction of the Canadian Northern (now Canadian National) Railway.

CLEMRETTA, N. shore of François L. (G-6). After two cows, Clementine and Henrietta, belonging to the first postmaster here. He coined the name in desperation after all the names that he had suggested were rejected by Ottawa since they duplicated the names of post offices elsewhere in Canada.

CLEVELAND DAM, North Vancouver (B-8). Named in 1954 after E.A. Cleveland, first chief commissioner of the Greater Vancouver Water District.

CLINTON, N. of Cache Creek (D-9). Originally known as Cut-Off Valley or 47 Mile House. In 1863, upon completion of the Cariboo Road, it was given its present name in honour of Henry Pelham Clinton, fifth Duke of Newcastle (and Colonial Secretary from 1859 to 1864).

CLOAK BAY, Langara I., QCI (G-3). So named in June 1787 by Captain Dixon because of the many cloaks, made of sea otter or beaver skins, that he received here in trade with the Indians.

CLO-OOSE, NW of Carmanah Point (A-7). Derived from the Nitinaht Indian word meaning 'campsite beach.' Before the village was built, the site was a favourite resting spot for halibut fishermen.

In the years before World War I, a Victoria real estate promoter sold nearby land to well-to-do English families, some of whom built fine houses. Evidently not happy with the Indian name Clo-oose for their postal address, they petitioned unsuccessfully to have the name changed to Clovelly.

CLOVER POINT, Victoria (A-8). James Douglas appears to have landed here from SS *Beaver* in March 1843 when he arrived to found Fort Victoria. Douglas himself gave the point this name because he found an abundance of red clover growing here.

CLOVERDALE, W. of Langley (B-8). Originally named Clover Valley by William Shannon, a pioneer settler who later wrote: 'A few days after arriving in the valley in the year 1875 I had occasion to write a letter and was in doubt as to how to head it. I looked out at the wild clover which grew luxuriantly everywhere and at once thought of "Clover Valley."'

CLUCULZ LAKE, W. of Prince George (F-8). Based on the Carrier Indian word meaning 'carp lake,' or 'place of big whitefish.'

CLUSKO RIVER, flows S. into Chilcotin R. (E-7). From a Southern Carrier word meaning 'mud river.'

CLUXEWE RIVER, W. of Port McNeill (C-6). According to Chief James Sewid, *cluxewe* is a Kwakwala Indian word meaning 'delta, or sand bar.'

COAL HARBOUR, Quatsino Sound (C-6). The coal deposits here were first investigated in 1883-5. When a coal mine was subsequently established here, it was soon closed because of the poor quality of the coal.

COAL HARBOUR, Vancouver (B-8). Takes its name from the discovery of veins of coal here in 1859. The veins ranged from four to fifteen inches in thickness.

COALMONT, NW of Princeton (B-9). So named because it was believed that there was a mountain of coal here which could be strip-mined.

COBBLE HILL, SE of Duncan, VI (A-8). There are no cobblestones on Cobble Hill, the formation here being of limestone. Two accounts of the naming of Cobble Hill have their rival champions locally. One says that the hill was named after a Lieutenant Cobble, RN, 'probably an officer on one of the gunboats which made occasional visits to Cowichan Bay when the settlement was apprehensive of Indian trouble.' Unfortunately no Lieutenant Cobble can be found in the navy lists. The other story is that, when the Esquimalt and Nanaimo Railway was being built, a visiting Englishwoman said that the place reminded her of a Cobble Hill in England, and so the name was adopted. It has also been claimed that the name comes from gravel hills in the vicinity.

COCCOLA, MOUNT, W. of Bear L. (I-6). After the fiery Corsican priest Father Nicholas Coccola, OMI, who – from 1880 until his retirement in 1934 – laboured in the mission field in British Columbia, first in the Kamloops area, then in East Kootenay, and finally in the region of Fort St. James. He died in 1943.

COCKBURN, CAPE, S. of Jervis Inlet (B-7). After Admiral Sir George Cockburn (1772-1853).

COFFEE CREEK, flows E. into Kootenay L. (B-11). From the muddy water resulting from an early mining operation on the creek.

COGLISTIKO RIVER, flows NE into Baezako R. (F-8). From a Carrier Indian name meaning 'stream coming from small jack-pine windfalls.'

COLDSTREAM, SE of Vernon (C-10). The land along this creek was originally preempted in 1863 by Colonel Charles F. Houghton of the 20th Regiment of Foot. (Trutch's map of 1871 shows 'Houghton's Coldstream.') In 1869 the land passed to the Vernon brothers, and in 1891 Forbes George Vernon sold the estate to Lord Aberdeen, who planted here the first orchards in the Okanagan.

G.M. Dawson, a visitor in 1877, reported that on 8 July of that year he found the temperature of the springwater at the head of Coldstream Creek to be 48.5°F.

COLES BAY, W. side of Saanich Pen. (A-8). After John Coles, who first came to British Columbia in 1851 as a midshipman on HMS *Thetis*. He left the navy after the Crimean War and returned to British Columbia, living on his farm near Coles Bay from 1857 to 1866 and representing Saanich in the first Legislative Assembly. Back in England he was the curator of the Royal Geographical Society for many years.

The name of the Saanich Indian reserve at Coles Bay, Paquechin, means 'drop off.'

COLLEY, MOUNT, N. of François L. (G-6). Also COLLEY-MOUNT PARK. After Edward Pomeroy Colley, who surveyed land in this area in 1905 and 1906. He perished on the *Titanic*.

COLLINGWOOD CHANNEL, Howe Sound (B-8). After Vice-Admiral Lord Collingwood. While still a captain, Collingwood commanded one of Lord Howe's warships at the battle of 'The Glorious First of June,' 1794. A personal friend of Lord Nelson, Collingwood was his second in command at the Battle of Trafalgar and took over when Nelson died, receiving his peerage for his part in the great victory. He is buried by the side of Nelson in St. Paul's Cathedral.

COLLINSON POINT, SW tip of Galiano I. (A-8). Named after William Tomkins Collinson, JP, who arrived in British Columbia from England in 1858. After spending some time in the Cariboo and the Fraser Valley, he moved to Mayne Island in 1871 with Mary, his Indian wife. In 1880 he became the first postmaster on the island.

COLLISON BAY, E. coast, Moresby I. (E-4). After Archdeacon William Henry Collison, the Anglican apostle to the Haidas, who began his missionary endeavours among them in 1876. Author of *In the Wake of the War Canoe*.

COLONEL FOSTER, MOUNT, Strathcona Park (B-7). After Colonel, later Major-General, William W. Foster (1875-1954). He came to British Columbia from England in 1894 and rose to become Deputy Minister of Public Works before serving with distinction in World War I (DSO with two bars). He was president of Pacific Engineers Ltd. in 1935 when a reform mayor made him Vancouver's chief of police, charged with reorganizing the force. In World War II, he was special commander for defence projects in northwest Canada.

In 1913 'Billy' Foster, an ardent alpinist, along with A.H. MacCarthy, made the first undisputed ascent of Mount Robson. From 1922 to 1924, he was president of the Alpine Club of Canada. (See *Foster Peak.*)

COLQUITZ RIVER, flows into Portage Inlet, Victoria (A-8). From the Straits Salish word meaning 'waterfall.'

COLUMBIA GARDENS, SE of Trail (B-11). So named because of a large orchard planted here to show that smoke from the Trail smelter would not harm the trees.

COLUMBIA RIVER, SE BC. First named the Rio de San Roque by the Spaniards after Bruno de Hezeta discovered the river's mouth in 1775. It was rediscovered in 1792 by Captain Robert Gray, an American, who named it after his ship, the *Columbia.*

COLWOOD, W. of Victoria (A-8). In 1851, when Captain E.E. Langford arrived from England to manage the Esquimalt farm of the Puget Sound Agricultural Company, a subsidiary of the HBC, he named the farmhouse Colwood after his home in Sussex.

COMIAKEN, N. of Duncan (A-8). From Comiaken Hill, *comiaken* being a Cowichan Indian word probably meaning 'bare, or void of vegetation.'

COMMERELL POINT, NW coast of VI (C-5). In 1862 Captain G.H. Richards gave the most northerly tip of Vancouver Island the name of Cape Commerell, after Admiral of the Fleet Sir John Edmund Commerell, GCB, VC. In 1905 the Geographic Board of Canada restored to this promontory the earlier name, Sutil Point, given to it by the Spaniards in 1792, and made the southern tip of Raft Cove Commerell Point.

COMMITTEE PUNCH BOWL, Athabasca Pass (E-10). On the old trail by which the fur-company brigades once travelled across the Rockies.

Governor Simpson of the HBC tells how he named the lake when he passed that way in 1824:

> At the very top of the pass or height of Land is a small circular Lake or Basin of water which empties itself in opposite directions and may be said to be the source of the Columbia & Athabasca Rivers as it bestows its favors on both

these prodigious Streams ... That this basin should send its Waters to each side of the Continent and give birth to two of the principal Rivers in North America is no less strange than true both the Dr. [John McLoughlin] & myself having examined the currents flowing from it East & West and the circumstance appearing remarkable I thought it should be honored by a distinguishing title and it was forthwith named the 'Committee's Punch Bowl.'

The 'committee' was presumably the managing committee of the HBC.

According to one account, the officers in charge of HBC parties customarily served their men with punch when they reached the top of the Great Divide here.

COMMONAGE, THE, S. of Vernon (C-10). In 1876 a joint Dominion-Provincial Indian Reserve Commission set aside this land, between Okanagan Lake and Kalamalka Lake, for pasturage to be shared by Indians and whites in common. In 1893, after the creation of an Indian reserve on the west side of Okanagan Lake, the province offered the land for sale to individuals. There were other 'commonage' reserves in British Columbia.

COMO LAKE, S. of Port Moody (B-8). Various origins have been suggested for this name: Como, a Sandwich Islander or Hawaiian who belonged to the party that founded Fort Langley in 1827; Lake Como in northern Italy; or even a compound of 'Co' for Coquitlam and 'mo' for Port Moody.

COMOX, SE of Courtenay (B-7). From a Kwakwala Indian word meaning 'place of plenty,' with reference to the abundant game and berries in the Comox Valley. Comox Harbour was once known as Port Augusta.

CONKLE LAKE, W. of Westbridge (B-10). After William H. Conkle, who settled in the Kettle Valley in the 1890s.

CONSTANCE COVE, Esquimalt Harbour (A-8). After HMS *Constance,* which in 1848 became the first Royal Navy ship to use Esquimalt as her base.

CONTACT CREEK, flows S. into Liard R. (L-6). Here the US Army engineers building the Alcan (Alaska) Highway from the east met those coming from the west.

CONUMA PEAK, NE of Nootka Sound (B-6). *Conuma* is said to be a Nootka Indian word meaning 'high, rocky peak.'

COOK, CAPE, NW coast of VI (C-6). After Captain James Cook, RN, who in 1778 became the first British navigator to enter these waters and the first white man to set foot in British Columbia. Cook named the promontory Woody Point, but in 1860 it was renamed Cape Cook by Captain G.H. Richards, RN.

COOMBS, S. of Qualicum Beach (B-7). Named after Captain Coombs of the

Salvation Army, which in 1910-11 established a settlement here for a number of working-class Englishmen and their families from Leeds, Yorkshire.

COPLEY LAKE, E. of Cheslatta L. (F-7). Named after George V. Copley, faithful aide to Frank Swannell, prince of surveyors, when their party was working in the area. (Also MOUNT COPLEY, N. of Trembleur L.)

COPPER ISLAND, Shuswap L. (C-10). So named since Walter Moberly found traces of copper on its southwest corner in 1865. The island's Shuswap Indian name, anglicized as 'Hum-tat-qua,' means 'something sitting in the water.'

COQUALEETZA, S. of Chilliwack (B-9). From the Halkomelem word for 'place of beating of blankets' (to get them clean).

COQUIHALLA RIVER, flows SW into Fraser R. (B-9). From the Halkomelem word meaning 'stingy container [of fish].' Dr. Brent Galloway supplies the following explanation: according to a legend, black water pygmies used to grab the fish spears of Indians trying to catch suckers in a certain stretch of the river and thus reduce their catch. Early diaries refer to this river as the Qua-que-alla.

COQUITLAM, W. of Pitt R. (B-8). From the Halkomelem word meaning 'stinking of fish slime.' During a great winter famine in traditional times, the Coquitlam people sold themselves into slavery to the more numerous and prosperous Kwantlen band (whose name means 'tireless runners'). The Coquitlams, while butchering salmon for their masters, became covered with fish slime – hence the name.

CORBIN, S. of Crowsnest Pass (B-12). After Daniel Chase Corbin of Spokane, Washington, capitalist and railway builder, owner of the Corbin Coal and Coke Company.

CORDERO CHANNEL, between E. Thurlow I. and mainland (C-7). After José Cordero, draughtsman on Galiano's expedition of 1792.

CORDOVA BAY, N. of Victoria (A-8). In 1790 Sublieutenant Manuel Quimper of the Spanish Navy gave Esquimalt Harbour the name of Puerto de Cordova, in honour of the forty-sixth Viceroy of New Spain. About 1842 the HBC transferred the name to this bay. For many years previous to 1906, Cordova Bay was known as Cormorant Bay.

The Saanich Indian name for the bay means 'white colour,' referring to the lush growth of waxberry/snowberry bushes on its shoreline.

CORMORANT ISLAND, E. of Port McNeill (C-6). After HMS *Cormorant*, a paddle sloop that was in these waters in 1846.

CORNWALL CREEK, flows SE into Thompson R. near Ashcroft (C-9). After Clement Francis and Henry Pennant Cornwall. (See *Ashcroft*.)

CORTES ISLAND, E. of Quadra I. (C-7). This and Hernando Island to the south were named in 1792 by Galiano after Hernando Cortes, the conqueror of Mexico.

COSENS BAY, Kalamalka L. (C-10). After Cornelius Cosens (1837-1930), a pioneer settler who sold his property in the area to the Earl of Aberdeen.

COSTE ISLAND, S. of Kitimat (F-5). After Louis Coste, chief engineer in the federal Department of Public Works, who in 1898 examined the heads of various inlets in quest of a suitable terminus for a proposed railway to the Yukon.

COURCELETTE PEAK, headwaters of Fording R. (C-12). After a village in France captured by Canadian troops in September 1917.

COURTENAY, E. coast, VI (B-7). When the town was laid out in 1891, it was named after the nearby Courtenay River, which had been named about 1860 after Captain George William Courtenay of HMS *Constance*, which was on the Pacific Station in 1846-9.

COUTLEE, NW of Merritt (C-9). Tiring of seeking for gold, Alexander Coutlee preempted at Boston Bar in 1863, moved to the Nicola Valley in 1873, and here ran a store, a hotel, and a ranch.

COW BAY, Prince Rupert (G-4). The first cows in the area were unloaded here in 1906.

COWAN, POINT, Bowen I. (B-8). After G.H. Cowan, KC (1858-1935), a Vancouver lawyer who had a summer home here.

COWARD'S COVE, off Wreck Beach, Point Grey (B-8). This small cove with its arresting name is simply a dredged area within the outer end of a breakwater, thus providing a mooring ground for fishing vessels. The name mocks those who remain in shelter here when wind and waves make rather risky further progress out of the mouth of the Fraser River.

COWICHAN LAKE, W. of Duncan (A-7). From an Island Halkomelem word meaning 'warm country' or 'land warmed by the sun.' The name originated because of a huge rock formation, on the side of Mount Tzuhalem, supposedly resembling a frog basking in the sun. Originally both Cowichan Lake and the settlement were known as Kaatza, Cowichan for 'big lake.'
    The Cowichan Indians are a large group who formerly ranged widely over the southern Strait of Georgia and had fishing rights in certain places on the mainland, including the mouth of the Fraser River.

CRACROFT ISLANDS, N. of Johnstone Strait (C-6). These islands and nearby Sophia Island are named after Sophia Cracroft, niece of the ill-fated Arctic explorer Sir John Franklin. With Lady Franklin she visited British Columbia in 1861.

CRAIGELLACHIE, W. of Revelstoke (C-10). It was here that the last spike of the CPR transcontinental line was driven in 1885.

Craigellachie is the name of a high rock in Moray, Scotland. There in olden days the beacon fire was lit to summon Clan Grant in time of battle. The battle cry of the Grants was 'Stand fast, Craigellachie!' In 1884, when the finances of the CPR were desperate, George Stephen (later Lord Mount Stephen) raised £50,000 by guaranteeing that he, Donald Smith (later Lord Strathcona), and R.B. Angus would be personally responsible if the railway defaulted. Stephen and Smith, who were cousins, had grown up close to the crag of Craigellachie and knew the old war cry. After completion of the loan, Stephen's message cabled from London to Smith in Montreal read, 'Stand fast, Craigellachie.'

CRAIGFLOWER, Esquimalt (A-8). After adjacent Craigflower Farm, founded by the HBC in 1853. This farm was named after Craigflower Farm in England, owned by Andrew Colvile, Governor of the HBC from 1852 to 1856.

Craigflower School, still surviving along with the old manor house, was one of the earliest schools built in the province. An entry in the diary of Robert Melrose, dated 24 September 1854, notes, 'School-house frame erected, whole company in general notoriously drunk.'

CRANBROOK, E. Kootenay (B-12). The area around Cranbrook was formerly known as Joseph's Prairie, and here the Kootenay Indian village of A'Qkis ga'ktlect (meaning 'two streams going along together') once stood. An early colonist, Colonel James Baker, sometime provincial Minister of Education and later Minister of Mines, settled here in 1885 and named his estate Cranbrook Farm, after the little Kentish town of Cranbrook from which he came. When a townsite was laid out in 1897, the name of Cranbrook was taken for the new settlement. Cranbrook became important as a divisional point after the opening of the CPR's Crowsnest line in 1898, and it was incorporated as a city in 1905.

CRAWFORD BAY, E. side of Kootenay L. (B-11). After James Crawford ('White Man Jim'). An old-time prospector and trapper in the district, he died in 1914.

CREASE ISLAND, off Knight Inlet (C-6). After Sir Henry Pering Pellew Crease (1823-1905), for ten years Attorney-General and for twenty-five years a judge of the Supreme Court of British Columbia.

CRESTON, S. end of Kootenay L. (B-11). Known originally as Seventh Siding, during the construction of the CPR's Crowsnest line, 1897-8. When the CPR sought to buy land from Fred Little for a townsite here, he stipulated that the new town be named Creston after his hometown in Iowa.

CRICK CREEK, flows E. into Kettle R. (B-10). After a local landowner, James H. Crick. A very minor stream, it is noted here only because of its delightful name. Authentic western pronunciation would, of course, make it 'Crick Crick.'

CRICKMER, MOUNT, W. of Stave L. (B-8). Named in 1859 after the Reverend W. Burton Crickmer, the Anglican clergyman at Fort Langley. His diary is in the Vancouver City Archives. For almost a century, all memory of the naming of this mountain seemed lost. Only in 1957 was it identified from a sketch in the diary and, no new name having become attached to it in the interim, the old name was restored.

CRIDGE PASSAGE, S. of Grenville Channel (F-5). After the Reverend Edward Cridge (1817-1913), HBC chaplain in Victoria 1856-74. After a magnificent row with Bishop Hills, he joined the Reformed Episcopal Church, taking most of the old HBC men in Victoria with him.

CRISS CREEK, flows SW into Deadman R. (D-9). Corrupted form of Chris Creek. After Christopher Pumpmaker, who ranched here from 1869 until his death in 1876.

CROFTON, NE of Duncan, VI (A-8). After Henry Croft, a mining engineer who was a brother-in-law of James Dunsmuir, the coal and lumber magnate. In 1900 Croft, manager of the Lenora, Mt. Sicker Copper Mining Company, bought land on Osborn Bay on which to build a smelter. In 1902, when a village was built to house the 400 men working at the smelter, it was named Crofton.

CROSS RIVER, flows SW into upper Kootenay R. (C-12). Translation of the Stoney Indian name, which alludes to Father De Smet's raising of a cross in 1845 at the pass at the head of the watershed.

CROWSNEST PASS, Alberta-BC boundary (B-12). In very early days, a band of Crow Indians, making a horse-stealing raid into Blackfoot territory, camped here. The Blackfeet, catching them by surprise in their 'nest,' massacred them. A Palliser Expedition map of 1859 shows 'lodge des Corbeaux' at the head of the 'Crow River.' Another map of the same expedition in 1860 shows 'Crow Nest' river and pass. Captain Palliser, the first white man to learn of the pass, referred to it at times as the 'British Kutanie pass.' Crowsnest became a CPR divisional point during the building of the railway's Crowsnest line in 1897.

CRUICKSHANK RIVER, flows into Comox L. (B-7). After George Cruickshank, secretary of the committee that launched the Vancouver Island Exploration Expedition of 1864.

CRYSDALE, on the BCR N. of Hixon (F-8). After C.R. Crysdale, in charge of the survey of the natural resources of the Cariboo and upper Peace River areas made by the PGE Railway in 1929-30.

CULTUS LAKE, S. of Chilliwack (B-9). Chinook jargon for anything bad, worthless, or foul. The anglicized form of the Halkomelem word for the lake is Sweltzer, now the name of the river draining the lake. The latter word means 'unclear liquid that warns secretly.' The Indians believed that dreaded supernatural creatures lived in Cultus Lake, often manifesting themselves as dirty swirlings in the water.

CUMBERLAND, S. of Courtenay, VI (B-7). The original settlement was named Union after the Union Coal Company. In 1898 Union as a post office address was replaced by Cumberland. Many of the miners had come from Cumberland in England. (See *Union Bay*.)

CUMSHEWA INLET, Moresby I. (F-4). Named about 1788 after the Indian chief here. In 1794 Cumshewa and his Haidas massacred the crew of the American trading vessel *Resolution*. Cumshewa is derived from the Kwakwala Indian word meaning 'rich at the mouth [of the river].'

CUNNINGHAM CREEK, flows E. into Cariboo R. (E-9). After William Cunningham, from Kentucky, who found gold on this creek during the Cariboo rush.

CUNNINGHAM ISLAND, E. of Seaforth Channel (E-6). After Thomas Cunningham, an Ulsterman who came to Victoria in 1859. When a member of Vancouver Island's House of Assembly, he voted for union with the mainland colony of British Columbia.

CURRIE, MOUNT. See *Mount Currie*.

CURZON, W. of Yahk (B-11). After the Marquess Curzon of Kedleston. Curzon (1859-1925) held various cabinet positions in Britain and, from 1899 to 1905, was Viceroy of India.

D

DAER, MOUNT, Kootenay NP (C-12). From Lord Daer, the title of the son and heir of the Earl of Selkirk. The last Lord Daer became the sixth Earl of Selkirk in 1820. (See *Selkirk Mountains*.)

DAHL LAKE, W. of Prince George (F-8). For the naming of this and nearby Norman Lake, see *Norman Lake*.

DAIRY CREEK, flows SE into N. Thompson R. (C-9). Takes its name from the dairy farm established here by the HBC. The company received a free grant of the land in 1869.

DAIS GLACIER, SW of Mt. Waddington (D-7). This is the 'platform' from which Mount Waddington rises. An example of the romantic names that Don Munday sprinkled around the area, to the amusement of his fellow alpinists. Others are Dauntless Mtn., Vigilant Mtn., and Combatant Mtn.

DAISY LAKE, expansion of Cheakamus R. (B-8). J.W. McKay of the HBC was this way in 1859 and, according to Lieutenant Mayne of the Royal Navy, who visited the lake in 1861, gave the name. Whether McKay's interest was botanical, familial, or amatory remains unknown.

DAK RIVER, flows SW into Kitsault R. (H-5). From the Nisgha Indian word meaning 'lake.'

DALA RIVER, flows W. into Kildala Arm (F-5). Either from the Haisla Indian word meaning 'in hand,' or from one meaning 'long way ahead.'

DALLAS, E. of Kamloops (C-9). This Dallas is named after neither Dallas, Texas, nor Alexander Grant Dallas, a notable HBC agent and director. It takes its name, instead, from Dallas Johnston (1887-1964), who farmed six miles east of Kamloops from 1902 until after World War II, when much of his land was purchased for Veterans' Land Act settlement. The name was chosen in 1955 by the settler veterans at a meeting called to decide upon a name for the community.

DAM MOUNTAIN, N. of North Vancouver (B-8). So named in 1894 for the view it afforded of the waterworks dam on Capilano River.

DAMFINO CREEK, flows SE into Kettle R. (B-10). William Fleet Robertson (provincial mineralogist 1897-1925), when crossing this stream with a pack train, asked one of the men what its name was. 'Damned if I know,' said the man, and onto the map it went as Damfino Creek.

DARBY CHANNEL, Rivers Inlet (D-6). After Dr. George Darby, for many years the much loved and respected medical missionary at Bella Bella. He used this channel when travelling to and from the summer hospital he maintained on Rivers Inlet.

D'ARCY, S. of Anderson L. (C-8). This BCR station is named after D'Arcy Tate, vice-president and general counsel 1912-18 for the PGE.

D'ARCY ISLAND, Haro Strait (A-8). After John D'Arcy, a mate on HMS *Herald* when she was in these waters in 1846. The Saanich Indian name for the island means 'arrive,' this being where the salmon first arrive when they are heading for the Fraser River to spawn.

DARFIELD, N. Thompson R. (D-9). When a post office was opened here in 1937, it was given this name at the suggestion of Hubert Janning as an anglicized version of Darfeld, his home in Germany, but at least partly because it echoed Darlington Creek nearby.

DARKE LAKE PARK, NW of Summerland (B-10). After Robert Darke, who settled in the area in the 1890s.

DARWIN SOUND, W. of Lyell I., QCI (E-4). Named by G.M. Dawson after Charles Darwin, the famous naturalist. Dawson named various features in the area after Victorian scientists: for example, Lyell Island, Sedgwick Bay, Huxley Island.

DAVIDSON, MOUNT, Garibaldi Park (B-8). After Professor John Davidson of UBC, who did the first botanizing in Garibaldi Park around 1912. In 1918 he founded the Vancouver Natural History Society, which he headed until 1937.

DAWN TREADER MOUNTAIN, S. of Franklyn Arm of Chilko L. (D-7). Named after the boat in C.S. Lewis's *The Voyage of the Dawn Treader*.

DAWSON CREEK, Peace R. district (H-9). After Dr. George Mercer Dawson (1849-1901), the famous geologist who became the head of the Geological Survey of Canada in 1895. A diminutive hunchback, Dawson showed heroic endurance in his surveys, which took him up and down the length of British Columbia. His excellent reports show a continual interest in place names.

In the summer of 1879, Dawson was travelling in this area and met Henry A.F. MacLeod, exploration engineer for the CPR. Although 'Dawson's Brook' appears on Dawson's subsequent map, it may have been MacLeod who gave the name to this tributary of the Pouce Coupé River.

DAWSON RANGE and MOUNT DAWSON in Glacier National Park are also named after G.M. Dawson.

DAWSON FALLS, Murtle R., Wells Gray Park (D-9). After George Herbert Dawson, Surveyor-General of British Columbia 1912-17.

DAWSONS LANDING [DAWSON'S LANDING], Rivers Inlet (D-6). After Jimmy Dawson, who ran a gas station and general store here with his wife, Jean, for many years. 'A short man but a lovable person who got on well with everybody.'

DEADMAN ISLAND, Coal Harbour, Vancouver (B-8). This was formerly an Indian burial ground.

DEADMAN RIVER, flows S. into Thompson R. near Savona (C-9). After Pierre Chivrette, or Charette, killed here in 1817 in a quarrel over the choice of a campsite.

DEADMANS [DEADMAN'S] ISLAND PARK, Burns L. (G-7). When the Grand Trunk Pacific Railway was being built, at least one man, killed during blasting, was buried here.

DEAN CHANNEL, Central Coast (E-6). Also DEAN RIVER, which flows into it. Named by Captain Vancouver after the Very Reverend James King, Dean of Raphoe in Ireland. Vancouver had served under Dean King's son James on Captain Cook's final voyage.

DEAS ISLAND, mouth of Fraser R. (B-8). After John Sullivan Deas, a mulatto tinsmith who was foreman at a cannery here in the 1870s. He originally came to British Columbia as a member of the Royal Engineers.

DEASE LAKE, Cassiar (K-4). Also DEASE RIVER. Named in 1834 by John McLeod, Chief Trader for the HBC. Born in 1788 Peter Warren Dease entered the fur trade as an employee of the XY Company in 1801. Upon the union of the NWC and the HBC in 1821, he became a Chief Trader, and in 1828 he became a Chief Factor. In 1825 Governor Simpson wrote of him, a trifle incoherently: 'This Gentleman is one of our best voyageurs, of a strong robust habit of body, possessing such firmness of mind joined to a great suavity of manners, and who from his great experience in the country – would be a most valuable acquisition.' From 1831 to 1835, based at Fort St. James, Dease was in charge of New Caledonia.

Dease was on Franklin's Arctic expedition of 1825-7 and on a second Arctic expedition in 1837-9. Because of his explorations, Queen Victoria granted him a Civil List pension of £100 per annum.

DEATH RAPIDS, N. of Revelstoke (D-10). The old Dalles des Morts of the fur traders, one of the most dangerous stretches of the Columbia River route from Athabasca Pass. The worst of the many disasters here occurred in 1838 when twelve people drowned, among them the young botanist Robert Wallace and his bride, a daughter of Governor Simpson of the HBC.

DECEPTION CREEK, flows S. into Mahood L. (D-9). When E.W. Jarvis reached Mahood Lake in the autumn of 1873, with directions to explore the Clearwater River, he mistook this creek for the river and travelled some nineteen kilometres before discovering his error. Although some coarse gold was taken from this stream, miners never found the large deposits that they had expected to find upstream.

DECKER LAKE, NW of Burns L. (G-7). After Decker, a foreman with the Collins Overland Telegraph party here in 1866.

DE COURCY ISLAND, S. of Gabriola I. (B-8). After Captain (later Vice-Admiral) Michael de Courcy of HMS *Pylades,* on the Pacific Station in 1859-61.

DEEKS CREEK, flows SW into Howe Sound (B-8). After John F. Deeks, whose Deeks Sand and Gravel Company began working here in 1908.

DEER PARK, Lower Arrow L. (B-10). In 1889, when G.M. Dawson published his geological survey of the area, he noted: 'The most attractive and park-like portion of this country is commonly named the "Deer Park," and is frequented by great numbers of deer, when in winter their higher pastures in the mountains become covered with snow.'

DE HORSEY ISLAND, mouth of Skeena R. (G-4). After Rear-Admiral Algernon F.R. de Horsey, RN, commanding the Pacific Station 1876-9.

DEKA LAKE, S. of Canim L. (D-9). A surveyor's error for 'Decker,' the name of a family from the Canim Lake Indian Reserve that used to fish here.

DEL CREEK, E. side of Finlay R. (J-7). After 'Del' Miller, who had a cabin here.

DELLA FALLS, NW of Great Central L. (B-7). Also DELLA LAKE and DRINKWATER CREEK. These falls, really a series of cascades descending 1,443 feet (440 metres), have been claimed to be the highest in Canada. They were named by Joe Drinkwater, a prospector, after his wife, Della, whom he married in 1899.

DELPHINE CREEK, flows E. into Toby Cr. (C-11). Named after the wife of pioneer settler George Starke.

DELTA, Lower Mainland (B-8). This municipality is on the delta of the Fraser River.

DEMPSEY LAKE, E. of Lac la Hache (D-9). In 1919 a boxing enthusiast in the Cariboo named this lake after Jack Dempsey, heavyweight champion of the world.

DENMAN ISLAND, SE of Comox (B-7). After Rear-Admiral Joseph Denman, FRS (1810-74), Commander-in-Chief, Pacific Station 1864-6.

DENNY ISLAND, Central Coast (E-5 and 6). After Lieutenant D'Arcy A. Denny, RN, commanding the gunboat *Forward* 1866-8.

DEPARTURE BAY, Nanaimo (B-8). So named by J.D. Pemberton in 1852 because it was the site of an abandoned Indian village. During his survey Pemberton also spotted that here was a good exit for ships leaving Nanaimo harbour.

DE PENCIER BLUFFS, Mt. Seymour, North Vancouver (B-8). After that redoubtable mountaineer the Most Reverend A.U. De Pencier (1866-1949), Anglican Archbishop of New Westminster.

DEPOT CREEK, Chilliwack L. (B-9). So named because the Royal Engineers had a base camp here in 1858-9 during their survey of the international boundary.

DERBY REACH, lower Fraser R. (B-8). This stretch of the Fraser provides the only survival of the name Derby, applied to old Fort Langley in 1858 when it was expected to become the capital of British Columbia. The Earl of Derby was Prime Minister of Britain at that time.

DEROCHE, E. of Mission (B-8). After Joseph Deroche, the first settler here. He arrived in British Columbia from California in 1860, became a teamster in the Cariboo, acquired a farm here, and died in 1922 at the age of ninety-nine.

DESCANSO BAY, Gabriola I. (B-8). Spanish for 'rest' or 'ease.' Galiano and Valdes left a record of how they came to give this name in June 1792 to this haven on Gabriola Island: 'We called this roadstead "Cala del Descanso" from our need of rest and our appreciation in finding it on this occasion.'

DESERTERS CANYON, Finlay R. (I-7). Here on the night of 27 May 1824, when Samuel Black of the HBC was making his great journey of exploration into northern British Columbia, two of his men, Jean Marie Bouche and Louis Ossin, deserted.

DESOLATION SOUND, E. of Cortes I. (C-7). So called by Captain Vancouver because of both the gloomy aspect of the coast and the lack of fish and berries.

DEVEREUX CREEK, flows S. into Klinaklini R. (D-7). For F.A. Devereux, the surveyor who in 1895 tried to find a practicable route inland from the head of Knight Inlet.

DEWDNEY, E. of Mission (B-8). After the Honourable Edgar Dewdney (1835-1916). He arrived in British Columbia in 1859 and the next year, with Walter Moberly, built the famous Dewdney Trail from Fort Hope to Rock Creek (later continued to Wild Horse Creek). He served as both a member of the Legislative Council and a Member of Parliament. In 1881 Dewdney was appointed Lieutenant-Governor of the North West Territory, serving through the Riel Rebellion. He was federal Minister of the Interior 1888-92 and Lieutenant-Governor of British Columbia 1892-7.

DICK BOOTH CREEK, SW of Port Hardy (C-6). Dick Booth was a land surveyor, killed in World War I.

DICKENS POINT, Portland Canal (H-4). After Sublieutenant Sydney Smith Haldemand Dickens, RN, son of novelist Charles Dickens. He served on the Pacific Station 1868-70.

DIGBY ISLAND, Prince Rupert (G-4). After Henry Almarus Digby, second lieutenant aboard HMS *Malacca* when she was on the Pacific Station in 1866-7.

DILWORTH, MOUNT, NE of Kelowna (B-10). After John Dilworth (1850-1917), a Manitoba farmer who served in the Red River Rebellion, later farmed near here, and became a Victoria alderman.

DISCOVERY PASSAGE, between VI and Quadra I. (C-7). Named by Captain Kellett, RN, in 1847. Apparently Kellett was mindful that Captain George Vancouver's HMS *Discovery*, 340 tons, mounting ten four-pounder cannon and ten swivel guns, had sailed through this passage during Vancouver's circumnavigation of Vancouver Island in 1792. Also DISCOVERY ISLAND off Victoria – though Vancouver's ship was never there.

DIXON ENTRANCE, N. of QCI (G-3). In 1788, after returning to England from the voyage on which he had named the Queen Charlotte Islands, Captain George Dixon showed his map of these waters to Sir Joseph Banks, president of the Royal Society. Dixon left an account of this interview. In it he mentions that, 'When I laid my manuscript chart before Sir Joseph Banks for his approbation, I, at the same time, requested him to name such places as I had not filled up, and he did me the honour to insert mine in the place you find it on the chart.'

DOC ENGLISH GULCH, E. of Riske Creek (D-8). After Benjamin Franklin ('Doc') English. Born in St. Louis in 1841, he travelled overland to the Pacific with his father on the Oregon Trail when he was a boy of five. Far from having an MD, he had a minimal formal education. 'Doc' English arrived in the Cariboo in 1860, driving cattle that he sold to the miners for meat. Thereafter he had a varied career as a packer, storekeeper, and stock raiser in the Chilcotin, at Ashcroft, and in Venables Valley. An acute horse trader, he had a passion for horse racing.

DOCTORS POINT [DOCTOR'S POINT], W. side of Harrison L. (B-9). Here is the effigy of 'The Doctor,' formerly known as 'Shay,' the local Indian god of the weather. When passing the image, one ensures good weather by tossing into the lake a coin for 'The Doctor.' The original doctor whose name is given to the point may have been Charles Forbes, the medical officer of HMS *Topaz*, who, in the report on his geological survey of the area in 1860, mentions seeing Shay's effigy. Another explanation is that Shay was not a deity but an Indian shaman or witch doctor who was here turned to stone by the 'Transformer,' who figures in so many Indian legends.

DODD NARROWS, between VI and Mudge I. (B-8) After Captain Charles Dodd, who came to the Pacific coast in 1835 as second officer on the HBC's steamboat *Beaver,* which he commanded from 1843 to 1852. Subsequently he became a Chief Factor in the Company's service.

DODGE COVE, Prince Rupert (G-4). After G.B. Dodge, DLS, who surveyed Prince Rupert Harbour in 1906.

DOGWOOD VALLEY, Fraser Valley, N. of Hope (B-9). In 1972 local residents successfully petitioned to have the name of Choate changed to Dogwood Valley. Dogwood trees are found in the area, and the dogwood flower is the floral emblem of British Columbia. The CPR continues to call its nearby station Choate (q.v.).

DOLLARTON, Burrard Inlet (B-8). After Captain Robert Dollar, founder of the Dollar line of steamships. He once owned a lumber mill here.

DOMETT POINT, Anvil I., Howe Sound (B-8). After William Domett, RN, flag captain to Admiral Sir Alexander Hood on HMS *Royal George* at the battle called 'The Glorious First of June,' 1794. He later became Admiral Sir William Domett.

DOMINIC LAKE, SW of Kamloops (C-9). Avosti Domenico settled here in the 1880s.

DONALD, NW of Golden (D-11). After Donald A. Smith, who became Lord Strathcona and Mount Royal, director of the CPR during and after construction. (For further biographical detail, see *Strathcona Park*.)

Donald was originally known as First Crossing since it was here that the CPR first crossed the Columbia River. Second Crossing (later Revelstoke) was on the west side of the Big Bend of the Columbia.

DONEGAL HEAD, Malcolm I. (C-6). This cape commemorates HMS *Donegal,* commanded by Captain Pulteney Malcolm at the time of the Battle of Trafalgar.

DOOGIE DOWLER, MOUNT, E. of Ramsay Arm (C-7). After Doogie Dowler, from 1949 to 1983 the well-known and widely respected operator of the store at Heriot Bay. The mountain can be seen from the store.

DOT, Nicola Valley (C-9). This little station was given the nickname of Dalton

P. Marpole, who owned land nearby. This Marpole was the son of Richard Marpole, formerly BC general superintendent of the CPR.

DOUGLAS, E. of White Rock (B-8). This border-crossing point is named after Benjamin Douglas, a pioneer settler.

DOUGLAS CHANNEL, approach to Kitimat (F-5). After Sir James Douglas. (See *Douglas, Mount.*)

DOUGLAS ISLAND, at confluence of Pitt R. and Fraser R. (B-8). Originally Manson Island after Donald Manson, a member of the expedition that founded Fort Langley in 1827. After Governor Douglas purchased the island, he gave it to his daughter Cecilia, and it became known as Douglas Island.

DOUGLAS LAKE, E. of Nicola L. (C-9). After the first preemptor of the Douglas Lake ranch, John Douglas. This Douglas, no relation of Governor Sir James Douglas, came to Nicola Valley hoping that the dry climate would help his lung condition. He acquired much land between 1872 and 1888. He died in California in 1889, apparently of tuberculosis.

Rather late in life, John Douglas married Mrs. Julia Cross, keeper of a boardinghouse in San Francisco. She took violent exception to life in a ranch house, so old John set her up with a boardinghouse in Victoria and returned to his bachelor ways.

DOUGLAS, MOUNT, Saanich (A-8). Originally known as Cedar Hill, since the cedar palisades around Fort Victoria were cut here. The original name still lingers in Cedar Hill Road and Cedar Hill Crossroad.

In a letter of 24 March 1859 to the Hydrographer of the Royal Navy, Captain G.H. Richards told how Cedar *Hill* became *Mount* Douglas: 'it has been much the fashion here to give the term Mountain to elevations which are by no means entitled to that distinction. I have taken the liberty of reducing all under 1000 feet to Hills, except Mount Douglas, which I have retained as a mountain altho 690 feet, partly from not wishing to *lower* the Governor, and partly because Douglas Hill does not sound well.' Actually, as early as 1851, the captain of HMS *Daphne* had noted in his remark book 'an abrupt but not high hill called Cedar Hill (or Mount Douglas).'

Sir James Douglas (1803-77) was born in Demerara (Guyana) of Scottish and Creole parentage. Educated in Scotland, he entered the service of the NWC in 1819. In 1825 the Hudson's Bay Company stationed him at Fort St. James. Here in 1829 he made a 'fur trade marriage' with Amelia, a mixed-blood daughter of Chief Factor Connolly. This union was confirmed by a Church of England marriage nine years later. In 1829, when the Indians seized Fort St. James and threatened the captive Douglas with death, his young wife saved his life by quickly handing over trade goods until the Indians agreed that sufficient ransom had been paid. Because of continuing trouble with the Indians, Douglas was transferred to Fort Vancouver in 1830. In 1832 Governor Simpson characterized Douglas thus:

> A stout powerful active man of good conduct and respectable abilities; – tolerably well Educated, expresses himself clearly on paper, understands our Counting House business and is an excellent Trader. Well qualified for any Service requiring bodily exertion, firmness of mind and the exercise of sound judgement, but furiously violent when roused. Has every reason to look forward to early promotion and is a likely man to fill a place at our Council board in course of time.

In 1834 Douglas was promoted to Chief Trader and in 1839 to Chief Factor. In 1842, after inspecting the site, Douglas recommended the founding of Fort Victoria. In 1849, Fort Victoria having become the new headquarters of the Columbia Department, Douglas took up residence there to supervise the management of that district. We are given an interesting glimpse of him in Victoria in 1850:

> I suppose there must have been more than twenty people in the [dining] room, when Mr. Douglas made his appearance – a handsome specimen of nature's gentlemen, tall, stout, broad-shouldered, muscular, with a grave bronzed face, yet kindly withal. After the usual greetings he took the head of the table ... I was informed that no frivolous conversation was ever allowed at table, but that Mr. Douglas, as a rule, came primed with some intellectual and scientific subject, and thus he educated his clerks. All had to go to church every Sunday.

After the resignation of Governor Blanshard in 1851, James Douglas became the second Governor of the Crown Colony of Vancouver Island. In 1858 he also became Governor of the mainland colony of British Columbia and resigned his HBC positions. During the years that followed, Governor Douglas's energy, sagacity, and courage saved the two colonies from annexation by the United States during the turbulent period of the Fraser River and Cariboo gold rushes. His task would have been impossible without the modest naval and military forces supplied by Britain and the small group of excellent civil servants provided by Sir Edward Bulwer-Lytton, the Colonial Secretary. It was Douglas's vision that provided British Columbia first with the tote road from Port Douglas at the head of Harrison Lake, via Anderson and Seton Lakes, to Lillooet; and then with the famous Cariboo Wagon Road from Yale, through the Fraser Canyon, and on, via Cache Creek and Clinton, to the goldfields around Barkerville.

In 1863, the year before Douglas handed over his two governorships to his successors, he was created a knight commander of the Order of the Bath. Few of the proconsuls of the British Empire better deserved knighthood. He died in Victoria in 1877. If any man may be called 'The Father of British Columbia,' Sir James Douglas is that man.

DOWNIE CREEK, flows W. into Columbia R. (D-10). After 'Major' William Downie, who also had Downieville, California, named after him. In 1858 Governor Douglas, who thought very highly of Downie, employed him with J.W. McKay to explore the route south from Pemberton to Howe Sound. The next year Downie explored around Jervis and Bute Inlets, led an unsuccessful expedition to the Queen Charlotte Islands, but rehabilitated himself with a tremendous journey up the Skeena, then via Babine Lake to Fort St. James, and then by way of the Fraser to the Cariboo goldfields. In 1865 he led a party of gold-seekers from Washington Territory up the Columbia River to the Big Bend country north of modern Revelstoke.

DOWNING PARK, Kelly Lake (D-9). After Claire S. Downing, who in 1970 presented the people of British Columbia with ninety-seven acres of land, including 2,400 feet of frontage on Kelly Lake, to provide the nucleus for this park.

DOWNTON LAKE, W. of Carpenter L. (C-8). Standing atop Mission Mountain in December 1912, Geoffrey M. Downton, BCLS, a ship's officer turned surveyor, suddenly realized that the Bridge River valley to the north was 1,200 feet higher than Seton Lake to the south. Let a two-mile tunnel be driven through the mountain on which Downton stood and a fantastic source of hydroelectric power would come into being. Only after the two world wars, in the first of which Downton won the Military Cross, was the Bridge River power finally developed and brought south to Vancouver. Downton, one of the original syndicate that obtained the rights, realized only $2,000 from the sale of his shares, but in 1948 the BC Electric Company chose him to push the button that set the current surging south. Ultimately British Columbia would get from here more power than Ontario gets from Niagara Falls.

A journalist who visited Downton and his artist wife, both natives of Norfolk, found him 'an urbane scholarly man like a retired professor or an Anglican parson.'

DRABBLE, MOUNT, W. of Courtenay (B-7). After George Fawcett Drabble, CE, pioneer surveyor, who arrived in the Comox area around 1868. Drabble was a bachelor and a misogynist. When complaints were made about his fierce dogs, he retorted, 'Don't worry, my dogs never bite anybody that is any good.'

DRAGON LAKE, SE of Quesnel (E-8). On the BCR. After Dick Dragon, the first settler here.

DRAKE ISLAND, Quatsino Sound (C-6). After Montague W. Tyrwhitt-Drake (1830-1908), who served as a justice on the Supreme Court of British Columbia from 1889 to 1904.

DRANEY INLET, Rivers Inlet (D-6). After Robert Draney, who in 1876 built the Aberdeen Cannery on the Skeena, in 1882 the Rivers Inlet Cannery, and in 1893 a large cannery at Namu.

DREWRY LAKE, S. of Canim L. (D-9). After the surveyor William Stewart Drewry, who was active in the area about 1916.

DRURY INLET, NW of Broughton I. (C-6). After Captain Byron Drury of HMS *Pandora,* on this coast in 1846-8.

DRYAD POINT, Campbell I. (F-5) After the HBC brig *Dryad* (Captain Kipling) in these waters in 1833.

DRYNOCH, S. of Spences Bridge (C-9). During the building of the CPR, a construction engineer, H.A.F. Macleod, named the station here after his home, a hamlet on the island of Skye.

DRYSDALE, MOUNT, W. boundary of Kootenay NP (D-11). After Dr. Charles W. Drysdale, an outstanding young geologist who, with his assistant, William Gray, drowned in the Kootenay River in 1917 while working for the Geological Survey of Canada.

DUCHESNAY, MOUNT, SE of Field (D-11). After E.J. Duchesnay, CE, assistant general superintendent of the Pacific Division of the CPR, killed by a rockfall in a tunnel near Spuzzum in 1901.

DUCHESS PEAK, headwaters of Findlay Cr. (C-11). After the *Duchess,* the first steamboat on the upper Columbia River. The boat was described as 'resembling an old canal-boat into which a travelling gipsy's van had been hastily crammed without regard for its position or safety.'

DUCK RANGE, E. of Monte Cr. (C-10). See *Monte Creek.*

DUFF, MOUNT, Alaska-BC boundary (L-0). After Sir Lyman Duff, Chief Justice of the Supreme Court of Canada 1933-44. He appeared as a counsel before the Alaskan Boundary Tribunal in 1903.

DUFFERIN, Kamloops (C-9). This suburb takes its name from Dufferin Hill, so named since Lord Dufferin, the Governor-General, sketched the view from here when, with Lady Dufferin, he visited Kamloops in 1876.

DUFFERIN ISLAND on the central coast was named in the same year when Lord and Lady Dufferin sailed past it aboard HMS *Amethyst* during their northern cruise.

DUFFEY LAKE, NE of Pemberton (C-8). After Sapper James Duffy or Duffey (he spelled his name both ways) of the Royal Engineers. He passed this way in 1860 while exploring the route from Lillooet to Pemberton.

DUFFY CREEK, flows N. into Kamloops L. (C-9). Patrick Duffy from California, after participating in the Cariboo gold rush and running a bar in Lillooet, settled here in the 1870s.

DUKE POINT, E. of Nanaimo (B-8). So named because it faces Northumberland Channel, named after the Duke of Northumberland in 1852.

DUNCAN, southern VI (A-8). After William Chalmers Duncan, born 1836 in Sarnia, Ontario. He arrived in Victoria in May 1862, and in August of that year, he was one of the party of 100 settlers that Governor Douglas took to Cowichan Bay. After going off on several gold rushes, Duncan settled close to the present city of Duncan. He married in 1876, and his son Kenneth became the first mayor of Duncan.

Duncan's farm was named Alderlea, and this was the first name of the adjacent settlement. In August 1886 the Esquimalt and Nanaimo Railway was opened. No stop had been scheduled at Alderlea for the inaugural train bearing Sir John A. Macdonald and Robert Dunsmuir. However, at Duncan's Crossing, the level crossing nearest to Alderlea, a crowd of 2,000 had assembled around a decorated arch, and the train came to an unplanned halt.

Children from the local school sang 'Welcome to You All,' and the local farmers petitioned Dunsmuir, the builder of the railway, for a station here. At the last moment, as the train pulled out, 'Old Bob' shouted, 'You will have a station there, boys!' When built, it was called 'Duncan's Station,' which was shortened first to 'Duncan's' and then to 'Duncan.' The City of Duncan was incorporated in 1912.

DUNCAN BAY, N. of Campbell River (C-7). Named by Captain G.H. Richards, RN, about 1860 after Captain Alexander Duncan, who entered the HBC maritime service in 1838, initially commanding the barque *Vancouver* and later the famous steamboat *Beaver*. A contemporary described Duncan as an accomplished seaman, 'an old British tar with a heart full of generosity for his friends and a fist full of bones for his enemies.'

DUNCAN LAKE, N. of Kootenay L. (C-11). This lake and the major dam at its lower end are named after 'old Jack Duncan,' a late-nineteenth-century prospector.

DUNDARAVE, West Vancouver (B-8). Russell E. Macnaghten (d. 1918), Assistant Professor of Greek at UBC, was one of the persons who purchased and subdivided land here in 1909. It was he who named the area after Dundarave Castle, Loch Fyne, Scotland, the ancestral home of Clan Macnaghten. Dundarave Castle was reached by rowboat, and *dundarave* (which should rhyme with 'have') is a Gaelic word meaning 'the castle of the two oars.'

DUNDAS ISLAND, Dixon Entrance (G-4). Named by Captain Vancouver after Henry Dundas (1742-1811), Treasurer of the Royal Navy. Dundas was created Viscount Melville in 1802.

DUNN LAKE, E. of Little Fort (D-9). Also DUNN PEAK. After James Dunn, who mined gold in the area around 1888.

DUNNS NOOK [DUNN'S NOOK], Esquimalt (A-8). This quaintly named cove commemorates Thomas Russell Dunn, MD, surgeon aboard HMS *Fisgard* when she was here in 1846-7.

DUNSMUIR, W. of Qualicum Beach (B-7). After Robert Dunsmuir (1825-89), British Columbia's first industrial magnate. Born in Kilmarnock, Scotland, Dunsmuir married Joanna (Joan) White in 1847. Promising that one day he would build her a castle on the Pacific, he persuaded her in 1850 to accompany him with their three children to Fort Rupert on Vancouver Island, where he worked for the HBC as a coal miner, soon being transferred to Nanaimo. After leaving the service of the HBC, Dunsmuir engaged in various rather unsuccessful ventures in coal mining until he discovered and obtained complete possession of the very rich deposits around Wellington. He also entered the lumber business, acquired an ironworks, built the Esquimalt and Nanaimo Railway, and became a very wealthy man. During the last years of his life, he made good on the promise given years earlier to his wife, building Craigdarroch Castle in Victoria, though it was not completed until a year after his death.

DUNSMUIR ISLANDS, Ladysmith Harbour (A-8). After James Dunsmuir (1851-1920), founder of Ladysmith, Premier of British Columbia (1900-2) and its Lieutenant-Governor (1906-9). Son of Robert Dunsmuir (see preceding entry). If Craigdarroch Castle in Victoria was Robert's monument architecturally, Hatley Park west of Esquimalt was James's. Hatley Park was the location of the Royal Roads Military College, now Royal Roads University.

DUNTZE HEAD, Esquimalt (A-8). After Captain J.A. Duntze of HMS *Fisgard*, here in 1846-7. The former British naval base had its first buildings on this point.

DURIEU, N. of Hatzic (B-8). After Father Paul Durieu, OMI. Appointed a coadjutor bishop in 1875, Durieu became Bishop of New Westminster in 1890. He died in 1899.

DUTCH CREEK, N. of Columbia L. (C-12). After 'Dutchy,' a placer miner.

DUTCH LAKE, near Clearwater (D-9). 'Dutch' is a corrupted form of 'Deutsch' and refers to an early German settler.

E

EAGLE PASS, E. of Sicamous (C-10). Walter Moberly, who 'rediscovered' this pass (shown on Archibald McDonald's map of 1827), has left us the story of his naming of it in *The Rocks and Rivers of British Columbia*:

> In the summer of 1865, I was exploring the Gold range of mountains for the Government of British Columbia, to see if there was any pass through them. I arrived at the Eagle River, and on top of a tree near its mouth I saw a nest full of eaglets, and the two old birds on a limb of the same tree. I had nothing but a small revolver in the shape of firearms; this I discharged eight or ten times at the nest, but could not knock it down. The two old birds, after circling around the nest, flew up the valley of the river; it struck me then, if I followed them, I might find the much wished-for pass. I explored the valley two or three weeks afterwards, and having been successful in finding a good pass, I thought the most appropriate name I could give it was the 'Eagle Pass.' (p. 39)

EALUE LAKE, E. of Eddontenajon L. (J-5). From a Tahltan Indian word meaning 'sky fish.'

EARL GREY PASS, NE of Kootenay L. (C-11). In 1907 Earl Grey, Governor-General of Canada, travelled on horseback over this pass between Kootenay Lake and the headwaters of the Columbia. Actually Fred Wells (see *Wells*) discovered this pass, which was originally named after him, and Lord Grey is said to have objected to the change in name.

EARLE RANGE, N. of Sechelt Inlet (B-8). After Lieutenant Wallace Sinclair Earle, DLS, BCLS, killed in action in France in 1916. He surveyed in this area in 1910.

EARLS COVE [EARL'S COVE], Sechelt Pen. (B-7). After a Mr. Earl, who pioneered in the area with his family.

EBURNE, N. Arm of Fraser R. (B-8). W.H. Eburne arrived in British Columbia in 1875 and, in 1881, after attempts at farming, he opened a store on the mainland, opposite Sea Island. In 1885 he moved, taking over Sexsmith's store on Lulu Island and becoming postmaster of the North Arm post office, housed in the store. In 1892 he moved his premises to Sea Island, close to the Marpole Bridge, taking with him the post office, which was renamed Eburne. The mainland portion of Eburne became Marpole in 1916.

ECHO BAY MARINE PARK, Gilford I. (C-6). Echo Bay was originally Echo Cove, so named because of reverberations of sound from the cliff on one side of the bay.

ECSTALL RIVER, flows NW into Skeena R. at Port Essington (G-5). From the Tsimshian Indian word meaning 'a tributary,' or 'something from the side.'

EDASP LAKE, E. of Tuya L. (L-4). After Ed Asp, owner of a trapline here.

EDDONTENAJON LAKE, headwaters of Iskut R. (J-5). From a Tahltan Indian phrase meaning 'a little boy drowned.' According to legend a small boy was standing by the shore, trying to imitate the cry of a nearby loon. His mother rebuked the boy, warning him that if he copied the bird he would go into the water. The child persisted, fell into the lake, and drowned.

EDENSAW, CAPE, N. coast of Graham I. (G-3). After Charlie Edensaw (1810?-94), last of the great chiefs of the Masset Haidas. He became a chief upon the death of his uncle in 1842, but only after slaying a rival cousin in combat. A British naval officer who got to know him in 1853 wrote: 'He is decidedly an interesting character, and an example of what splendid abilities are only waiting culture among these Indians. He would make a Peter the Great or Napoleon, with their opportunities. He has great good sense and judgment, very quick, and as subtle and cunning as the serpent.'

EDGEWATER, NW of Radium (C-11). Edges the Columbia River.

EDINBURGH MOUNTAIN, N. of Port Renfrew (A-7). Named by a party of Scots who staked mineral claims in the area.

EDMONDS, Burnaby (B-8). After Henry Valentine Edmonds, born in Ireland in 1837. Edmonds was one of the original promoters of the Vancouver-New Westminster interurban railway.

EDZIZA, MOUNT, SE of Telegraph Cr. (J-4). This peak is named after the Edzertza family, a well-known Indian family living in the area.

EFFINGHAM ISLAND, Barkley Sound (A-7). In 1788 Captain Meares named a harbour here Port Effingham after Thomas Howard, third Earl of Effingham. In 1905 the island itself, hitherto named Village Island, was renamed Effingham Island.

EGERIA MOUNTAIN, Porcher I. (F-4). After HMS *Egeria*. The last ship of the Royal Navy to be stationed in BC waters, she was engaged in surveys from 1898 to 1910.

EGMONT, Jervis Inlet (B-8). After nearby Egmont Point, named after HMS *Egmont*, which served under Rear-Admiral Sir John Jervis (who later became Earl St. Vincent) at the Battle of St. Vincent, 14 February 1797.

EHAHCEZETLE MOUNTAIN, E. of Eddontenajon L. (J-5). From the Tahltan Indian word meaning 'sky mountain,' or 'mountain reaches the sky.'

EHOLT, NW of Grand Forks (B-10). After Louis Eholt, who began ranching in the area around 1890. Eholt was a busy rail centre during the mining activity at Phoenix and elsewhere early in the twentieth century.

EKINS POINT, Gambier I. (B-8). After Admiral Sir Charles Ekins, RN (1768-1855).

ELEPHANT CROSSING, near Holberg, VI (C-5). Years ago a warrant officer from eastern Canada was transferred to the Canadian Armed Forces base at Holberg. Crossing an access road to that base was another road, one used by the trucks of the Western Forest Products Company. Loggers, after they have unloaded a logging truck, commonly hoist its trailer portion up behind the driver's cab and carry it piggyback for its return journey, with the 'reach' (the long connecting bar) jutting out above the cab. The warrant officer, seeing this strange sight for the first time, remarked that the empty trucks looked like elephants holding up their trunks. Thereafter this intersection became known as Elephant Crossing, a sign here was embellished with pink elephants, and the name found its way into the *Gazetteer of Canada*.

ELIZA DOME, Esperanza Inlet (B-6). Also ELIZA EARS. After Francisco de Eliza, who in 1790 reestablished the Spanish base at Nootka abandoned the previous year. In 1791, with Narvaez, he conducted explorations on the west coast of Vancouver Island and in the Strait of Georgia.

ELK LAKE, N. of Victoria (A-8). Elk were once numerous in this area. The Saanich Indian name for the lake means 'drifts from place to place' and referred to a floating island of weeds. The Indians thought this was a monster, the sight of which brought bad luck.

ELK RIVER, flows SW into Kootenay R. (B-12). This is David Thompson's 'Stag River.'

ELKIN LAKE, N. of Taseko L. (D-8). After Ed Elkins, an Englishman who in 1890 became the first white settler in the Nemaia Valley, operating a trading post with his Indian wife. He left about 1900 after his brother, who had a post north of Tatlayoko Lake, was murdered by an Indian.

ELKO, SW of Fernie (B-12). Founded in 1905 and named after the Elk River, on whose bank it stands.

ELLICE, POINT, Victoria (A-8). After Edward Ellice (1781-1863). Ellice joined the NWC in 1805 and was largely responsible for its decision to unite with the HBC in 1821. He entered politics in Britain and became Secretary to the Treasury (1830-2) and Secretary for War (1832-4). He was Deputy Governor of the HBC from 1858 to 1863.

ELLIS CREEK, Penticton (B-10). After Tom Ellis (1845-1918), a young Irishman who arrived in the Penticton area in 1865 and became the first settler here. At the peak of his career, Ellis was a cattle baron owning more than 30,000 acres extending from Okanagan Lake to Osoyoos Lake. He worked his ranch hands hard but was a good boss. He and Mrs. Ellis were hospitable hosts, and their home, where Penticton now stands, was a favourite stopping-place for early travellers

ELLISON LAKE, N. of Kelowna (B-10). After Price Ellison (1852-1932), pioneer Okanagan rancher and farmer. Ellison came to British Columbia from Manchester in 1876. He became an MLA in 1898 and provincial Minister of Finance and Agriculture in 1910.

ELPHINSTONE, MOUNT, N. of Gibsons (B-8). This mountain on Howe Sound was presumably named after Captain J. Elphinstone, who commanded HMS *Glory* in Howe's famous naval victory in 1794 known as 'The Glorious First of June.'

ELSJE POINT, English Bay (B-8). Pronounced 'El-shuh.' This point, at the end of the Vancouver Maritime Museum's breakwater, commemorates Mrs. W.M. Armstrong, née Elsje De Ridder (1918-81), a former chairwoman of the Museum's board of trustees. Daughter of a former conductor of the Vancouver Symphony Orchestra, she was a warm-hearted and accomplished lady who, among many contributions to the cultural life of Vancouver, brought to prominence the city's Community Music School, now the Vancouver Academy of Music.

EMILY CARR INLET, Princess Royal I., Central Coast (E-5). After the Victoria eccentric (1871-1945) who, at her best, painted with genius BC landscapes and Indian scenes.

EMORY CREEK, S. of Yale (B-9). After one Emory, who washed for gold on nearby Emory Bar in the Fraser River before 1859.

EMPEROR FALLS, Mount Robson Park (F-10). Of these magnificent falls, A.O. Wheeler wrote:

> The total drop is 145 ft ... At a distance of 60 feet from the crest, the full volume of the water strikes a ledge and bounds outwards for 30 feet, creating a splendid rocket which gives the idea of a giant leap. There is such a feeling of majesty and power inspired by the spectacle that I christened it, 'The Emperor Falls,' and the rocket, 'The Emperor's Leap.'

ENDAKO RIVER, flows E. toward Fraser L. (G-7). Derived from a Carrier Indian name meaning 'ancient monster river.'

ENDERBY, SE of Salmon Arm (C-10). Originally known as Spallumcheen, Steamboat Landing, Fortune's Landing, Lambly's Landing, or Belvidere. One afternoon in 1887, Mrs. George R. Lawes had some friends in for tea. The Shuswap (then the Spallumcheen) River was in flood, moving one of the ladies to recite a poem by Jean Ingelow, 'The High Tide on the Coast of Lincolnshire.' It begins:

> The old mayor climbed the belfry tower,
> The ringers rang by two, by three;
> 'Pull, if ye never pulled before;
> Good ringers, pull your best,' quoth he.
> 'Play uppe, play uppe, O Boston bells!
> Play all your changes, all your swells,
> Play uppe "The Brides of Enderby."'

The musical name of Enderby so enchanted the ladies that they decided to make that the name of their settlement. Ottawa acquiesced, and Enderby post office opened on 1 November 1887.

ENGINEERS POINT, N. of Salmon Arm (C-10). During construction of the CPR, some of the supervising engineers had their camp on Shuswap Lake at this point.

ENGLEFIELD BAY, W. coast of Moresby I. (E-3). Named by Captain Vancouver in 1793 after his 'much esteemed friend' Sir Henry Charles Englefield, FRS, FSA (1752-1822).

ENGLISH BAY, Vancouver (B-8). This and nearby Spanish Bank commemorate the meeting of the English (under Captain Vancouver) and the Spanish (under Galiano and Valdes) in this area in June 1792. (See *Spanish Bank*.)

ENGLISH BLUFF, Tsawwassen (B-8). So named because it was on the British side of the nearby international boundary.

ENGLISH COVE, E. side of Christina L. (B-10). After a group of English remittance men who led a colourful life here at the end of the nineteenth century.

ENGLISHMAN RIVER, flows N. near Parksville (B-7). Named, at least as early as 1883, in commemoration of an otherwise forgotten Englishman who drowned trying to cross this stream.

ENOS LAKE, N. of Nanoose (B-7). After John Enos, the name adopted by a Portuguese who preempted here in 1864. Captain Walbran has a story that, when the Indians began to harass Enos, he flitted around them, dressed in white, playing a ghost in the night, and so scared them away.

ENTERPRISE, NW of Lac la Hache (D-9). In 1916 the Enterprise Cattle Company was formed and, acquiring four ranches, merged them into the Enterprise Ranch, after which the PGE (now the BC Railway) named its station here.

ENTIAKO LAKE, S. of Natalkuz L. (F-7). According to Father Ouilette, this Carrier Indian name means 'lake with a brown-coloured creek.'

ERICKSON, E. of Creston (B-11). After E.G. Erickson (1857-1927), CPR superintendent at Cranbrook from 1904 to 1908.

ERMATINGER, MOUNT, Hamber Park (E-10). After Edward Ermatinger, who, with his younger brother Francis, entered the service of the HBC as a clerk in 1818. They were posted to the Columbia Department in 1825.

ERRINGTON, W. of Parksville (B-7). An early resident, Duncan McMillan, is presumed to have taken this name from a passage in Sir Walter Scott's 'Jock of Hazeldean.' This portion runs: 'Young Frank is chief of Errington / And Lord of Langley-dale ...' This Errington is a small village in Northumberland, much involved in the old border wars with the Scots.

ERROCK, LAKE, SW of Harrison Hot Springs (B-8). Originally Squakum Lake, but this inelegant name, for a kind of spring salmon, was changed to Loch Erroch for the post office maintained here from 1892 to 1896. (The name was apparently borrowed by A.W. Ross, local rancher and MP, from Loch Ericht in Inverness-shire, Scotland.) When the post office was reestablished in 1940, it was as Lake Errock, but Victoria insisted upon calling the lake Squakum Lake for about thirty more years.

ESOWISTA PENINSULA, S. of Meares I. (B-6). Named after a village on its west coast, whose Nootka Indian name can be translated as 'battle place.'

ESPERANZA INLET, W. coast of VI (B-6). In 1778 Captain Cook gave the name of Hope Bay to the stretch of coast between Estevan Point and Cape Cook. Malaspina translated the name into Spanish in 1791 and restricted it to this inlet.

ESPINOSA INLET, arm of Esperanza Inlet (B-6). After Lieutenant Josef de Espinosa, who explored this area under Malaspina's directions in 1791.

ESQUIMALT, W. of Victoria (A-8). From the Straits Salish word meaning 'place of gradually shoaling water,' referring to the mudflats at the mouth of Mill Stream at the head of the harbour.

In 1790 Sublieutenant Quimper named the harbour Puerto de Cordova, after the Mexican Viceroy. Later this name was transferred to a bay northeast of Victoria that had been known as Cormorant Bay, and the Indian name was restored to Esquimalt Harbour.

ESSONDALE, SW of Coquitlam (B-8). After Dr. Henry Esson Young, Provincial Secretary when the provincial mental hospital (now renamed Riverview) was established here. The Minister of Education from 1907 to 1916, he became the 'father' of UBC.

ESTEVAN POINT, S. of Nootka Sound (B-6). In 1774 the Spanish authorities, alarmed by reports of Russian penetration into the Pacific Northwest, sent north from Monterey Lieutenant-Commander Juan Perez with the *Santiago* and eighty-seven officers and men. He explored part of the coast of the Queen Charlotte Islands, then turned south. Bad weather kept him from sighting Vancouver Island until 7 August, when he moved inshore in the vicinity of Nootka Sound and, without landing, made contact with the local Indians.

Perez named Estevan Point after his second lieutenant, Estevan José Martinez, making this the first place in British Columbia to be named by a white man. The nearby Perez Rocks were named later.

Captain Cook, arriving four years after Perez but ignorant of his nomenclature, named this Breakers Point.

ETHELBERT, MOUNT, NW of Radium (C-11). Named after the first nun to ascend the upper Columbia River. She died aboard Captain Armstrong's riverboat, the *Ptarmigan,* and was buried as Sister Ethelbert (a surprising name for a nun – one wonders if Armstrong got it wrong). In any event Armstrong named the mountain after her.

ETHEL F. WILSON PARK, S. of Babine L. (G-7). Although childless, Ethel Wilson and her husband Walter, a BC Forest Service ranger, did much for the children of the Burns Lake district, among other things holding summer camps for the forestry-oriented Young Ranger Band.

EUCHINIKO RIVER, flows E. into West Road R. (F-8). Two meanings have been advanced for this Carrier Indian word: 'blueberry creek,' and 'good feeding place,' which are not mutually exclusive!

EUCHU REACH, Natalkuz L. (F-7). This Carrier Indian name means 'first lake.'

EUTSUK LAKE, Tweedsmuir Park (F-6). According to Father Morice, the first white man to see this lake, his Indian companions named it after him. The authorities in Victoria refused to accept Morice's 'Morice Lake' and renamed it Eutsuk, an almost untranslatable name, used in the sense that this lake was very far off (from Ootsa Lake).

EVA LAKE, Revelstoke NP (D-10). After Eva Hobbs, an enthusiastic member of the Revelstoke Mountaineering Club in the years before World War I.

EWEN SLOUGH, Fraser R. mouth (B-8). After Alexander Ewen, an early Scottish cannery owner, so dour that he was reputed never to have laughed.

EWING, W. side of Okanagan L. (C-10). Originally Ewing's Landing after R.L. Ewing, who settled here in the early 1900s.

EXACT POINT, Skidegate Channel, QCI (F-3). After the American vessel *Exact,* here in 1851.

EXETER, W. of 100 Mile House (D-9). This BCR station is named after the Marquess of Exeter, who bought a large tract of land before World War I around the 100 Mile House and established the Bridge Creek Cattle Ranch.

EXLOU, N. Thompson R. (D-9). A curious name coined to show that this settlement is just outside the village of Louis Creek.

EXTENSION, S. of Nanaimo (B-8). Having widely advertised the excellence of its Wellington coal, the Wellington Collieries Co., when it opened a mine here, thirteen miles from Wellington, called the place Wellington Extension. The name was abbreviated when a post office was established. Earlier the locality was known as Southfield because of its position relative to Nanaimo.

F

FACE LAKE, SW of Kamloops (C-9). Translation of the Shuswap Indian name Ski-kloosha, meaning 'face,' with possible reference to the mirror effect of the clear water.

FAIRMONT HOT SPRINGS, S. of Windermere (C-12). Indians and white men passing through the country bathed in the hot springs here long before Mr. and Mrs. Brewer, around 1888, built their Fairmont Hotel, a log house within half a mile of the springs.

FAIRVIEW, Vancouver (B-8). This district was named by L.A. Hamilton, the CPR land commissioner. A newspaper advertisement announced in 1891 that 'the clearing of the Land being nearly completed in that new and beautifully situated part of the VANCOUVER TOWNSITE known as FAIRVIEW,' the CPR was now offering lots for sale.

FAIRWEATHER MOUNTAIN, Alaska-BC boundary (K-l). Named Mount Fair Weather in 1778 by Captain Cook, who was enjoying good weather when he discovered it. Its peak is the highest point in British Columbia (15,300 feet, or 4,663 metres).

FALKLAND, NW of Vernon (C-10). This little settlement was named in compliment to Colonel Falkland G.E. Warren, CMG, CB. Retired from the Royal Horse Artillery, he made his home here in 1893 and proved a good friend to the other settlers.

FALSE CREEK, Vancouver (B-8). So named by Captain G.H. Richards, RN, in the late 1850s presumably because, despite its promising entrance, this small inlet soon ended in mudflats. (In England the word 'creek' applies to any narrow indentation in a coast.) Galiano's name for Boundary Bay, Ensenada del Engeño (Mistake Bay), is very similar.

FANNIN RANGE, N. of Vancouver (B-8). These mountains are named after John Fannin (1839-1904), one of the Overlanders of 1862. He was the first curator of the Provincial Museum in Victoria.

FANNY BAY, S. of Courtenay (B-7). None of the various explanations – comical, romantic, local, or historical – for this name can be regarded without scepticism. Fanny Bay and Henry Bay (Denman Island) six miles to the north first appeared in the 1864 edition of the *Vancouver Island Pilot*. If Captain G.H. Richards, RN, whose surveys provided the basis for that volume, knew who Fanny and Henry were, he apparently left that information unrecorded.

FARNHAM, MOUNT, W. of Invermere (C-11). This fine peak, over 11,000 feet or 3,352 metres high, is named after Paulding Farnham from New York. Between 1898 and 1904, he exhausted his considerable private means while trying to make a success of the nearby Red Line (Ptarmigan) mines. His wife was a distinguished sculptor. Of him it was written: 'Mr. Farnham stands out like his mountain among mining men in this – he lost his fortune like a man and paid every cent he owed (an unusual thing with defunct mining companies).'

FAUQUIER, Lower Arrow L. (B-10). After F.G. Fauquier, pioneer rancher and fruit grower. In the 1890s he served as government agent, mining recorder, customs officer, and first policeman at Nakusp.

FERGUSON POINT, Vancouver (B-8). This promontory in Stanley Park is named after Alfred Graham Ferguson, a Vancouver building contractor devoted to the development of Stanley Park. In 1888 he became the first chairman of the Vancouver Parks Board – since he was an American, the swearing-in ceremony was quietly dispensed with. He died in San Francisco in 1903.

FERNIE, E. Kootenay (B-12). Originally Coal Creek. Named after William Fernie (1837-1921). An Englishman by birth, Fernie travelled in Australia, New Zealand, and South America before arriving in British Columbia. In 1861 he prospected for gold near Revelstoke, later moved on to the Cariboo, and in 1864 arrived at the Wild Horse Creek gold camp. The next year he was foreman on the extension of the Dewdney Trail to Wild Horse Creek. Fernie was government agent and mining recorder for the Kootenay district from 1876 to 1882.

Prospecting in 1887, Fernie discovered coal on one of the tributaries of Michel Creek. Further explorations over the next two years by Fernie and his brother Peter resulted in the discovery of the vast coalfields in the Elk River valley and the adjacent areas. William Fernie became wealthy through his subsequent association with the Crows Nest Pass Coal Company. According to his own recollections, 'In 1898 the townsite of Fernie was cleared and the site surveyed into streets and lots. The town was named after me as being the discoverer of the coal, a director of the company and having had control of all the field work until the work got too big for one man to handle.' Such was the beginning of the Crowsnest coal industry.

FEUZ PEAK, Glacier NP (D-11). This and nearby Hasler Peak, both part of Mount Dawson, were named by Professor Charles Fay and Herschel C. Parker of the Appalachian Mountain Club after the guides who accompanied them in 1899 on the first ascent of Dawson. Edouard Feuz Sr. and Christian Hasler were two of the Swiss guides brought out by the CPR to encourage wealthy alpinists to make expeditions to Glacier, Yoho, and Banff National Parks. These men lived in Edelweiss, a 'Swiss village' just outside Golden.

FIELD, Yoho NP (D-11). After Cyrus Field (1819-92) of Atlantic cable fame, who visited the area in 1884.

FINDLAY CREEK, flows E. into Kootenay R. (C-12). Findlay (or more correctly Finlay) was a part-Indian son of Jaco Finlay, an associate of David Thompson. A prospector and fur trader, he found gold on this stream in 1863.

FINLAY RIVER, flows SE into Williston L. (J-7). After John Finlay of the NWC, who ascended the river (possibly as far as its junction with the Ingenika River) in 1797. The lower part of Finlay River, including Finlay Forks (where the Finlay and the Parsnip came together to form the Peace), has been swallowed up in Williston Lake as a result of the building of the W.A.C. Bennett Dam.

FINLAYSON ARM, Saanich Inlet (A-8). After Roderick Finlayson (1818-92). A native of Ross-shire, he arrived in Canada in 1837 and entered the service of the HBC. He was second in command to Chief Trader Charles Ross when the latter was placed in charge of Fort Victoria in 1843. Upon Ross's death in 1844, Finlayson took over command of the fort, and it is he who must be considered the real father of Victoria. Finlayson took a leading part in the political life of the infant colony of Vancouver Island. He became a Chief Factor in 1859 and retired in 1872. Also FINLAYSON CHANNEL, Milbanke Sound.

FINTRY, W. side of Okanagan L. (C-10). The land here was bought in 1909 by Captain James Cameron Dun-Waters, who named his estate Fintry after his family home in Stirlingshire.

FIRESIDE, near confluence of Kechika R. and Liard R. (L-6). After the Fireside Inn on the Alaska Highway.

FISGARD ISLAND, entrance to Esquimalt harbour (A-8). After HMS *Fisgard*, forty-two guns, in these waters in 1846. Various points and islands in Esquimalt harbour are named after the *Fisgard*'s officers. In fact, just as Victoria's early names are almost all reminiscent of the HBC, so Esquimalt's remind us of the Royal Navy.

FISHEM LAKE, W. of Taseko L. (D-8). Obviously good fishing here. The Chilcotin Indian name is Nach'eẑnadinlin, meaning 'river going off into lake.' A very short river empties Fishem Lake into Upper Taseko Lake.

FISHER CHANNEL, leads into Dean Channel (E-6). Named by Captain Vancouver after his 'much respected friend,' the Reverend John Fisher, DD, vicar of Stowey.

FISHER, MOUNT, E. of Fort Steele (B-12). After Jack Fisher, who discovered gold on Wild Horse Creek in 1863. The mining camp that sprang up here after this discovery was named Fisherville.

FISHERMANS COVE [FISHERMAN'S COVE], entrance to Howe Sound (B-8). Got its name around 1888 when some Newfoundland fishermen and their families settled here.

FITZ HUGH SOUND, Central Coast (D-6). Named 'Fitzhugh Sound' in 1786 by Captain James Hanna, who, sailing from China, became the first man to come to the British Columbia coast to trade for furs. His Fitzhugh was probably William Fitzhugh, a partner of Captain John Meares in the latter's expedition of the same year.

FITZSTUBBS CREEK, N. of Slocan L. (C-11). After Captain Napoleon Fitzstubbs, appointed gold commissioner at Nelson in 1891. His superior English manner caused a lot of resentment among the locals.

FITZWILLIAM, W. of Yellowhead L. (E-10). This settlement takes its name from nearby Mount Fitzwilliam, named in 1863 by Viscount Milton after his family's senior title. (His father was the sixth Earl of Fitzwilliam.)

FLAMINGO INLET, SW coast of Moresby I. (E-4). After the trawler *Flamingo* (Captain Freeman).

FLATHEAD RIVER, E. Kootenay (B-12). In northwestern Montana Flathead Range, Flathead Lake, and Flathead River (whose upper stretches are in British Columbia) bear witness to the fact that this was the homeland of the Flathead Indians. The origin of the band's name is uncertain. Some believe that at a remote time certain members of the band, like Indians on the Pacific coast, artificially deformed the skulls of their infants. Others believe the name arose from the Indian sign language that indicated members of this band by holding the hands tight on either side of the head. The two explanations are not necessarily contradictory.

FLEMING ISLAND, Barkley Sound (A-7). This and nearby Sandford Island are named after Sir Sandford Fleming. (See *Sir Sandford Range*.)

FLEWIN POINT, entrance to Port Simpson (G-4). Commemorates a family whose history goes back to British Columbia's roots. Thomas Flewin, who had joined the HBC in 1852, arrived in Victoria in 1853. His son, John (born in 1857), was the first government agent, stipendiary magistrate, and gold commissioner for the Skeena district, and this point was named for him.

FLOODS [FLOOD'S], W. of Hope (B-9). After W.L. Flood, early builder and sawmill operator.

FLORENCIA BAY, Pacific Rim NP (A-7). In December 1860 HMS *Forward* was towing the badly damaged Peruvian brig *Florencia* from Nootka to Victoria when bad weather and boiler trouble forced the gunboat to cast off the tow. As a result the *Florencia* was wrecked here.

FLORES ISLAND, Clayoquot Sound (B-6). Named in 1791 after Don Manuel Antonio Flores, fifty-first Viceroy of Mexico (1787-9).

FLOURMILL CREEK, flows SE into Clearwater R. (D-9). Since this creek enters the Clearwater River four miles north of its confluence with Mahood River, the name originally proposed for it was Four Mile Creek. The province already having plenty of Four Mile Creeks, it was decided in Victoria to juggle the letters a bit and come up with a more interesting and distinctive name. Thus, it became Flourmill Creek, though there has never been any flourmill here.

FLY HILL, W. of Salmon Arm (C-10). Flies and mosquitoes are numerous here.

FOGHORN MOUNTAIN, SE of Clearwater (D-10). 'Foghorn' seems to have been a popular name for mineral claims – there were Foghorn claims here, near Greenwood, and near Ymir, and this fact casts doubt on the picturesque local story about a foghorn being used to help sheepherders lost in the mountain mists.

FONTAS RIVER, flows W. into Sekani Chief R. (K-9). Fontas or Fantasque was the chief of a Sekani Indian band that had traditional fishing rights on this river.

FORBIDDEN PLATEAU, W. of Courtenay (B-7). So named because the Indians believed it to be inhabited by evil spirits.

FORD PASS, NW of Spatsizi Park (J-5). The policeman at Telegraph Creek told old man Ford that, because of his age, he was not to go trapping any more. Ford was not going to be stopped by any policeman – he went out on his line and was never seen again.

FORDING RIVER, flows S. into Elk R. (B-12). Given this name by G.M. Dawson in 1884 because the trail frequently crossed and recrossed the stream, requiring much fording.

FOREST GROVE, NE of 100 Mile House (D-9). Not after the town in Oregon, but after a grove here. The name was suggested by a sister of E.C. Phillips, the local storekeeper, at a public meeting held to choose a name for the post office.

FORSTER CREEK, flows E. into Columbia R. near Radium (C-11). After Harold E. Forster, alpinist, rancher, mining speculator. Born in Ontario and orphaned at the age of one, he was sent to England and educated at Eton and Cambridge. He came to British Columbia in 1890. In 1898 he acquired Firlands Ranch on this creek, where he built a spacious manor house, complete with tennis court and croquet lawn. He served a term as MLA for the riding. In 1940, living apart from his wife, with most of the house closed off, he and a friend spent an evening drinking with a young Indian, who shot them and burned down the house in an attempt to conceal his crime.

FORT FRASER, E. of Fraser L. (G-7). After Simon Fraser, who established the post here in 1806.

FORT LANGLEY, Fraser Valley (B-8). After Thomas Langley, HBC director (1800-30). In October 1824 Governor Simpson of the HBC, under the misapprehension that the Fraser River provided a navigable route to the Interior, wrote: 'I imparted to Mr. McMillan my views in regard to extending the trade to the Northward of Fort George [Astoria] and pointed out to him the importance of having an Establishment at the mouth of Frazer's River ...' Later the same year, Chief Trader McMillan made a reconnaissance of the lower Fraser Valley and in 1827 founded Fort Langley. In 1839 the original fort was abandoned and a new establishment (present Fort Langley) built several miles farther up the river, where the land was better for agriculture. The colony of British Columbia was proclaimed here in 1858, but the capital was soon after moved to New Westminster. Fort Langley then went into a slow decline. The HBC closed its store here in 1896, and the old fort began to rot away. Restoration of the fort began in 1957-8 as part of the centennial of British Columbia, though by then only one of the original buildings was still standing. Fort Langley is now a National Historic Park administered by the federal government.

FORT NELSON, near confluence of Muskwa R. and Fort Nelson R. (K-8). Apparently named after Horatio, Lord Nelson (1758-1805), the British naval hero.

Both Fort Nelson and a post at McLeod Lake seem to have been founded in 1805, and the two places compete for the title of 'oldest white settlement in British Columbia.' (See, however, *Fort St. John.*)

FORT RODD, Esquimalt (A-8). After John Rashleigh Rodd, first lieutenant on HMS *Fisgard,* on the Pacific Station from 1844 to 1847. Guns were installed here in 1864 to protect Esquimalt harbour.

FORT RUPERT, E. of Port Hardy (C-6). Named after Prince Rupert, the first Governor of the HBC, Fort Rupert was founded in 1849 when the company decided to work the coal seams discovered in the vicinity some years earlier (see *Beaver Harbour*). Commander Mayne, who visited the post around 1860, wrote: 'Fort Rupert is the newest and best built station of the Hudson [sic] Bay Company I have seen, and the gardens are very nicely laid out. Of course, like all the rest, it is stockaded and has its gallery and bastions. It stands almost in the middle of the Indian village.' All that now remains of the HBC fort is a single chimney.

FORT ST. JAMES, Stuart L. (G-7). Founded by Simon Fraser in 1806, it was referred to as the Stuart Lake post until 1822, when it became Fort St. James.

The reason for the new name is not known. Governor Simpson, visiting here in 1828, described the post as 'the capital of Western Caledonia.' It was in fact the administrative centre for the HBC's department of New Caledonia. The original buildings have disappeared, though five dating from the late nineteenth century survive. The fort has been restored by the National and Historic Parks Branch.

The Indian name for the site, 'Na'-Kra-ztli' (Nakasleh) means 'arrows floating by' and refers to a legendary battle with dwarfs that left the Stuart River full of arrows where it leaves the lake.

FORT ST. JOHN, Peace R. district (I-9). Many of the early fur-trading forts were rebuilt on different locations, sometimes with changes of name. Rocky Mountain House, built in 1799 some six miles upstream from modern Fort St. John, can be regarded as the original fort. In this case, Fort St. John is the oldest white settlement on the mainland of British Columbia.

FORT STEELE, NE of Cranbrook (B-12). Originally named Galbraith's Ferry, after R.L.T. Galbraith, who ran a ferry across the Kootenay River here. In 1887, in consequence of an appeal from Colonel Baker, the local magistrate, who anticipated trouble with Chief Isadore's band of Indians, Major Steele (later Major-General Sir Samuel Steele) arrived here with a force of two inspectors and seventy-five NCOs and men of the North West Mounted Police. In his memoirs, *Forty Years in Canada,* Steele has left the following account of the building of the NWMP post here:

> I asked Galbraith for permission to build on a point of land which was in an angle formed by the confluence of Wild Horse Creek and the Kootenay. He very kindly gave me a lease of the ground for as long as we should require it for the modest sum of one dollar. The site was an ideal one. It commanded the trails to Tobacco Plains, the Crow's Nest Pass, Moyea [Moyie], and the Columbia Lakes, and was the most central situation from which to communicate with the Indians and give protection to the whites. (p. 248)

Major Steele soon established good relations with Chief Isadore, who surrendered an escaped prisoner whom he had been harbouring. With the end of the crisis, the BC government declined to pay for the continuance of the NWMP post here, and the force was withdrawn. Just before Steele and his men left, the name of the adjacent settlement was changed, courtesy of Galbraith, to Fort Steele. Visitors to the modern 'reconstruction' of Fort Steele may be interested to know that the NWMP 'fort' had no stockade.

Steele was born in Medonte, Ontario, in 1849, the fourth son of Captain Elmes Steele, RN. At the age of sixteen, he joined the militia during the Fenian Raids, served against the Métis in the Red River Rebellion of 1870, and joined the NWMP as sergeant major in 1873. He served with the force in the Riel Rebellion and in the Yukon gold rush. After service in the Boer War, he became general officer commanding Military District 13 (Calgary). He died in 1919.

FORTUNE CREEK, S. of Enderby (C-10). After Alexander Leslie Fortune, one of the Overlanders of 1862. In 1866 he became the first settler in the North Okanagan, where he lived until his death in 1915. A monument to him in Enderby (formerly Fortune's Landing) describes him as 'A Friend of all Classes and Creeds – Indian and White. A Gracious Gentleman.'

FORWARD INLET, Quatsino Sound (C-5). After the Royal Navy's little gun-boat *Forward*, two guns, 233 tons, which did sterling service in BC waters from 1860 to 1869.

FOSSLI PARK, Sproat L. (B-7). Takes its name from Fossli, east of Eidfjord, Norway. *Fossli* is the Norwegian word for a waterfall.

FOSTER PEAK, Kootenay NP (D-11). After Colonel (later Major-General) W.W. Foster. (See *Colonel Foster, Mount.*)

FOSTHALL, W. side of Upper Arrow L. (C-11). Gets its name from the clerk in charge of the post maintained here in the mid-nineteenth century by the HBC.

FOUL BAY, Victoria (A-8). See *Gonzales Bay.*

FOUNTAIN, N. of Lillooet (C-9). Originally 'The Fountain.' Commander Mayne, who was here in 1859, gives us the explanation of the name: 'Fountain is a flat at a sharp turn in the [Fraser] river 12 miles below Pavillon [*sic*] and derives its name from a small natural fountain spouting out in the middle of it.'

FOX RIVER, flows SE into Finlay R. (J-7). After William Fox, manager of the HBC trading post at Fort Graham, south of here, around the end of the nineteenth century. He was 'a well informed man, great reader, and very obliging.'

FRANÇOIS LAKE, W. of Fraser L. (F- and G-6 and 7). According to Father Morice, the Indians called this Nitapoen (meaning 'lip lake') because of its shape. Due to confusion with the Indian word *neto*, meaning 'white man' (most white men seen here being French-Canadian voyageurs), the Indian name was mistranslated Lac des Français (French Lake), which became François Lake.

FRANKLIN RIVER, flows W. into Alberni Inlet (B-7). Named in 1864 for Selim Franklin, auctioneer, real estate dealer, political figure in Victoria, and chair-man of the Exploration Committee, which backed Dr. Robert Brown's Vancouver Island Exploration Expedition of that year.

FRANKLIN RIVER, flows S. into Knight Inlet (D-7). After Benjamin Franklin, a rancher who explored a Klinaklini route from Tatla Lake to Knight Inlet in 1892, hoping to find a way of driving his cattle down to the coast.

FRASER LAKE, E. of François L. (G-7). After Simon Fraser, who spent the winter of 1806-7 here. The Indian name for Fraser Lake was Nat-leh, which means 'they return,' referring to the salmon.

FRASER RIVER. After Simon Fraser (1776?-1862). Fraser was born in Bennington, Vermont. At the end of the American Revolution, his Loyalist family moved to Canada, where he joined the NWC in 1792 and was admitted as a partner in 1802. In 1805 he was chosen to open to the company's trade the country west of the Rocky Mountains. Between May and July 1808, with a party of twenty-three others (two clerks, two Indians, and nineteen French-Canadian voyageurs), he made his great journey down the Fraser River from modern Prince George to the river's mouth south of modern Vancouver. It was a bitter disappointment for Fraser when he reached the ocean to discover that the river down which he had travelled could not be the Columbia.

The Fraser River was first discovered by Sir Alexander Mackenzie during his journey to the Pacific in 1793. In the map printed with his *Voyages* in 1801, Mackenzie called the river 'Tacoutche Tesse or Columbia River.' Father Morice said 'Tacoutche Tesse' came from *tacoutche*, a corrupted form of a Carrier word meaning 'one river within another' (possibly referring to the confluence of the Fraser and Nechako), and *tesse* or *desse*, meaning 'river' in the language of Mackenzie's eastern Indian companions. The coastal Indians called the Fraser the Cowichans' River. The Spaniards never found their way up the mouth of the Fraser, but in 1791, finding evidence that they were near the mouth of a major river, they named it the Rio Floridablanca in honour of the Prime Minister of Spain. The Fraser was also known in early days as the New Caledonia River and as the Jackanet River. It was named after Simon Fraser in 1813 by David Thompson. By a pleasing coincidence, the Thompson River was given its name by Fraser.

FRENCH CREEK, N. of Revelstoke (D-10). During the Big Bend gold rush of 1865, some French-Canadians, former engagés of the HBC, worked this stream.

FRENCH BEACH PARK, W. of Sooke (A-8). After James George French, who settled here around 1890. A pioneer conservationist, he maintained wildlife protection areas in the Victoria and Sheringham Point areas. He was a brother of Sir John French, later Earl of Ypres, the British Commander-in-Chief during the first part of World War I.

FRESHFIELD, MOUNT, Alberta-BC boundary (D-11). Named in 1899 after Sir Douglas Freshfield, president of the Alpine Club of Great Britain.

FREYA, MOUNT, Valhalla Range, W. of Slocan L. (B-11). After the Norse goddess of fertility and sexual love. Other peaks in the range take their names from Scandinavian mythology.

FRIENDLY COVE, Nootka Sound (B-6). Although visited by Captain Cook while he was at Nootka in 1778, the cove did not receive its name until James Strange put in here in 1786. The local Indians proved decidedly unfriendly during the Cook Bicentennial of 1978, not permitting white visitors to come ashore. The Indian name for the village here, Yuquot, means 'village exposed to the wind.'

FROMME, MOUNT, N. of Vancouver (B-8). After J.M. Fromme, 'Father of Lynn Valley,' the lumber camp foreman who in 1899 built the first house in the valley. From 1924 to 1929, he was reeve of North Vancouver District. He died, aged eighty-three, in 1941.

FULFORD HARBOUR, Saltspring I. (A-8). Named after Captain John Fulford, commanding HMS *Ganges,* which served on this coast in 1858-60.

FURRY CREEK, flows W. into Howe Sound (B-8). After Oliver Furry, trapper and prospector, who in 1898 staked claims to the rich copper deposits later worked at Britannia Beach.

G

GABRIOLA ISLAND, E. of Nanaimo (B-8). Henry Wagner maintains – in his authoritative *Cartography of the Northwest Coast of America* (2:389) – that the name originally given by the Spaniards in 1791 to the southeastern part of this island was Punta de Gaviota (Cape Seagull) but that this was corrupted first into Gaviola by the Spaniards themselves and then into Gabriola.

GALBRAITH CREEK, flows S. into Bull R. (B-12). After R.L.T. Galbraith, who came to the Cranbrook area in 1872 and was the Indian Agent here for many years. (For Galbraith's Ferry, see *Fort Steele.*)

GALENA BAY, Upper Arrow L. (C-11). Galena is lead sulphide ore, the principal source of lead.

GALIANO ISLAND, Gulf Islands (A-8). After Dionisio Alcala Galiano, captain of the Spanish vessel *Sutil*, which explored this area in 1792. Galiano was killed, fighting against the British, in the Battle of Trafalgar in 1805.

GALLOWS POINT, Protection I., Nanaimo (B-8). Two Indian murderers were hanged here in 1853. (See also *Chase River.*)

GAMBIER ISLAND, Howe Sound (B-8). After Admiral of the Fleet James, Lord Gambier (1756-1833). As captain of HMS *Defence*, he took part in the victory of 'The Glorious First of June' in 1794, in which he demonstrated that the puritanical regime that he imposed on his crew did not keep them from fighting effectively. In 1807 he became Commander-in-Chief of the Baltic Fleet and bombarded Copenhagen into surrender in the same year. In 1809 Gambier tarnished his honour at the Battle of Aix Roads by refusing to bring his big ships of the line into action after Lord Cochrane, finest of frigate captains, had scattered and demoralized the French in a preliminary attack with fireships.

GANG RANCH, S. of Williams L. (D-8). The first ranch in the Interior of British Columbia to use a double-furrowed gang plough.

GANGES HARBOUR, Saltspring I. (A-8). Earlier known as Admiralty Bay, it takes its name from HMS *Ganges*, eighty-four guns, flagship on the RN's Pacific Station 1857-60. The *Ganges*, built in 1821, was the last British sailing battleship commissioned for service outside home waters.

GARDNER CANAL, E. of Hawkesbury I. (F-5). Named by Captain Vancouver after his friend and former commander, Captain Alan Gardner (later Admiral Lord Gardner, Commander-in-Chief, Channel Fleet). Gardner had strongly recommended Vancouver for the command of his famous expedition. In

1860 Captain Richards, RN, named MOUNT GARDNER on Bowen Island after this admiral.

GARIBALDI, MOUNT, N. of Squamish (B-8). Also GARIBALDI PARK. After the Italian patriot and soldier (1807-82). A colourful story has it that Mount Garibaldi was named by an Italian serving as a sailor on a survey ship, the mountain being in view on Garibaldi's birthday. All we can say with certainty is that the name, whatever its origins, was officially conferred by Captain Richards of HM survey ship *Plumper* sometime around 1860. Garibaldi Park was created in 1927. Garibaldi station on the BCR was formerly Daisy Lake station.

GARRY POINT, SW end of Lulu I. (B-8). After Nicholas Garry, Deputy Governor of the HBC 1822-35. Named by Captain Aemilius Simpson of the HBC when he brought the schooner *Cadboro* into the mouth of the Fraser in 1827, the first vessel to enter the river.

GASPARD CREEK, flows E. into Fraser R. (D-8). After one of the early settlers in the Cariboo, Isidore Versepuche, also known as Vespuois Gaspard, who preempted on nearby Dog Creek in 1861 and built a flour mill, bringing in millstones from California.

GASTOWN, Vancouver (B-8). Here in 1867 'Gassy Jack' Deighton built his saloon, which became the centre of a settlement later to be named Vancouver (q.v.).

GATES RIVER, flows NE into Anderson L. (C-8). Also GATES STATION on the BCR. After James Gates, who settled here around 1900.

GEDDES, MOUNT, N. of Mount Waddington (D-7). After M.D. Geddes of Calgary, an outstanding alpinist who was killed on Mount Lefroy in 1927.

GELLATLY, W. side of Okanagan L. (B-10). After David Erskine Gellatly, who arrived in the Okanagan in 1893, settling near Peachland. At one time 'Tomato King of the Okanagan,' he pioneered in the growing of market produce. On his trips to Calgary, where he had a wholesale fruit business, 'he dressed up very well with a waxed moustache, as if he were a laird.'

GENELLE, N. of Trail (B-11). After the Genelle brothers, active in the Kootenay lumber industry as early as 1890.

GENOA BAY, N. side of Cowichan Bay (A-8). Since an Italian, Giovanni Baptiste Ordano, came here in 1858, it seems likely that Genoa Bay commemorates his birthplace. Later that year Ordano opened the first store in the district at Tzuhalem.

GEORGIA, STRAIT OF, (A-8, B-7 and 8). This is the euphonious 'Gran Canal de Nuestra Senora del Rosario la Marinera' of the Spanish explorers. Captain Vancouver tells of his naming of it:

> On Monday, the 4th [of June 1792], they [his seamen] were served as good a dinner as we were able to provide for them, with a double allowance of grog to drink the King's health, it being the anniversary of His Majesty's birth; on which auspicious day, I had long since designed to take formal possession of all the countries we had lately been employed in exploring, in the name of, and for His Britannic Majesty, his heirs and successors.
>
> To execute this purpose, accompanied by Mr. Broughton and some of the officers, I went ashore about one o'clock, pursuing the usual formalities which are generally observed on such occasions, and under the discharge of a royal salute from the vessels, took possession of the coast, from that part of New Albion, in the latitude of 39°20' north and longitude 236°26' east, to the entrance of this inlet of the sea, said to be the supposed straits of Juan de Fuca: as likewise all the coast islands, etc. within the said straits, as well on the northern as on the southern shores; together with those situated in the interior sea we had discovered, extending from the said straits, in various directions, between the northwest, north, east, and southern quarters; which interior sea I have honored with the name of THE GULF OF GEORGIA.

On 27 November 1858, Captain G.H. Richards wrote to the Hydrographer of the Royal Navy: 'You will observe that I have called the Gulf of Georgia a Strait in the Chart; it seemed a more appropriate name and I thought it better to alter it at once.'

GEORGINA POINT, Mayne I. (A-8). After Lady Georgina Mary Seymour, wife of Rear-Admiral Sir George F. Seymour, RN, in command of the Pacific Station 1844-6. The discovery of an English penny of 1784 on this point supports the theory that one of Captain Vancouver's boat parties camped here on 12 June 1792.

GERMANSEN LANDING, Omineca (H-7). After James Germansen, born in St. Paul, Minnesota, who discovered gold on Germansen River in 1870.

GERRARD, S. end of Trout L. (C-11). A locality name, often spelled Gerard, is all that remains of the northern terminus of the Lardeau-Gerrard line once operated by the Kootenay and Arrowhead Railway, an alias of the CPR. Named after G.B. Gerrard, at one time manager of the Kaslo branch of the Bank of British North America.

GIBSON PASS, Manning Park (B-9). After Luke Gibson, who arrived in Hope around 1912 and took pack trains between Hope and Princeton. He maintained stables at both places and at Chilliwack.

GIBSONS [GIBSON'S], Howe Sound (B-8). Originally 'Gibson's Landing,' but in 1947 local businessmen persuaded the post office to drop the 'Landing.'

The Gibson here immortalized was George William Gibson (1829-1913), who arrived on this coast with his two sons when he was in his mid-fifties. J.W. Bell of the Nanaimo Saw Mill remembered how he had sold Gibson rough lumber to build himself *The Swamp Angel,* a thirty-foot, double-ender sloop. Bell further recalled.

> One day he asked me if I knew of any place on the coast where he could take up a piece of land – preferably on the mainland – not isolated – reasonably near some town. 'I have a family; I would like to make a home for them some place. I have not many more years ahead of me. I have not money enough to buy a place. What I would like is enough ground to raise vegetables, keep a cow, some chickens and where there is good fishing so I can make enough money selling fish to buy necessities.'

Bell declared that 'Gibson was a fine character, good citizen. His path was not strewn with roses – it had many thorns and rough spots, but he did make a home for his family.' (See Leslie R. Peterson, *The Gibson's Landing Story* [1962].) Bell recommended to Gibson a small sheltered bay just inside Howe Sound, and there Gibson and his sons staked their claims to the land in May 1886. Gibson lived to a ripe old age, becoming in time magistrate and postmaster at Gibson's Landing.

GIL ISLAND, Central Coast (F-5). Named by Caamaño in 1792. There is no record as to what Gil he intended to honour, but Walbran notes that a Juan Gil was ensign bearer on the Duke of Medina-Sidonia's flagship when the Spanish Armada sailed for England.

GILFORD ISLAND, mouth of Knight Inlet (C-6). After Richard James Meade, Lord Gilford (or Gillford), who commanded HMS *Tribune* when she was on this coast in 1864-5. A few years later, he married the first daughter of Sir Arthur Kennedy, who had been Governor of Vancouver Island at the time of Lord Gilford's visit. In 1879 Gilford became Earl of Clanwilliam. He retired from the Royal Navy in 1902 with the rank of Admiral of the Fleet and died in 1907.

GILLATT ARM, Cumshewa Inlet, QCI (F-4). After Captain J.B. Gillatt, a retired Indian Army officer who settled at Sandspit in 1911.

GILLIES BAY, W. side of Texada I. (B-7). Local tradition has it that Gillies was a very mean captain whose crew threw him into the bay.

GILLIS CREEK, S. of Manson Cr. (H-7). A lonely grave in the wilderness here reminds the traveller of the tragedy of Hugh Gillis. Successful in his mining, he left Manson Creek with his gold in August 1872, en route to his home on Prince Edward Island. On the trail he met an Indian bringing in the mail from Fort St. James. In it was a letter from the girl in PEI whom he was going to marry – she had written to tell him that she had married another man. Gillis shot himself on the spot.

GILPIN, E. of Grand Forks (B-10). After Ralph Gilpin, first customs officer in Grand Forks, 1884, and local farmer.

GILTTOYEES INLET, N. from Douglas Channel (F-5). From the Haisla Indian word meaning a 'long and narrow stretch of water leading outward.'

GINGIT CREEK, flows W. into Tseax R. (H-5). From the Nisgha Indian word meaning 'place of spawning sockeye.'

GINLULAK CREEK, flows W. into Nass R. (H-5). From a Nisgha Indian word meaning 'place of corpses,' referring to the finding of a large number of bones in the creek.

GISCOME, NE of Prince George (G-8). After John Robert Giscome (c. 1832-1907). In 1863, during a prospecting expedition to the Peace River country, Giscome and Harry McDame, another black West Indian, became the first non-Indians to use the Giscome Portage route from the Fraser River to Summit Lake. (See Linda Eversole, 'John Robert Giscome, Jamaican Miner and Explorer,' *British Columbia Historical News* 18,3 [1985]:11-15.)

GITNADOIX RIVER, flows N. into Skeena R. (G-5). From the Tsimshian Indian word meaning 'people of the swift water.'

GITZYON CREEK, S. of Aiyansh (H-5). From a Nisgha Indian word meaning 'people of the glacier.'

GLADWIN, E. of Lytton (C-9). After Walter D. Gladwin, who operated a roadhouse here. A resident of Yale as early as 1858, Gladwin later forwarded freight on the old Cariboo Wagon Road.

GLADYS LAKE, W. of Teslin L. (L-3). When the future Commissioner Perry of the RNWMP was in charge of this district, he named the lake after his youngest daughter.

GLENANNAN, François L. (G-7). After 'Sandy' Annan, who had a farm and stopping house here.

GLENEMMA, N. of Okanagan L. (C-10). After Emma, wife of Kenneth Sweet, first postmaster here in 1895.

GLENORA, SW of Telegraph Cr. (J-4). This name is said to have been compounded from the Gaelic *glen* ('valley') and the Spanish *oro* ('gold'), producing 'The Valley of Gold.' Glenora became important during the Cassiar gold excitement of the 1870s.

GLUNDEBERY CREEK, flows N. into Teslin R. (L-4). Derived from the Tahltan Indian word meaning 'hungry mice.'

GNAWED MOUNTAIN, E. of Spences Bridge (C-9). A translation of the Shuswap Indian name 'Skutl-hĕh'-tl,' literally meaning 'eaten to the bone.'

GOAT MOUNTAIN, N. of Vancouver (B-8). So named in 1894 by a party of hunters who had shot two mountain goats here.

GOD'S POCKET MARINE PARK, near Port Hardy (C-6). Whereas other safe havens known to mariners as God's Pocket have had their names bowdlerized, this one survives unspoiled. (See Introduction.)

GOGIT PASSAGE, E. of Lyell I., QCI (E-4). After a Haida legendary creature possessing an evil spirit and usually intending harm.

GOLD BRIDGE, Bridge R. (C-8). The road from Lillooet to the gold country around Bralorne and Pioneer here crosses the Bridge River.

GOLD RIVER, flows S. into Muchalat Inlet (B-6). Chinese were taking gold out of the area in the 1860s. The name appears on the Trutch map of 1871.

GOLDEN, upper Columbia R. (D-11). Known in earlier days as The Cache or Kicking Horse Flats. Around 1883, when the CPR's end of steel was reaching west of Banff, a syndicate of crooked mining promoters with very rich samples of silver ore from Montana began showing them around Calgary, saying that they were samples from their claims near Castle Mountain. They succeeded in starting a rush, and a small mining town named Silver City came into existence where there had been only a railway construction camp. When the men at the Kicking Horse Flats camp learned of the new city, they were determined not to be outdone and, on the suggestion of F.W. Aylmer, renamed their settlement Golden City. After a time the little village dropped the 'City.'

GOLDEN EARS, N. of Haney (B-8). This name for the peaks on Mount Blanshard (q.v.) is found as early as 1862, when Commander Mayne makes mention of 'the beautiful peaks known as the Golden Ears.'

GOLDEN HINDE, W. of Buttle L. (B-7). The highest mountain (7,219 feet or 2,200 metres) on Vancouver Island. It is named after the ship in which Sir Francis Drake circumnavigated the world. (See also *Sir Francis Drake, Mount.*) Among alpinists it has the unofficial name of The Rooster's Comb.

GOLDSTREAM RIVER, flows into Finlayson Arm (A-8). In 1863-4 there was a small gold rush here, with up to 300 men working on the stream. Its Saanich Indian name is anglicized as Sawluctus and means 'our fishing ground tucked inside the arm.'

GOLDSTREAM RIVER, flows W. into Columbia R. (D-10). As the Gold River, it was important during the Big Bend gold rush of 1865-6.

GOLETAS CHANNEL, S. of Hope I. and Nigei I. (C-6). *Goletas* is Spanish for 'schooners.' In 1792 Galiano and Valdes named this passage after their two goletas, *Sutil* and *Mexicana*.

GONZALES BAY, Victoria (A-8). Adjacent Gonzales Point was named by Manuel Quimper in 1790 after his first mate on the *Princesa Real,* Gonzalo Lopez de Haro. This bay is better known as Foul Bay, signifying a poor anchorage.

GOOCH ISLAND, Haro Strait (A-8). After Thomas Sherlock Gooch, second lieutenant on HMS *Satellite* when she was on this coast from 1857 to 1860.

GOODFELLOW CREEK, Manning Park (B-9). After the Reverend John C. Goodfellow (1890-1968), for many years the United Church minister in Princeton. Keenly interested in British Columbia's history, he tried to preserve the Indian pictographs in the Princeton area using protective coatings of lacquer.

GOODSIR, MOUNT, Yoho NP (D-11). Named in 1858 by Dr. Hector after his former teacher John Goodsir (1814-67), Professor of Anatomy at Edinburgh University.

GOODWIN FALLS, Mahood R. (D-9). Named after Walter Goodwin of Spokane, Washington, a dental technician. He was a friend of A.W. ('Wells') Gray, Minister of Lands, who in 1941 put him on a list of his personal friends whose names he wanted attached to features in Wells Gray Park.

GORDON HEAD, N. of Oak Bay (A-8). After Captain the Hon. John Gordon, commanding HMS *America* when she was in the North Pacific in 1845. Disappointed in his hunting and fishing here, Gordon gave a very poor account of the country to his brother, the fourth Earl of Aberdeen, the British Foreign Secretary. There is no evidence to support the theory that Gordon's unfavourable report contributed to the British decision to let the Americans have Washington and Oregon rather than risk a war.

GORDON RIVER, flows S. near Port Renfrew (A-7). After Commander (later Admiral) George Thomas Gordon, on this coast with HMS *Cormorant* in 1846 and HMS *Driver* in 1850.

GORE ISLAND, Nootka Sound (B-6). After John Gore, first lieutenant on HMS *Resolution* when Captain Cook put in at Nootka in 1778. After the deaths of Cook and his successor, Captain Charles Clerke, Gore took over command of the expedition.

GOSNELL, near confluence of N. Thompson R. and Albreda R. (E-10). After R. Edward Gosnell (1860-1931), who came to British Columbia as a newspaperman in 1888. He was the first Provincial Librarian (1893-8) and the first Provincial Archivist (1908-10). He was a prolific writer on British Columbia and served the provincial government in a variety of positions.

GOTT CREEK, SW of Lillooet (C-8). After Frank Gott, a Lillooet Indian, hunter, guide, rancher, and prospector. In World War I, he dyed his white hair so that he could serve overseas with the 102nd Battalion (North British Columbians). Although an excellent sniper, he was returned to Canada as overaged in 1917. In October 1932, in the climax to a feud with a game warden, Gott shot him in the back when the warden sought to arrest him for having an untagged deer. Pursuing wardens and police caught up with Gott finally and demanded that he surrender. 'I am a soldier, and I never surrender,' shouted Gott, continuing his flight. An officer opened fire, and Gott was wounded in the leg. He died in hospital in Lytton, more from advanced tuberculosis, exposure, and lack of nourishment than from his wound. Opinion was largely sympathetic to Gott, and he was given a public military funeral.

GOTTFRIEDSEN, MOUNT, headwaters of Trépanier Cr. (B-10). After November Gottfriedsen, of Danish and Indian ancestry, who settled in the district around 1910. He said he was originally named September since he came after his brother August but that, since he was always late, people took to calling him November.

GOVERNOR ROCK, centre of Trincomali Channel (A-8). Sardonically named by Captain G.H. Richards in 1859 when, steaming at full speed toward Victoria in response to an urgent message from Governor Douglas, he ran HMS *Plumper* on this previously unknown rock.

GOWER POINT, entrance of Howe Sound (B-8). Although Captain Vancouver gives no indication as to whom he had in mind when he gave this name, Captain Walbran and Professor Meany agree that he was thinking of Admiral Sir Erasmus Gower (1742?-1814), who was knighted in 1792 shortly before sailing with HMS *Lion* to carry Lord Macartney, the first British ambassador to China, to his new post.

GOWGAIA BAY, W. coast of Moresby I. (E-4). From the Haida word meaning 'inlet.'

GOWLLAND HARBOUR, W. side of Quadra I. (C-7). After John Thomas Gowlland, second master, first of HMS *Plumper* and then of HMS *Hecate*.

GRAHAM ISLAND, QCI (F-3). After Sir James Robert Graham (1792-1861). A principal member of the reform party led by Lord Grey, Graham held various cabinet posts, among them that of First Lord of the Admiralty from 1852 to 1855. He was a polished speaker and an able administrator.

GRAHAM RIVER, flows E. into Halfway R. (I-8). After Lieutenant John R. Graham, MC, BCLS, who made surveys in this area. He was killed in France in World War I.

GRAINGER CREEK, flows into Skaist R., Manning Park (B-9). After Martin Allerdale Grainger (1874-1941). After graduating from Cambridge he came to British Columbia and started the strenuous work of hand logging – about which he wrote in *Woodsmen of the West*, a BC classic. He remained in forestry, both in the provincial service (he was Chief Forester 1916-20) and later in industry. He loved the outdoors and spent much time riding horseback and camping in the area around Princeton.

GRANBY RIVER, flows S. into Kettle R. at Grand Forks (B-10). After the Granby Company, which established a smelter on this river. The founder of this company had come from Granby, Quebec. Earlier this river was known simply as the North Fork of the Kettle River. Also GRANBY BAY on Observatory Inlet, site of the ghost town of Anyox.

GRAND FORKS, Boundary district (B-10). Originally known as Grande Prairie. It takes its present name from the junction here of the Kettle River and the Granby River.

GRANISLE, W. side of Babine L. (G-6). A combination of 'Gran' for the Granby Company, the original owner of the mine, and 'isle' for the island in the lake where the mine was located.

GRANT BROOK, Mt. Robson Park (E-10). After George Monro Grant (1835-1902), who in 1872 accompanied Sir Sandford Fleming on an overland journey to the Pacific. He later published an account of this expedition in his *Ocean to Ocean*. He became Principal Grant of Queen's University.

GRANT ROCKS, Sooke Inlet (A-8). After Captain Walter Colquhoun Grant of the 2nd Dragoons (Scots Greys), who arrived on Vancouver Island in 1849 as the first independent colonist. With him he brought eight men to work on his farm. He also brought impedimenta such as carriage harness and cricket sets, only to find that there were no carriages on Vancouver Island and that the HBC men at Fort Victoria did not play cricket. Finding that the HBC owned all the good land around Victoria, Grant established his farm at Sooke and built a sawmill here. In 1853, admitting failure, he returned to England. He died eight years later, serving in the Indian Mutiny. Although Grant had a very attractive personality, he lacked a business sense. He was the first of many hundreds of young gentlemen from the British Isles who came out to British Columbia with scant knowledge of what they would find here and eventually lost a good deal of money. He planted the first broom (*Cytisus scuparius*) to grow on Vancouver Island.

GRANTHAMS LANDING [GRANTHAM'S LANDING], Howe Sound (B-8). After Frederick Charles Grantham (1871-1954), a Vancouver lime juice manufacturer and 'a very kindly quiet gentleman.' In 1909 he arrived at Gibson's Landing looking for a good location for a summer cottage. Gibson's son-in-law Glassford took him around to see 'the prettiest spot on Howe Sound.' Taken with the place, Grantham bought District Lot 687, seventy-five acres and 800 feet of waterfront. Since the property was bigger than Grantham needed for his cottage, he subdivided it, putting in roads and sidewalks, a wharf and a water system, and so Grantham's Landing was born.

GRANVILLE ISLAND, Vancouver (B-8). In 1870 Gastown (the future Vancouver) was renamed Granville in honour of George Leveson-Gower, Earl Granville, Britain's Secretary of State for the Colonies. Granville Street and Granville Island are derived from this naming.

GRASMERE, E. of Lake Koocanusa (B-12). Earlier named McGuire after Howard McGuire, who arrived in the district in 1898. In 1922, when a new school was built, every pupil submitted a new name for the settlement. A draw was held, and the entry picked was that of young Warren Lancaster, who had proposed Grasmere.

GRAVE CREEK, flows W. into Elk R. (B-12). So named because of two Indian graves near the mouth of the creek.

GRAY CREEK, Crawford Bay, Kootenay L. (B-11). After John Hamilton Gray (1853-1941), civil engineer and surveyor, active in railway construction.

GRAY, MOUNT, W. boundary, Kootenay NP (D-11). After William J. (Billy) Gray, a UBC student who drowned in 1917 in the Kootenay River during a summer geological survey. A founding member of the BC Mountaineering Club, and at one time its president, Gray also has GRAY PASS in Garibaldi Park named after him. (See *Drysdale, Mount.*)

GRAYMALKIN LAKE, expansion of Birnam Cr. (C-11). After the familiar spirit of one of the witches in Shakespeare's *Macbeth*. There are other *Macbeth* names in the area.

GREAT CENTRAL LAKE, NW of Port Alberni (B-7). This is not only a large lake but is also the deepest on Vancouver Island (1,100 feet).

GREATA, W. side of Okanagan L. (B-10). After George Greata, pronounced 'Greeta,' who looked 'exactly like Edward VII.' Around 1900 he developed a large orchard here.

GREEN LAKE, N. of Alta Lake (C-8). This lake on the BCR was named by

Lieutenant R.C. Mayne, RN, when he visited it in 1861. He wrote, 'Finding the Indians knew no name for it, I called it "Green lake" from the remarkably green colour of the water.'

GREEN LAKE, S. of 100 Mile House (D-9). This lake, approximately thirty miles long, has no outlet other than underground drainage. The consequent accumulation of soda, salt, sulphur, and other elements has caused its notable green colour. The early anthropologist James Teit wrote:

> Exchange of various goods took place at Green Lake, where great numbers from all divisions of the Shuswap tribe congregated once a year to have sports and to trap trout etc ... Here the Lake Division sold some dried trout, nets, carrying-bags, some cedar-root, very dark marmot-skins, and a few baskets, principally to the Fraser River bands, for dried salmon, salmon oil, and shells. The North Thompson band brought hazel nuts to sell.

GREENSTONE MOUNTAIN, W. of Kamloops (C-9). A translation of the Shuswap Indian name, 'Kwil-āl-kwila' ('green stone').

GREENVILLE, lower Nass R. (H-5). After the Reverend A.E. Green, Methodist missionary in the area from 1876 to 1889. Its Nisgha Indian name is 'Lach Al Zap,' meaning 'at the village,' or 'the place of the village.'

GREENWOOD, W. of Grand Forks (B-10). Writing to the Chief Geographer of Canada in 1905, R.A. Brown reported that Greenwood was named after another mining camp, Greenwood, Colorado. This agrees with the account of R.W. Haggen, a local historian and MLA, that Greenwood is one of those places 'named after camps in the American Cordillera.' He added: 'Most of the miners and companies which operated in the Boundary [district] were from the USA.' The townsite for Greenwood was laid out in 1895. Greenwood acquired a post office in 1896 and was incorporated as a city in 1897.

GRENVILLE CHANNEL, North Coast (F-5). Named by Captain Vancouver after William Wyndham Grenville, Baron Grenville (1759-1834). Grenville was an able politician, at one time Speaker of the House of Commons and later Home Secretary. Grenville's brother-in-law, the Hon. Thomas Pitt, who inherited the title of Lord Camelford, was one of Vancouver's midshipmen. He proved so unsatisfactory that Vancouver had to discharge him at Hawaii in 1794.

GREY, POINT, Vancouver (B-8). Named by Captain Vancouver 'in compliment to my friend Captain George Grey of the Navy.' Captain Grey (1767-1828) commanded HMS *Victory,* Jervis's flagship at the Battle of St. Vincent in 1797. Later he commanded HMS *Ville de Paris,* the flagship of Jervis (now Earl St. Vincent) during his blockade of the French fleet in Brest. Grey subsequently commanded the royal yacht *Amelia.* In 1814, when commissioner of Portsmouth dockyard, he was knighted by the Prince Regent.

The Spaniards, here a year before Vancouver, named Point Grey Punta de Langara, in honour of Admiral Don Juan de Langara.

GRIBBELL ISLAND, N. of Princess Royal I. (F-5). Named in 1867 by Lieutenant (later Captain) D. Pender, RN, after his brother-in-law, the Reverend Francis B. Gribbell. Gribbell lived in Victoria from 1865 to 1875, serving first as rector of St. John's, Victoria, then of St. Paul's, Esquimalt, and finally as principal of the Collegiate School, Victoria.

GRICE BAY, Tofino Inlet (B-7). After John Grice (1841-1931), who in 1888 became Tofino's first white settler and later the local justice of the peace. He was 'a well-educated man, a lover of Shakespeare and the classics, and something of an astronomer.'

GRINDER CREEK, flows E. into Fraser R. (D-8). After Philip Grinder, who settled in the area in the 1860s.

GRINDROD, SE of Salmon Arm (C-10). After Edmund Holden Grindrod, first CPR inspector of telegraphs in British Columbia (1886) and later a farmer near Kamloops.

GROHMAN CREEK, N. of Nelson (B-11). Also MOUNT GROHMAN. After W.A. Baillie-Grohman. (See also *Canal Flats.*)

GROUSE MOUNTAIN, N. of Vancouver (B-8). Received its name in 1894 when it was climbed by a party including E.A. Cleveland, who became chief commissioner of the Greater Vancouver Water District many years later. They named it Grouse Mountain because of the blue grouse they shot on it.

GUICHON CREEK, flows S. into Nicola R. (C-9). After Joseph, Pierre, and Laurent Guichon, 'Old Country French,' not French-Canadians, who settled in the district in 1873. Joseph later founded the Guichon Cattle Co. Ltd. at Quilchena and became one of British Columbia's cattle kings. (For Laurent, see *Port Guichon.*)

GUILDFORD, Surrey District Municipality (B-8). When the original developers of Guildford Town Centre, the directors of Grosvenor-Laing, came to name it, they borrowed the name of Guildford in the county of Surrey in England, where their headquarters are located.

GUISACHAN, Kelowna (B-10). When Kelowna townsite was laid out, an agent for the Earl of Aberdeen bought an adjacent farm for him. Aberdeen named his new estate Guisachan, which in Gaelic means 'place of the firs.'

GUN LAKE, W. of Carpenter L. (C-8). Apparently a gun was lost when a packhorse drowned in nearby Gun Creek.

GUNANOOT LAKE, NW of Babine L. (H-6). Named after Simon Gunanoot, a Kispiox Indian, object of the most famous manhunt in the history of British Columbia. A skilful hunter and a crack shot, Gunanoot eluded the police for thirteen years after he was accused of murdering two men partly Indian. He finally gave himself up, stood trial, and was acquitted.

GUNBOAT PASSAGE, N. of Denny I. (E-6). Named in 1867 after the two little gunboats, HMS *Forward* and HMS *Grappler,* that gave sterling service on this coast for years.

GUNN VALLEY, W. of Taseko L. (D-8). After a Chilcotin Indian named Ganin, who lived in the area late in the nineteenth century.

GWILLIM RIVER, flows N. into Murray R. (H-9). After John Cole Gwillim (1868-1920), Professor of Mining at Queen's University, who made surveys in the area in 1919.

H

HADDINGTON ISLAND, NE of Port McNeill (C-6). After Thomas Hamilton, ninth Earl of Haddington (1780-1858), a mediocrity whom fortune loved and made, successively, Lord Lieutenant of Ireland and First Lord of the Admiralty. (He declined to serve as Viceroy of India.) Haddington Island has only two claims to attention. The stone that faces the provincial legislative buildings came from here, and a provincial ferry, the *Queen of Prince Rupert,* ran on the rocks here in 1967.

HADDO LAKE, SE of Vernon (C-10). After George, Lord Haddo, eldest son of the seventh Earl (and first Marquess) of Aberdeen. (See *Aberdeen Hills.*)

HAFFNER, MOUNT, Kootenay NP (D-11). After Lieutenant Henry John Haffner, CE (1880-1916). He made the first survey of the Banff-Windermere road and was an engineer during its early construction. In World War I, he gave splendid service with the 8th Field Company of the Canadian Engineers until killed by a sniper.

HAGAN ARM, Babine L. (G-6). After Fred Hagan, who, with his fellow prospector Hughie McDonald, discovered the copper deposits at Granisle.

HAGENSBORG, E. of Bella Coola (E-6). Hagen B. Christenson, first postmaster here, writing in 1906 to James White, Chief Geographer, explained the name thus: 'The Hagensborg P.O. was established in 1900, and is situated on the Bella Coola River 10 miles up the valley. To my first name, Hagen, was added "Borg," which means in Norwegian a fortified place where chiefs used to live.'

HAGWILGET CANYON, near junction of Bulkley R. and Skeena R. (H-6). From the Gitksan Indian word meaning 'peaceful, deliberate people.'

HAHAS LAKE, NE of Kimberley (B-12). From the Kootenay Indian word meaning 'skunk.'

HAIG, N. of Hope (B-9). This station on the CPR mainline was named Hope until the CPR transferred the name to the station of its Kettle Valley line on the other side of the Fraser. This station was then renamed Haig, in honour of the British Commander-in-Chief on the western front in the later years of World War I. Also MOUNT SIR DOUGLAS and HAIG GLACIER in the Rocky Mountains.

HAIG, MOUNT, Alberta-BC boundary (B-12). After Captain R.W. Haig, RA, astronomer with the British Boundary Commission, which, with its American counterpart, established the international boundary from the Rockies to the Pacific from 1858 to 1862.

HAIG-BROWN, MOUNT, W. of N. end of Buttle L. (B-7). Commemorates Roderick Haig-Brown (1908-76), noted author and conservationist, and his wife, Anne (1908-90). He also has RODERICK HAIG-BROWN PARK (where the famous salmon-producing Adams River empties into Shuswap Lake) named after him.

HAIHTE RANGE, SW of Woss L. (C-6). Formerly called Rugged Range. When a new name was needed, an Indian vocabulary list was consulted, and *haihte*, Kwakwala for 'fish head,' was chosen.

HAISLA, S. of Kitimat (F-5). Means 'people who live on the seashore' (in contrast to Interior people), or 'living downstream.'

HAKAI PASSAGE, N. of Hecate I. (D-5). The name of this well-known fishing area comes from a Heiltsuk Indian word said to mean 'wide passage.'

HALFMOON BAY, W. of Sechelt (B-8). Formerly known as Priestland Bay, after the Priestland family who preempted land here in 1889, this bay takes its present name from its shape.

HALFWAY RIVER, flows SE into Peace R. (I-9). So named since it is approximately halfway between Fort St. John and Portage Mountain.

HALLOWELL, MOUNT, Sechelt Pen. (B-8). After Admiral Sir Benjamin Hallowell (1760-1834), captain of HMS *Swiftsure* under Lord Nelson at the Battle of the Nile. After the victory he sent Nelson a decidedly original gift made out of floating wreckage from one of the French battleships. With it he sent the following note: 'My Lord – Herewith I send you a coffin made of part of *L'Orient*'s mainmast, that when you are tired of this life you may be buried in one of your own trophies; but may that period be far distant is the sincere wish of your obedient and much obliged servant. – Ben Hallowell.' He is said to have been a giant of a man, with tremendous muscular strength. After his inheritance in 1828 of the Carew estates, he changed his name to Benjamin Hallowell Carew.

HAMBER PARK, W. side of Jasper NP (E-11). This provincial park (now greatly reduced from its original 3,800 square miles) was established in 1941 and named in honour of Eric Werge Hamber, Lieutenant-Governor of British Columbia from 1936 to 1941. British Columbia has been singularly fortunate in the calibre of the men who have been the sovereign's representative, and E.W. Hamber was among the best.

HAMILL, MOUNT, NE of Kootenay L. (C-11). After Thomas Hammil (note correct spelling), a young Cornish prospector. For the story of his murder, see *Sproule Creek*.

HAMMETT CREEK, flows NE into McLeod L. (G-8). After Tommy Hammett, long the HBC trader at McLeod Lake. Since company policy would not permit him to accept the cheques tendered by surveyors who came to his post for much-needed provisions, Hammett would put up the money himself.

HAMMOND. See *Port Hammond.*

HAMMOND BAY, N. of Nanaimo (B-8). After George Crispin Hammond, RN, navigating sublieutenant aboard the early steamship *Beaver* when, as a hired surveying vessel, she was active in these waters in 1867-70.

HANCEVILLE, SW of Williams L. (D-8). After Thomas Orlando (Tom) Hance (1844-1910), who came to the Chilcotin country from Illinois in 1869. He opened a trading post in an Indian village where Hanceville now stands and later founded the T-H Ranch.

HANEY, E. of Pitt Meadows (B-8). After Thomas Haney (1841-1916), who settled here in 1876. The settlement was originally known as Haney's Landing and later as Port Haney. With the increasing importance of roads, the centre of Haney gravitated away from the river and up to the highway.

HANINGTON, MOUNT, NW of Jasper NP (G-9). After C.F. Hanington, who passed this way in 1875 with E.W. Jarvis, CE, using the Jarvis Pass through the Rockies.

HANKIN RANGE, E. of Nimpkish L. (C-6). After Philip James Hankin, who first arrived on this coast as a mate on HMS *Plumper,* then served as a lieutenant on HMS *Hecate.* Back in England in 1864, he left the navy in order to return to British Columbia, and in that year he became superintendent of police for Vancouver Island. In 1869 he became Colonial Secretary of British Columbia.

HANNA CHANNEL, Nootka Sound (B-6). After Captain James Hanna, who put in at Nootka in 1785 with his fur-trading vessel *Harmon,* in which he had sailed from China. Hanna, the first white visitor to British Columbia since Cook in 1778, came back for more furs in 1786, this time with the *Sea Otter.*

HANNA CREEK, N. of Trail (B-11). After Frank Hanna, who in 1890, in association with 'Colonel' Eugene Sayre Topping, preempted over 300 acres at the mouth of Trail Creek, the future townsite of Trail. Later he quarrelled with his wife and with Topping, moved down to Texas, and died. Topping and Hanna's widow were married in 1906.

HANNAH MOUNTAIN, W. of Alberni Inlet (B-7). From the Nootka Indian word meaning 'naked'; in other words, this is 'Bare Mountain.'

HANSEN BAY, Cape Scott Park (C-5). After Rasmus Hansen, who landed here in 1895 with another Danish fisherman named Jensen and discovered many acres of fertile meadow lying inland. The ultimate consequence was the establishment by Hansen and Jensen of the Danish settlement, long since abandoned, in the Cape Scott area.

HANSON ISLAND, W. end of Johnstone Strait (C-6). After James Hanson, who arrived on this coast in 1792 as lieutenant aboard HMS *Chatham* and later that year was transferred by Captain Vancouver to the command of his supply ship, the *Daedalus.*

HAPPY VALLEY, NW of Metchosin (A-8). Around 1860 some blacks who had emigrated from the United States settled here. According to some accounts, their singing gave the place its name of Happy Valley. However, I.G. Walker, postmaster at Happy Valley in 1905, reported that it was one of the blacks, Isaac Mull (who lived to be more than 100), who gave Happy Valley its name, possibly with reference to joy at becoming a free man under the British flag.

HAPUSH MOUNTAIN, NW of Schoen L. (C-6). From the Kwakwala Indian word for 'boy.' Originally a surveyor proposed that this mountain be named Stung Mountain because, when he tried to make his way to a certain locality, he was stopped by this large mountain. However, a bureaucrat in Ottawa judged the name 'not very appropriate.'

HARDWICKE ISLAND, Johnstone Strait (C-7). After Philip Yorke, Earl of Hardwicke (1757-1834), who seems to have taken an interest in the career of Spelman Swaine, master's mate on HMS *Discovery* during Vancouver's voyage.

HARDY ISLAND, Jervis Inlet (B-7). After Vice-Admiral Sir Thomas Masterman Hardy, baronet, RN. Hardy was captain of HMS *Victory* at the Battle of Trafalgar, in which Lord Nelson died in his arms. (See also *Port Hardy.*)

HAREWOOD, S. of Nanaimo (B-8). Takes its name indirectly from the third Earl of Harewood. In 1864 one of Lord Harewood's sons, Lieutenant the Hon. H.D. Lascelles, on this coast as captain of the gunboat *Forward,* set up the Harewood Coal Mining Company to work the deposits that he had acquired here.

HARKIN, MOUNT, SE boundary of Kootenay NP (C-12). After James B. Harkin, from 1911 to 1936 Canada's first commissioner of national parks. He has been called the 'Father of National Parks in Canada.'

HARMAC, SE of Nanaimo (B-8). After H.R. MacMillan (1885-1976), founder of the H.R. MacMillan Export Co. Ltd. (now part of MacMillan Bloedel Ltd.), which in 1948 built the pulp mill here. After taking his BSA from Toronto and his MSCF from Yale (1908), MacMillan joined the Dominion Forestry Service, leaving it in 1912 to become the first Chief Forester of British Columbia. After making a survey for the province of potential lumber markets abroad, he left the government's service in 1916 and entered business, determined to take full advantage of the opportunities that he had discovered. The following years saw H.R. MacMillan emerge as by far the most powerful industrial magnate in the province.

HARO STRAIT, NE of Victoria (A-8). Named in 1790 by Quimper after Gonzalo Lopez de Haro, his *primo piloto* (or first mate) on the *Princesa Real.*

HAROLD PRICE CREEK, flows NW into Suskwa R. (H-6). After Captain Harold Price, MC (1890-1916), a BC land surveyor killed in action while serving with the Northumberland Fusiliers.

HARPERS CREEK, flows W. into Fraser R. (D-8). After Jerome and Thaddeus Harper, brothers who arrived in British Columbia in 1859 or earlier and built an enormous cattle empire in the Cariboo. Jerome died in 1874. In 1888 Thaddeus sold 38,572 acres, including the Gang Ranch and land by the South Thompson River and Cache Creek, but he went bankrupt the next year. He died in 1898. Also MOUNT HARPER, NE of Kamloops.

HARRIS, MOUNT, Alaska-BC boundary (L-1). After Dennis R. Harris, CE (1851-1932). A son-in-law of Sir James Douglas, and one of British Columbia's best-known surveyors, he was City Engineer of Victoria in the 1880s, located the Malahat highway, and was engaged in the Alaska boundary survey of 1904.

HARRISON LAKE, N. of Fraser Valley (B-9). Named in 1828 by Governor Simpson after Benjamin Harrison, a philanthropic Quaker who was first a director and then, from 1835 to 1839, Deputy Governor of the HBC. For fifty years, without salary, he served as treasurer of Guy's Hospital, London.

Story has it that the hot springs were discovered accidentally one wintry day when a boat upset and its occupants, expecting to perish in the icy lake, were amazed to find themselves in warm water.

The Indian name for Harrison Lake is anglicized as Pook-pah-Kohtl ('many large spring salmon'); that for the hot springs can be translated as 'boiling water.'

HARROGATE, SE of Golden (C-11). After the fashionable resort in Yorkshire.

HARROP, West Arm, Kootenay L. (B-11). After Ernest Harrop, who settled here in 1905 and became postmaster in 1907.

HARVEY CREEK, flows E. into Flathead R. (B-12). After J.A. Harvey, a Cranbrook lawyer.

HARVEY, MOUNT, E. side of Howe Sound (B-8). After Captain John Harvey, who commanded HMS *Brunswick* at 'The Glorious First of June,' Howe's victory over the French. Severely wounded, Harvey died at Portsmouth on 30 June 1794.

HARWOOD ISLAND, W. of Powell River (B-7). Named by Vancouver, who gave no indication as to whom Harwood might be. Captain Walbran believed that the man in question was Edward Harwood, naval surgeon and numismatics enthusiast, who died in 1814. This Harwood was described as 'a benevolent friend and an elegant scholar.'

HARWOOD POINT PARK, Texada I. (B-7). After Commodore Henry Harwood, RN, who commanded the cruisers *Exeter, Ajax,* and *Achilles* in the action against the German battleship *Graf Spee* off Montevideo in December 1939.

HASCHEAK CREEK, flows N. into N. Thompson R. (D-9). Formerly Red Creek. *Hascheak* is based on the Shuswap Indian word for 'red.'

HASLAM LAKE, E. of Powell River (B-7). After Andrew Haslam, pioneer lumberman. He bought the Nanaimo Sawmill in 1892, served as mayor of Nanaimo in 1892-3, and was an MLA and MP. In 1906 he moved to Vancouver and set up the provincial government's log-scaling organization.

HASLER PEAK, Glacier NP (D-11). See *Feuz Peak.*

HASTINGS ARM, Observatory Inlet, N. coast (H-5). After Rear-Admiral the Hon. G.F. Hastings, Commander-in-Chief of the Pacific Station 1866-9.

HAT CREEK, flows NE into Bonaparte R. (C-9). G.M. Dawson has preserved for us the origin of this name in a Shuswap legend about their hero Kwil-ï-elï and two other supernatural men:

> A trial of strength was arranged, Kwil-ï-elï proposing that each should push his head against a rock and see which could make the deepest impression. Klesá and Took-im-in-ēlst tried first, and each managed to make a shallow impression, but Kwil-ï-elï followed and pressed his head in to the shoulders. This happened at a place near the mouth of Hat Creek, and the name of this stream ... is derived from this story, and from the circumstance that the impressions made in the rock at this time are still shown by the Indians. (*Transactions of the Royal Society of Canada,* Section 2 [1891]:32)

HATZIC, E. of Mission (B-8). This settlement was originally known as Wells' Landing, after an early settler. The meaning of the present name, derived from a Halkomelem word, may be either 'shore' or 'beach,' or perhaps is the name of an early spring edible plant that grew profusely in the area. (See *Katz.*)

HAVANNAH CHANNEL, S. of East Cracroft I. (C-6). After HMS *Havannah,* which in 1858 brought to British Columbia some seventy Royal Engineers to work on surveying the international boundary from the Pacific to the Rockies.

HAWKESBURY ISLAND, Douglas Channel, north coast (F-5). Named by Vancouver after Sir Charles Jenkinson, Baron Hawkesbury, Earl of Liverpool, President of the Board of Trade 1786-1804.

HAWKINS CREEK, flows NW into Moyie R. (B-11). After Lieutenant-Colonel John S. Hawkins, RE, commissioner, British Boundary Commission, 1858-62.

HAWKS CREEK, N. of Williams L. (E-8). Originally Deep Creek, but renamed after John F. Hawks, who in 1874 took over a number of preemptions in the area.

HAWORTH LAKE, Kwadacha Wilderness Park (J-7). After Paul Leland Haworth, an American Professor of History who made exploratory trips into the area in 1916 and 1919. He was the author of *On the Headwaters of Peace River.*

HAY RIVER, Alberta-BC boundary (K-9). Because of the prairie hay on the plain through which it winds.

HAYNES CREEK, flows SW into Osoyoos L. (B-10). Also HAYNES POINT PARK. After John Carmichael Haynes (1831-88), who arrived from Ireland in 1858 and enlisted in the colonial police. He served around Yale and Osoyoos before being sent as magistrate to the Wild Horse Creek goldfield. Arriving just after a free-for-all in which one man was killed and others injured, Haynes was informed that at least 20 per cent of the miners were criminals. Declaring himself 'horrified to think of such a thing happening in Her Majesty's Domain,' Haynes swiftly brought law and order to the camp. In 1866 he settled at Osoyoos, where he received a commission as a county court judge. He began buying land and ended up owning everything between the international border and the ranch of his fellow cattle baron, Tom Ellis, at Penticton.

A granddaughter has preserved a picture of the man for us: 'Judge Haynes was an expert horseman, and to him a good mount was one of the necessities of life ... On horseback he invariably appeared as if "riding in the Row" with his Irish tweed coat, riding breeches, and English riding boots. An army helmet was part of the picture in summer; a felt hat at other seasons – never a Stetson or "cowboy."'

HAYS, MOUNT, S. of Prince Rupert (G-4). After Charles Melville Hays, president of the Grand Trunk Pacific Railway (now part of the CNR), who perished in the sinking of the *Titanic* in 1912.

HAYWARD LAKE, S. of Stave L. (B-8). After R.F. Hayward, general manager of the Western Canada Power Co., who had much to do with planning the dams and power plants at Ruskin and Stave Falls.

HAZELMERE, E. of White Rock (B-8). Named by a pioneer homesteader, H.T.

Thrift – according to his own account not so much because he came from near Hazelmere, England, as because here, on his farm, he had both a thick growth of hazel brush and a small pond or 'mere.'

HAZELTON, Skeena R. (H-6). Soon after 1868 Thomas Hankin settled on the flats bordering the Skeena close to the Indian village of Gitenmaks or Kitanmaksh, meaning 'the place where people fished by torchlight.' When Hankin staked out a townsite here at Skeena Forks, he named it Hazelton because of the large amount of hazelnuts ripening at that time. Hazelton, at the confluence of the Skeena and Bulkley Rivers, was the head of navigation for the sternwheelers that used to come up the Skeena.

HEART CREEK, flows NW into Lower Arrow L. (B-10). From the shape of the cirque at the head of the creek.

HEBER RIVER, flows SW into Gold R. (B-7). After Heber DeVoe, a member of the party that in 1913 made the first survey of Strathcona Park.

HECATE STRAIT, E. of QCI (E- and F-4). After the Royal Navy's survey ship *Hecate,* commanded by Captain G.H. Richards, which worked on this coast in 1860-2. At least nine other places or features in British Columbia help to immortalize *Hecate.*

HECKMAN PASS, W. of Anahim L. (E-7). Named, on the recommendation of the Bella Coola Board of Trade, after Max Heckman, a longtime resident of the area.

HECTOR, Kicking Horse Pass (q.v.) (D-11). This station on the CPR mainline is named after Sir James Hector (1834-1907), who, as young Dr. Hector, served as geologist with the Palliser Expedition of 1857-60.

HEDLEY, E. of Princeton (B-9). Named about 1899 after Robert R. Hedley, manager of the Hall mines smelter in Nelson. Somewhat earlier Hedley grubstaked Peter Scott, who had staked some of the first mineral claims in the area.

HEFFLEY CREEK, flows W. into N. Thompson R. (C-9). After Adam P. Heffley, a pioneer rancher who died in 1871. In 1862, Heffley, an enterprising American, with Henry Ingram and Frank Laumeister, brought in some twenty camels from California and unsuccessfully attempted to use them to pack supplies to the Cariboo goldfields. The camels terrified the horses.

HEINZE, MOUNT, N. of Trail (B-11). After F. Augustus Heinze, American speculator and capitalist, who in 1895-6 built the first smelter at Trail.

HELA PEAK, W. of Slocan L. (B-11). In Norse mythology, Hela or Hel was goddess of the world of the dead.

HELEN POINT, Mayne I. (A-8). This point at the western entrance to Active Pass is named after the wife of Joseph W. McKay, Chief Trader, HBC. She came to Victoria from England in 1858 and married McKay in 1860.

HELL'S HALF ACRE, E. of Monkman L. (G-9). A slide area of immense boulders, some as big as a house, thrown helter-skelter, with many dangerous crevices. Named by men seeking a route for the Monkman Pass Highway.

HELMCKEN FALLS, Wells Gray Park (D-9). After Dr. John Sebastian Helmcken (1825-1920), who arrived in Victoria from England in 1850. After a dangerous and arduous assignment to Fort Rupert, he returned to Victoria, where in 1852 he married Cecilia, daughter of Governor Douglas. Helmcken took an active part in the political life of the young colony as a staunch supporter of his father-in-law. He was a member of the first Legislative Assembly of Vancouver Island in 1855, of which he subsequently served as Speaker. After first opposing BC's entry into the Canadian confederation, Dr. Helmcken came around to support it, and he was offered, but declined, a seat in the federal senate. He preferred to devote himself to his family and his medical practice.

In his memoirs Eric Duncan recalls the shack in the middle of a thoroughfare where 'rough but kindly Dr. Helmcken had his office and surgery, a queer place, crammed with old copies of *Blackwood's Magazine.*'

After R.H. Lee, BCLS, discovered these magnificent falls in 1913, he wrote to Premier McBride asking his permission to name them McBride Falls in his honour. McBride suggested that Lee 'call the falls you mention after the venerable J.S. Helmcken, whose name I believe has not been connected with anything on the mainland and who desires to have his name preserved in the geography of British Columbia for which he has done so much' (quoted in R. Neave, *Wells Gray Park,* pp. 52-3).

HELVEKER, MOUNT, W. of confluence of Chutine R. and Stikine R. (J-4). A thinly disguised naming for Helen Vicars Kerr, by her husband, Forrest A. Kerr, who mapped this area.

HEMMINGSEN CREEK, NE of Port Renfrew (A-7). After Matt Hemmingsen, a Wisconsin lumberman who came to British Columbia in 1906 and introduced high-lead logging on Vancouver Island.

HEMP CREEK, flows S. into Clearwater R. (D-9). Formerly Little Clearwater River, but now named for Indian hemp *(Apocynum cannabinum L.),* out of the stems of which the Indians wove rope and fishing lines. Since Indian hemp does not grow here, the name probably derives from a misidentification of the look-alike spreading dogbane *(Apocynum androsaemifolium L.).*

HENDERSON CREEK, N. of Greenwood (B-10). After Richard Arthur Henderson, BCLS, killed in action on 11 April 1917.

HENDERSON LAKE, W. of Alberni Inlet (B-7). After Captain John Henderson of the barquentine *Woodpecker,* who in 1860 brought from England the machinery for the Alberni Saw Mill Co. The *Woodpecker* soon became a total wreck off the mouth of the Columbia River, and for a few years Captain Henderson commanded coastal ships. This lake was named by Dr. Robert Brown, who first learned of its existence from Captain Henderson.

HENDRIX LAKE, N. of Canim L. (E-9). After John 'Slim' Hendrix (1870-1938), who lived at the foot of Canim Lake for many years and had a trapline north of it, in the area of HENDRIX CREEK.

Slim Hendrix was a tall, dark, gaunt man who wore a black patch over one eye. Like a number of men who dwelt in isolated parts of the province, Slim had a 'mail order' wife – Mrs. Hendrix was a most pleasant person.

HEN INGRAM LAKE, S. of Quesnel L. (E-9). After Henry Ingram, who discovered Horsefly River.

HENRETTA CREEK, flows SW into Fording R. (C-12). After C.M. Henretta, CPR mining engineer who had charge of the company's development of coalfields in the Fording River area.

HENRY, CAPE, W. coast of Moresby I. (E-3). Named by Vancouver, together with adjacent Englefield Bay (q.v.), in honour of his friend Sir Henry Englefield, the antiquary and scientist.

HENSON HILLS, E. of Ootsa L. (F-7). After George Henson, pioneer settler, whose son Frank became a well-known guide. Frank, according to a friend, 'was born in a log cabin there, and at the age of twelve he was out trapping and selling furs to the Indians, and it sure sharpened him up.'

HERALD PARK, N. of Salmon Arm (C-10). Dr. Dundas Herald acquired this fine lakefront property in 1906. In 1975 the provincial government purchased the land from Miss Jessie Herald but retained the family name for the park that it then established.

HERIOT BAY, Quadra I. (C-7). After F.L.M. Heriot, a relative of Rear-Admiral Sir Thomas Maitland, Commander-in-Chief of the Royal Navy's Pacific Station 1860-2.

HERNANDO ISLAND, E. of Campbell River (B-7). After Hernando Cortes, the conqueror of Mexico. Nearby Cortes Island was named by Galiano and Valdes in 1792.

HESQUIAT HARBOUR, SE of Nootka Sound (B-6). From the Nootka Indian word meaning 'people of the sound made by eating herring eggs off eel grass.' Captain Walbran gives an explanation: 'At Hesquiat village a saltwater grass called "segmo" drifts on shore in large quantities, especially at the time of the herring spawning, which the Indians are in the habit of tearing asunder with their teeth to disengage from the grass or weed the spawn, which is esteemed by them a great delicacy.'

HEWITT BOSTOCK, MOUNT, SW of Merritt (C-9). Commemorates Hewitt Bostock (1864-1930). An upper-class Englishman with a Cambridge degree and a training in law, Bostock emigrated to British Columbia in 1893, bought the Duck Ranch at Monte Creek, and plunged into politics. In 1896 he became an MP and in 1904 a senator. From 1921 to 1922, he was federal Minister of Public Works, and in the latter year he became Speaker of the Senate.

HIBBEN ISLAND, Englefield Bay, QCI (E-3). After the Victoria stationer Thomas Napier Hibben (1828-90), first agent for admiralty charts in British Columbia.

HIELLEN RIVER, N. coast of Graham I. (G-3). From the Haida name meaning 'river-by-Tow,' Tow Hill being the distinctive round hill not far from Rose Point.

HIHIUM LAKE, E. of Clinton (D-9). From a Shuswap Indian word meaning 'little big lake.'

HILL CREEK, flows W. into Upper Arrow L. (C-11). After Sam Hill, who had preempted at Galena Bay on a traditional Indian camping ground. When an Indian, Cultus Jim, made a threatening move, Hill shot him dead but was acquitted since an Indian woman testified that Cultus Jim had meant to kill him.

HILLIERS, SW of Qualicum Beach (B-7). Originally Hilliers Crossing, this settlement is named after Thomas Hellier (note the correct spelling), who settled here before 1912.

HILLS BAR [HILL'S BAR], S. of Yale (B-9). James Moore, who belonged to the first party of gold-hunters in the late spring of 1858, left the following account of the naming of Hill's Bar:

> we camped for lunch on a bar about ten miles from Hope to cook lunch, and while doing so one of our party noticed particles of gold in the moss that was growing on the rocks. He got a pan and washed a pan of this moss and got a good prospect, and after our gastric wants were satisfied we all prospected the bar and found it a rich bar of gold. With our crude mode of working with rockers we made on an average of fifty dollars per day to the man. We named this bar in honor of the man that washed the first pan of moss, Hill's Bar.

To the Halkomelem Indians, Hill's Bar was Qualark, meaning it was a good place to barbecue salmon heads.

HIPPA ISLAND, W. of Graham I. (F-3). Named by Captain George Dixon in 1787. The appearance of an Indian settlement here reminded him of a Maori 'hippa' or 'pa' (fortified village) in New Zealand.

HISNIT INLET, NW of Tlupana Inlet (B-6). From the Nootka Indian word meaning 'place of sockeye salmon.'

HITCHCOCK CREEK, flows W. into Atlin L. (I.-3). After Mary E. Hitchcock of New York, author of *Two Women in the Klondike*. A wealthy widow, she visited Atlin and grubstaked at least one miner in the area.

HIXON, S. of Prince George (F-8). After Joseph Foster Hixon, who sought gold here in 1866.

HKUSAM MOUNTAIN, S. of Johnstone Strait (C-7). From the Island Comox Indian word for 'having fat or oil,' referring to the Salmon River and the village at its mouth.

HOBITON LAKE, W. of Nitinaht L. (A-7). Derived from the Nitinaht Indian word meaning 'sound of [someone] snoring,' said to be descriptive of the sound made by the outlet stream where it flows into Nitinaht Lake.

HOBSON LAKE, Wells Gray Park (E-9). After John B. Hobson, at the end of the nineteenth century considered one of the world's best hydraulic mining engineers. He was responsible for the famous Bullion Pit at Hydraulic. He always travelled in style, and when he went up to the Cariboo in the summer, he hired the Dufferin fancy four-horse stagecoach.

HOGEM RANGES, N. of Takla L. (H-6). 'Hogem' (hog them) is a BC provincialism, going back to the gold rush days, used for a trader who exploited a monopoly by charging outrageous prices. By extension a post where exorbitant prices were charged became known as a 'hogem' post.

HOGUE, MOUNT, Wells Gray Park (E-10). After Henry and John Hogue, who trapped in this area and guided fishing and hunting parties from 1936 until the early 1950s.

HOHM ISLAND, off Port Alberni (B-7). From a Nootka Indian word meaning 'blue grouse.' The island is also known as Observatory Island.

HOIK ISLAND, off Port Alberni (B-7). From a Nootka Indian word meaning 'willow grouse.' The island is also known as Deadman Island.

HOLBERG, SE of Cape Scott (C-5). Danish settlers who came here in 1895 named their settlement after Baron Ludvig Holberg (1684-1754), the distinguished Danish historian and dramatist. Holberg was the first writer to use Danish as a literary language.

HOLDEN LAKE, SE of Nanaimo (B-8). After John Holden, a Yorkshireman who settled in the area in the 1870s and farmed by the lake. He founded the first orchestra in Nanaimo.

HOLDICH CREEK, flows S. into Columbia R. (D-10). After A.H. Holdich, a graduate of the Royal School of Mines, who ran an assay office in Revelstoke around 1893.

HOLE IN THE WALL, Sonora I. (C-7). The Mainland Comox Indian word for this small round cave in a cliff near high-tide line means 'Raven's chamberpot.'

HOLLYBURN, West Vancouver (B-8). When John Lawson, the first permanent white settler, moved here, he brought with him some holly trees from his former home in Vancouver. These hollies and the *burn* (Scots for 'stream') running across his new property inspired him to coin the name Hollyburn.

HOLMES RIVER, flows S. into Fraser R., E. of McBride (F-10). After A.W. Holmes, provincial forest ranger, who died in the 1920s.

HOMATHKO RIVER, flows S. into Bute Inlet (C-7). From the Mainland Comox Indian word meaning 'swift water.'

HOMFRAY CHANNEL, E. of Redonda Islands (C-7). After Robert Homfray, CE (1824-1902), a native of Worcestershire who arrived in Victoria from California in 1859. He was a very competent civil engineer and worked for a while on the CPR surveys. He was also an eccentric and, while still alive, had his tombstone erected in Ross Bay cemetery, Victoria.

HONEYMOON BAY, Cowichan L. (A-7). Many years ago a young bachelor lived on this bay. Tired of the loneliness of his life, he announced that he was going back to England to get a bride. Mrs. March, another early settler on the lake, suggested that he bring her back to live on the bay. The young Englishman never returned, and nobody knows if he married or not, but the name Honeymoon Bay remains on the map.

HOOD POINT, Bowen I. (B-8). After Sir Alexander Hood (1727-1814), Admiral of the Blue, second in command to Lord Howe at the latter's victory over the French in 1794, in the battle known as 'The Glorious First of June.' Later Hood rose to the dignity of Viscount Bridport.

HOODOO LAKES, NW of Prince George (G-8). So named by Forin Campbell in 1910 when his survey of the area was dogged by misfortunes: men with cut feet, lack of horses, and early snowstorms that left them stranded here until November.

HOOKER, MOUNT, W. of Hamber Park (E-10). This landmark by Athabasca Pass was named in 1827 by David Douglas, of Douglas-fir fame, 'in honour of my early patron the enlightened and learned Professor of Botany in the University of Glasgow.' This professor was Sir William Jackson Hooker (1785-1865), who became director of the Royal Botanical Gardens, Kew, in 1841.

HOOPER, MOUNT, NW of Cowichan L. (B-7). Named by Dr. Robert Brown after William Hooper, a 'pioneer and miner' in his Vancouver Island Exploration Expedition of 1864.

HOPE, Fraser Valley (B-9). Fort Hope was built in 1848-9 by Henry Newsham Peers, a clerk in the service of the HBC. A year earlier Peers had discovered a way through the mountains here (up the Coquihalla River and Peers Creek, over Fools Pass, along Podunk Creek, and across the Tulameen River). The HBC hoped that, with the building of a trail, this would prove a feasible all-British route by which their brigades could travel between Fort Kamloops and Fort Langley. (It was important that the brigades should not have to dip below the forty-ninth parallel into what had recently become American territory.) This hope, which was in fact realized, that a usable all-British route had been found led the HBC to name its new establishment Fort Hope.

The townsite at Hope was laid out in 1858 by O.J. Travaillot and Corporal William Fisher, RE.

The Halkomelem Indian name for the site of Hope means 'skinned rocks' (i.e., bare of moss).

HOPE BAY, E. side of N. Pender I. (A-8). After Rutherford Hope, who settled on the island around 1880.

HOPE ISLAND, NE of Cape Scott (C-6). After Vice-Admiral Sir James Hope, Commander-in-Chief of the North America and West Indies Station, RN, 1864-7.

HOPKINS LANDING, Howe Sound (B-8). After George Henderson Hopkins (1853-1931). After some years as an engineer on the ships of the British India Steam Navigation Co., Hopkins gave up the sea and became a partner in the engineering firm of Richards and Hopkins, Monmouthshire. He retired in 1906 because of ill health and came to British Columbia. Soon after arriving on the coast, he bought 160 acres on which to build his summer cottage, but he liked the place so well that he spent the rest of his life here.

HORETZKY CREEK, NE of Kemano (F-6). After Charles Horetzky (1839-1900), engineer, explorer, and author. Horetzky played an important part in the early surveys for the CPR between 1872 and 1880. He himself surveyed a northern route, which he fiercely championed, by way of the Peace River to Port Simpson, a short distance north of the modern Prince Rupert. Horetzky explored this creek in 1876.

HORNBY ISLAND, SE of Courtenay (B-7). According to Captain Walbran, named by the HBC about 1850 after Rear-Admiral Phipps Hornby (1785-1867), commanding the Pacific Station from 1847 to 1851.

HORNE LAKE, W. of Qualicum Beach (B-7). After Adam Horne (1831-1903), HBC storekeeper in Nanaimo, who discovered the lake in 1856. Horne, a man of tremendous personal courage, was perhaps the first white man to cross the waist of Vancouver Island.

HORSE LAKE, SE of 100 Mile House (D-9). This name appears on A.C. Anderson's map of 1867 with the notation 'Lac des Chevaux Noyés [Lake of the Drowned Horses] 1827.' The first HBC brigade trail from Kamloops to Alexandria passed by here.

HORSEFLY, NE of Williams L. (E-9). The settlement of Horsefly was originally called Harper's Camp after Thaddeus Harper, pioneer rancher and miner. The present name comes from the horseflies that can be a great nuisance here in the summer.

HORSERANCH LAKE, E. of Dease R. (L-5). There never was a ranch here, but packers found that the area around this lake provided good winter pasturage for their animals and took to calling it 'The Horse Ranch.'

HORSESHOE BAY, E. side of Howe Sound (B-8). So named because of its shape.

HORSETHIEF CREEK, flows into Columbia R. S. of Radium (C-11). Commemorates the capture here, in the 1880s, of a rogue who had stolen horses from some whisky pedlars.

HORSFALL ISLAND, NW of Campbell I. (E-5). Named by Captain Walbran, author of *British Columbia Coast Names 1592-1906*, after his maternal grandfather, the Reverend Thomas Horsfall (1795-1869), vicar of Cundall, Yorkshire.

HOSMER, NE of Fernie (B-12). After Charles R. Hosmer (1851-1927), at one time manager of the CPR telegraph system and later a director of the railway.

HOSPITAL CREEK, N. of Golden (D-11). During construction of the CPR, there was a hospital here.

HOTHAM SOUND, Jervis Inlet (B-7). After Admiral William Hotham, RN, Commander-in-Chief in the Mediterranean 1794-5. A courageous and conscientious subordinate, Hotham lacked the energy and force of character needed in the top command and was relieved in 1795 by Sir John Jervis, the future Earl St. Vincent.

HOTLESKLWA CREEK, flows N. into Laslui L. (J-6). From the Sekani Indian phrase meaning 'when the water is high, fish look like mud.'

HOT SPRINGS COVE, E. of Hesquiat (B-6). The Indian name for these springs was 'Mak-seh-kla-chuck,' meaning 'smoking water.'

HOTTAH LAKE, N. of Spatsizi Park (K-5). When E.C.W. Lamarque, the pioneer surveyor, came to this lake, he was told that the Tahltan Indians called it Ottah, meaning 'the same at both ends.' Lamarque rendered the name Hottah, making it almost a palindromic name for this lake whose ends look alike.

HOUDINI NEEDLES, Adamant Range (D-11). So named since only a contortionist like Houdini could ascend these peaks.

HOUSTON, SE of Smithers (G-6). In 1910, as a result of a newspaper contest for a suitable name, this settlement was named after John Houston, the first newspaperman in Prince Rupert. Earlier Houston had been the first mayor of Nelson, where he had edited the *Tribune*. He was a man with a rather dubious reputation.

HOUSTON STEWART CHANNEL, S. end of Moresby I. (E-4). Formerly Barrell Sound. Named in 1853 by Captain J.C. Prevost, commanding HMS *Virago,* in compliment to his predecessor. Stewart had been a most popular captain, he and his wife being roundly cheered by the *Virago*'s officers and crew when he went ashore after relinquishing command at Callao.

HOUSTOUN PASSAGE, N. of Saltspring I. (A-8). After Captain Wallace Houstoun, commanding HMS *Trincomalee* on the Pacific Station 1852-7.

HOWE SOUND, NW of Vancouver (B-8). Named by Captain Vancouver after Admiral the Rt. Hon. Richard Scrope, Earl Howe (1726-99). Howe won his most famous victory in 1794 in the battle that has gone down in the annals of the Royal Navy as 'The Glorious First of June.' On that day he not only defeated a superior French fleet but also captured seven ships of the line. When he returned from this great victory, the royal family visited him aboard his flagship, HMS *Queen Charlotte,* and the King presented him with a sword with a diamond-studded hilt valued at 3,000 guineas.

Because of the consideration that Lord Howe showed his crews, he was known as 'The Sailors' Friend.' Their nickname for him was 'Black Dick.' On the morning of his great victory, one of his seamen was heard to say, 'I think we shall have a fight today. Black Dick has been smiling.'

In 1859-60 Captain Richards, RN, named many of the islands, points, and mountains in Howe Sound after officers and ships that had seen service on 'The Glorious First of June.' 'Thus,' as Captain Walbran has noted, 'this sound is a record of the battle' (*British Columbia Coast Names,* p. 256).

HOWSE PASS, Alberta-BC boundary (D-11). Used by David Thompson in 1807 on his way to found Kootenae House on the upper Columbia River. The pass is named, however, after the next white man to go that way, Joseph Howse, who used the same route in 1810 when sent by the HBC to find out what the NWC was doing on the far side of the Rockies. Howse established a post near present-day Kalispell, Montana.

HOWSER, N. of Kootenay L. (C-11). This settlement was originally named Duncan City because of its location on the shore of Duncan Lake. After a post office (Duncan) was opened here in 1899, its mail was constantly being confused with that for Duncan's Station (now Duncan) on Vancouver Island. A change of name being ordered by Ottawa, the locals opted for Hauser, after a pioneer who had found gold in the area just a year before he and his partner went into the wilderness together, never to be seen again. In his letter to Ottawa, the postmaster misspelled the new name as Howser, and Howser it has remained.

HOWSON CREEK, flows N. into Telkwa R. (G-6). After Jack Howson, prospector and teamster at Telkwa after World War I.

HOYA PASSAGE, Darwin Sound, E. coast of Moresby I. (E-4). From the Haida name for 'raven,' but the Haidas' own name for the passage means 'channel behind.'

HOZAMEEN RANGE, SE of Hope (B-9). Around 1860 Dr. H. Bauerman, a geologist with the British Boundary Commission, wrote of 'two very remarkable peaks of black slate, which rise precipitately to a height of about 1800 feet above the watershed. They are called by the Indians "Hozamen", which name has been adopted for the pass and the ridge.'

Miss Annie York of Spuzzum, who spoke the Thompson Indian language, gave the meaning of *hozameen* as 'sharp, like a sharp knife.'

HUBER, MOUNT, NE of L. O'Hara, Yoho NP (D-11). After Emil Huber of the Swiss Alpine Club, one of the three climbers who in 1890 made the first ascent of Mount Donald.

HUBERT, SE of Telkwa (G-6). After the son of Sir Alfred Waldron Smithers, chairman of the Grand Trunk Pacific Railway (now part of the CNR).

HUDSON BAY MOUNTAIN, NW of Smithers (G-6). This impressive mountain takes its name from the ranch that the HBC maintained nearby, on Driftwood Creek, from 1899 to 1901.

HUDSON'S HOPE, Peace R. (I-9). This is the municipality's official spelling of its name even though the post office here, ever since it was opened in 1913, has used the name 'Hudson Hope.'

The post was known as early as 1873 as 'Hudson's Hope' or 'The Hope of Hudson.' The origin of the name is unknown. Sir William Butler, who passed this way in 1873, tells an anecdote in which it figures as an HBC outpost, thus giving support to those who maintain that the name was an ironic comment on the HBC's action in putting a seasonal trading post here, administered from Fort St. John. On the other hand, another school of thought maintains that the Hudson of 'The Hope of Hudson' was a hypothetical prospector who panned here for gold. A very old meaning of the word *hope* is 'a small inlet, valley, or haven.'

HUGH KEENLEYSIDE DAM, Lower Arrow L. (B-11). When the Arrow Lakes were flooded in the 1960s as part of the Columbia River project, this dam was named for one of the two co-chairmen of the BC Hydro and Power Authority. Earlier he had a distinguished career as a diplomat and as a federal civil servant.

HULATT, E. of Vanderhoof (F-8). After Henry Hulatt, manager of the Grand Trunk Pacific Telegraph Company.

HULLCAR, SW of Enderby (C-10). There seems to be agreement that this word is related to the nearby high rock bluff, but it is not clear whether it is derived from the Chinook jargon or the Okanagan Indian language, or what its meaning is.

HUNAKWA LAKE, N. of Anstey Arm, Shuswap L. (D-10). G.M. Dawson noted in 1877 that this name is derived from the Shuswap Indian word meaning 'one lake only.'

HUNGABEE MOUNTAIN, Yoho NP (D-11). From the Stoney Indian word meaning 'chiefs' or 'council.' This name is justified by the dominating appearance of this peak.

HUNLEN FALLS, on Hunlen Cr. W. of Lonesome L. (E-7). These, the third highest falls in British Columbia (1,300 feet or 396 metres), were discovered in 1911 by Walter and Frank Ratcliffe, who named them after the Indian whose trapline ran by them.

HUNTER CREEK, flows N. into the Fraser R. W. of Hope (B-9). After Harry Hunter (1825-1910), an English storekeeper who acquired land here in 1889.

HUNTINGDON, S. of Abbotsford (B-8). After the hometown of Benjamin Douglas, the original locator, who came from Huntingdon, Quebec. Douglas and a partner, Higginson, laid out the townsite and named it.

HUPEL, W. of Mabel L. (C-10). After Herman ('Mike') Hupel, a genial, cross-eyed German-American who became postmaster here in 1910.

HURD, MOUNT, Yoho NP (D-11). After Major M.F. Hurd, who was engaged in exploratory surveys during the building of the CPR.

HURLEY RIVER, flows N. into Bridge R. (C-8). After Daniel Hurley, miner and rancher, at one time a hotelkeeper in Lillooet, where he died in 1942.

HUSON LAKE, S. of Nimpkish L. (C-6). After Alden Westley Huson (1832-1913), a prospector who arrived in British Columbia from the California goldfields in 1858. He found his way to this lake, which he named after himself, and to the nearby caves. Huson is pronounced 'Whó son.'

HUXLEY ISLAND, E. coast of Moresby I. (E-4). Named by G.M. Dawson in 1878 after Thomas Henry Huxley (1825-95), the famous biologist.

HYDRAULIC, SW of Quesnel Forks (E-9). Takes its name from the hydraulic monitors that washed away the overlay to get at the gold deposits in the nearby Bullion Pit, a huge chasm excavated by water under high pressure. The operation was carried on by the Cariboo Hydraulic Company between 1892 and 1942. (See also *Hobson Lake.*)

HYLAND RIVER, flows S. into Liard R. (L-5). After John Hyland, pioneer trader and founder of one of the dynastic families in the Cassiar.

I

ICE RIVER, Yoho NP (D-11). A translation of the Stoney Indian name for this glacier-fed stream, 'Washmawapta' (literally 'deep snow river').

IKEDA COVE, E. coast of Moresby I. (E-4). After Arichika Ikeda, the Japanese who discovered copper here in 1906. Awaya, Ikeda and Co. subsequently operated a mine here.

IKNOUK RIVER, flows SW into Nass Bay (H-5). From a Nisgha Indian word having something to do with the old type of halibut hook.

ILLAHEE MOUNTAIN, E. of Horsefly L. (E-9). Chinook jargon meaning 'country, region, farm, or the place where one resides.'

ILLECILLEWAET RIVER, flows W. into Columbia R. near Revelstoke (C-10). From the Okanagan Indian word meaning 'big water.'

INCOMAPPLEUX RIVER, N. of Arrow Lakes (C-11). From the Okanagan Indian word meaning 'head end of lake' or 'end of the water.'

INDIAN ARM, Burrard Inlet (B-8). Explored by the Spanish in June 1792, who met a few Indians here. From them they learned that the Indians applied the name Sasamat either to Indian Arm or to the whole of Burrard Inlet – just which is not clear. Sasamat Lake (q.v.), near the mouth of Indian Arm, preserves this name. Indian River, at the head of Indian Arm, was officially Meslisloet River until 1921.

INDIAN PEAK, NW of Mt. Assiniboine (C-12). From a fancied resemblance to the head of an Indian in war regalia.

INGENIKA RIVER, flows E. into Williston L. (I-7). A Sekani Indian word indicating that the low-lying kinnikinnick shrub (bearberry) is found along this river.

INGRAM CREEK, flows S. into Kettle R. (B-10). After Jack Ingram, who had a roadhouse here during the Rock Creek gold rush of the early 1860s.

INGRAM CREEK, near Westwold (C-10). After Henry Ingram, who became the first settler in the area around 1865. According to tradition, he brought with him six of the camels that had proved unsatisfactory on the Cariboo Road. Ingram established a good ranch, started a popular race meet in 1873, and died in 1879.

INKANEEP CREEK, flows S. into Osoyoos L. (B-10). From the Okanagan Indian word meaning 'bottom end.'

INKLIN RIVER, flows NW into Taku R. (K-3). Apparently derived from the Tlingit Indian word meaning 'big river.'

INSKIP CHANNEL, E. of Englefield Bay, QCI (F-3). After George Hastings Inskip, RN (1823-1915), master of HMS *Virago* during her cruises in these waters in 1853.

INTERSECTION MOUNTAIN, Rocky Mtns. (F-9). This mountain stands where the 120th meridian (which becomes the BC-Alberta boundary north of here) meets the Continental Divide (which becomes the boundary south of here).

INVERMERE, N. end of Windermere L. (C-11). In 1890, in consequence of some local mining excitement, Edmund T. Johnson laid out a townsite here and named it Copper City. In 1900 the Canterbury Townsite Company took over the site and renamed it after the famous cathedral city in England. Finally the Columbia Valley Irrigated Fruit Lands Company acquired the site, and the name was changed to Invermere by the Hon. R. Randolph Bruce. The name Invermere means 'at the mouth of a lake.'

INVERNESS PASSAGE, mouth of Skeena R. (G-4). After the Inverness Cannery, which once stood on its shore. This cannery, established in 1876 but acquiring the Inverness name only in 1880, was the first on the northern coast of British Columbia. Earlier the site was known as Woodcock's Landing, William H. Woodcock having established, before 1870, an inn here for miners en route to and from the Omineca gold area.

INZANA LAKE, N. of Stuart L. (G-7). Said to be from the name of an Indian man, Inzanum.

IOCO, Burrard Inlet (B-8). From the initial letters of the Imperial Oil Company, which owns the large refinery here.

IONA ISLAND, mouth of Fraser R. (B-8). Originally Mole Island, then McMillan Island after Donald McMillan (1848-1901), a pioneer settler, and finally renamed by McMillan after Iona, the island in the Inner Hebrides where St. Columba in 563 AD began the christianizing of the Scots.

IRBY POINT, Anvil I., Howe Sound (B-8). After Rear-Admiral the Hon. Frederick Paul Irby, who served as a midshipman on HMS *Montagu* at Lord Howe's great victory, 'The Glorious First of June,' 1794.

IRELAND CREEK, S. of Mabel L. (C-10). After De Courcy Ireland, son of a Vernon magistrate, who preempted land here in 1895.

IRON MOUNTAIN, S. of Merritt (C-9). So named by G.M. Dawson because of a spectacular deposit of iron ore near the top of this mountain.

IROQUOIS CREEK, W. from McLeod L. (G-8). A number of Iroquois came west to British Columbia with the early fur traders as servants and as trappers. At one time there was a little Iroquois settlement in the Jasper area. This creek gets its name from the fact that an Iroquois, his wife, and two children were murdered here by two Carrier Indians who had warned him that he was poaching on their beaver territory.

IRVINE, N. of Vavenby (D-10). After John L. Irvine, divisional engineer, killed near here during location of the Canadian Northern Railway (now CNR) in 1911.

IRVINES LANDING [IRVINE'S LANDING], Sechelt Pen. (B-7). After Charles Irvine, who became the first postmaster here in 1897 and left in 1904.

ISAAC LAKE, Bowron L. Park (F-9). After George Isaac, an old-time prospector who led a party past here in 1886. He ran the Barkerville Club in Barkerville.

ISHKHEENICKH RIVER, flows N. into Nass R. (G-5). From the Nisgha Indian word meaning 'water out from among the pine tree(s).'

ISKUT RIVER, flows SW into Stikine R. (I-4). Possibly from the Nisgha Indian word meaning 'stinking,' otherwise of unknown origin.

ISOLILLOCK PEAK, S. of Hope (B-9). The name of this fine mountain, known also as Holy Cross Mountain, comes from the Halkomelem word meaning 'double heads,' describing its split double peaks.

ITALY CREEK, near Christina L. (B-10). So named because of a nearby Italian community.

IVY GREEN PARK, NW of Ladysmith (A-8). After Richard and Ellen Ivey, who settled in the Ladysmith district in 1903 and delivered the rural mail north of that town.

JACK POINT, Nanaimo (B-8). After Norwegian-born Jack Dolholt (1819-1905), a captain on coastal schooners until he settled here in the 1860s.

JACKASS MOUNTAIN, Fraser Canyon (C-9). The old Cariboo Road, narrow and without any protective parapet, came around a corner here with a drop of 500 feet to the Fraser below. According to an old story, a frightened woman passenger screamed at Steve Tingley, the famous driver of the Cariboo coach, 'What happens if we go over the edge, Mr. Tingley?' To which he replied imperturbably, 'Lady, that all depends on what sort of life you've been leading.'

Among those things that did go over the edge in the days before the trail was widened into a road was a jackass laden with miners' goods, presumably en route to the Cariboo goldfields.

JACK LEE CREEK, flows S. into Toodoggone R. (J-6). After Jack Lee, a popular guide, who has been described as 'slight and wiry with quick slogging strides.'

JACK OF CLUBS LAKE, NW of Barkerville (F-9). An item in the *Daily British Colonist* of 30 July 1861 gives the story of this name: 'a redskin had first struck the new creek and had communicated the discovery to an individual rejoicing in the sporting appelation of Jack-of-clubs.' Louis Le Bourdais (see *Le Bourdais Lake*) had a story that Jack of Clubs was an Italian gambler so nicknamed because he cut his beard like that of the jack of clubs playing card.

JACKSTONE CREEK, flows NE into Frog R., Cassiar (K-6). E.C.W. Lamarque, the noted surveyor, recalled Jack Stone, a Sekani Indian, as 'silent, inscrutable, efficient ... an excellent companion on the trail.'

JAFFRAY, SW of Fernie (B-12). After Senator Robert Jaffray of Toronto (1832-1914), president of the Crows Nest Pass Coal Co.

JAMES BAY, Victoria (A-8). Named in 1846 by Captain Kellett (HMS *Herald*) after Chief Factor (later Sir James) Douglas. In later years Douglas had his home on the south shore of the bay. The Straits Salish name for James Bay, before fill reclaimed the site for the Empress Hotel, was 'Whosaykum,' meaning 'muddy place.'

JAMES ISLAND, northeast of Victoria, and MOUNT DOUGLAS are also named after Douglas.

JAMIESON CREEK, flows SE into N. Thompson R. (C-9). After James Jamieson, sawmill operator, who preempted land here in August 1871.

JARVIS PASS, NW from Jasper NP (G-9). After E.W. Jarvis, CE, who used this pass in 1875 when exploring a route from Quesnel to Edmonton for the CPR. On this trip he almost starved to death.

JAWBONE CREEK, flows SW into Lightning Cr., Cariboo (F-9). 'Jawbone' was a word used in the gold camps for 'credit.' Presumably the prospectors on this creek were particularly adept at persuading others to finance them.

JAYEM, S. of Nanoose Bay (B-7). Jayem stands for J.M. Cameron, manager of the Esquimalt and Nanaimo Railway, 1928-35.

JEDWAY, Skincuttle Inlet, Moresby I. (E-4). Despite the appearance of this word, it is of Indian origin. From the Haida word *gigawai*, meaning the 'throat of a salmon trap' or a 'snare.'

JEMMY JONES ISLAND, off Oak Bay (A-8). After one of the most colourful characters in BC's early history, Captain James (Jemmy) Jones (1830-82), who lost a schooner, the *Caroline,* on the shore here. His most famous exploit occurred in 1865 after he had been imprisoned for debt in Victoria and his schooner, the *Jenny Jones,* had been seized at Olympia, Washington, by a US marshal. Disguised as a woman, Jemmy escaped from Victoria, went to Olympia, and sailed as a passenger on the *Jenny Jones* when the marshal took her to Seattle for sale. En route, while the marshal was ashore for a night, Jemmy repossessed his ship. After sundry adventures he reached Mexico aboard it, where he sold the ship. Later, tried for stealing the *Jenny Jones,* Jemmy was acquitted on the very reasonable ground that the marshal had left the ship, not the ship the marshal. Jones could neither read nor write, but he made do with acuteness and a good memory.

JENNINGS RIVER, flows into Teslin L. (L-3). After W.T. Jennings, CE, who made surveys in this area.

JERICHO BEACH, Vancouver (B-8). Takes its name from Jeremiah (Jerry) Rogers (1818-79). From his camp at Jerry's Cove ('Jericho'), Rogers sent axemen inland to fell the giant trees that once grew on Point Grey. A native of New Brunswick, he started logging Point Grey in either 1864 or 1865.

JERVIS INLET, S. coast (B- and C-8). Named by Captain Vancouver after Rear-Admiral Sir John Jervis (1735-1823), made Earl St. Vincent after his great victory over the Spaniards off Cape St. Vincent in 1797. Jervis was a formidable disciplinarian. When the crew of HMS *Marlborough* refused to hang one of their number as a mutineer, Jervis was ready to sink the ship with her crew if the condemned seaman was not hanging from a yardarm by the time that he had set. On the other hand, stories are also told of Jervis's kindness and generosity to his men.

JESMOND, NW of Clinton (D-9). After Ottawa refused to accept 'Mountain House' for the post office opened here in 1918 (presumably to avoid duplication of names), Mr. Coldwell, a former resident of Jesmond, a district in Newcastle-upon-Tyne in England, successfully proposed this name.

JESSOP ISLAND, mouth of Laredo Inlet (E-5). After John Jessop, born in Norwich, England, in 1829, who came to British Columbia in 1859, one of a party of eight men who walked overland from Fort Garry. A teacher and journalist, Jessop founded the first nonsectarian school on Vancouver Island. He was the province's first superintendent of education from 1871 until 1878 and later joined the federal immigration service.

JESTER PEAK, near Mt. Waddington (D-7). Don Munday (see *Munday, Mount*) saw this peak, close to Mount Waddington, as a jester 'at the emperor's feet.'

JEUNE LANDING, Neroutsos Inlet, Quatsino Sound (C-6). After the Jeune brothers, Channel Islanders from Jersey, who manufactured canvas products at Victoria from the time of the gold rushes. This was a drop-off point to which the firm shipped tents, tarpaulins, etc. ordered by surveyors, loggers, and others.

JEWITT COVE, Nootka Sound (B-6). After the central character in *The Narrative of the Adventures and Sufferings of John R. Jewitt, Only Survivor of the Crew of the Ship* Boston, *during a Captivity of Nearly Three Years among the Savages of Nootka Sound.* Actually young Jewitt was one of two survivors when Indians captured the *Boston* in 1803.

JIM SMITH LAKE, SW of Cranbrook (B-12). Presumably after the James Smith who was one of a party of five prospectors from Walla Walla, Washington, who arrived at Wild Horse in March 1864.

JOB, MOUNT, NW of Pemberton (C-8). Dr. Neal Carter, recalling a mountaineering holiday in 1932, noted, 'We named one mountain "Job Mountain" after the biblical character who was so afflicted with boils, because the mountain had many volcanic excrescences.'

JOBE, MOUNT, N. of McBride (F-10). After Mrs. Carl Akeley (née Mary Jobe), the wife of the African explorer. She herself mountaineered and explored in this area over a period of ten years.

JOE RICH CREEK, E. of Kelowna (B-10). After an American prospector who had a cabin here in the 1880s. He was subsequently killed in a mine accident.

JOFFRE, MOUNT, Alberta-BC boundary (C-12). After Marshal J.J.C. Joffre (1852-1931), Commander-in-Chief of the French Army on the western front 1914-16.

JOHN BENNETT CREEK, flows NE into Pine R. (H-8). Commemorates the tragic death of an eighteen-year-old English lad, John Noel Patch Bennett, heir to a fortune back home. In late October 1930, en route to Prince George, disregarding the advice of experienced woodsmen, he set out to cross the Pine Pass, wearing light shoes, having insufficient food and clothing, no guide, and only one packhorse. His diary, found with his remains the following spring, told of wet matches, frozen hands and feet, a trail lost in the snow, and all the other disasters that can befall a tenderfoot. Found in his packsack was a Latin-English dictionary.

JOHN BUCHAN ISLAND, Eutsuk L. (F-6). After John Buchan, Lord Tweedsmuir, Governor-General of Canada 1935-40 (see *Tweedsmuir Park*). The Carrier Indians knew this, the largest lake isle in British Columbia, as 'Ukwe-ses-nê-rc-thel-krêh-nu,' meaning 'the island where the black bear escapes us,' indicating that the island gives a bear space enough to elude pursuers.

JOHN DEAN PARK, N. Saanich (A-8). After John Dean (1850-1943), a fascinating eccentric. Born in Cheshire, he arrived in British Columbia from California in 1885. In 1903 he was mayor of Rossland. Subsequently he moved to Victoria, where he engaged for a while in the real estate business. Entering civic politics in Victoria as an ardent advocate of putting the city under a city manager, he was defeated in civic elections in 1928 and 1929.

An enthusiastic naturalist, Dean owned property high on Mount Newton, and here he spent much of each summer in a log cabin. With considerable misgivings he was persuaded to let the Boy Scouts camp on his property. His delighted amazement at finding how excellently they had cared for it may have had something to do with his decision in 1931 to turn over eighty acres for a provincial park to preserve the original flora and fauna of the area.

In 1936, a lifelong bachelor of eighty-six, Dean had his own tombstone erected in Ross Bay cemetery. On it he had engraved his verdict about life: 'It is a rotten world, artful politicians are its bane. Its saving grace is the artlessness of the young and the wonders of the sky.' He died some six years later.

JOHN SANDY CREEK, flows SW into Lillooet R. (C-8). After a Lillooet Indian who was captured by the Chilcotins when a boy. He drifted into the Cariboo, where he killed one of two men who had robbed him and was acquitted at the subsequent trial. He is said to have been given his name of John or Johnny Sandy by a us excise officer in the Port Simpson area. He died about 1937, regarded with awe by the Indians of the Pemberton area.

JOHNSTONE STRAIT, N. VI (C-6 and 7). Named by Captain Vancouver after James Johnstone, master of the *Chatham,* who, exploring with the *Chatham*'s cutter in July 1792, found for Vancouver this passage linking the Strait of

Georgia with the open Pacific to the north. Johnstone first visited this coast in 1787-8 on Colnett's trading voyage with the *Prince of Wales*. No doubt it was because of the experience that Johnstone had gained on this private venture that Vancouver secured him his post on the *Chatham*. Johnstone became a captain in the Royal Navy in 1806 and was later a commissioner of the navy at Bombay

JOLLY JACK CREEK, flows S. into Boundary Cr. (B-10). After 'Jolly Jack' Thornton, a prospector and the second white settler in the district.

JORDAN RIVER, W. of Sooke (A-7). Named by the Spaniards after Alejandro Jordan, the chaplain who accompanied Lieutenant Francisco Eliza to Nootka in 1790 and remained there for some time.

JOSEPHINE FLAT, Menzies Bay, VI (C-7). In 1895 the officers of HMS *Nymphe* attended an amateur performance of 'HMS *Pinafore*' at Nanaimo. Smitten by the charms of the girl who played the part of Josephine, the captain's daughter, they gave this name as they proceeded with their resurvey of the coast to the north.

JUAN DE FUCA, STRAIT OF, S. of VI (A-7 and 8). In *Purchas His Pilgrimes,* published in 1625, was printed a memoir of Michael Lok, an Englishman trading in the Near East. This memoir related how he met in Venice in 1596 'an old man, about threescore yeares of age, called commonly Juan de Fuca, but named properly Apostolos Valerianos, of Nation a Greeke, borne in the Iland Cefalonia, of Profession a Mariner, and an ancient Pilot of Shippes.' This Juan de Fuca told Lok that in 1592 the Viceroy of Mexico sent him north in the Pacific to seek the Straits of Anian, a supposed Northwest Passage. De Fuca told Lok specifically:

> he followed his course in that Voyage West and North-west in the South Sea, all alongst the coast of Nova Spania, and California, and the Indies, now called North America ... untill hee came to the Latitude of fortie seven degrees, and that there finding that the land trended North and North-east, with a broad inlet of Sea, between 47. and 48. degrees of Latitude: hee entred thereinto, sayling therein more than twentie dayes, and found that Land trending still sometime North-west and North-east, and North, and also East and South-eastward, and very much broader Sea then [than] was at the said entrance, and that hee passed by divers Ilands in that sayling. And that at the entrance of this said Strait, there is on the North-west coast thereof, a great Hedland or Iland, with an exceeding high Pinacle, or spired Rocke, like a piller thereupon.

In 1787 Captain Charles Barkley of the trading ship *Imperial Eagle,* finding an opening in the coast just where Juan de Fuca said there was one, named it after him.

In many ways de Fuca's account tallies with the location of the strait, the widening into the Strait of Georgia, and the numerous Gulf Islands. On the other hand, the inventions of an old scamp may have happened to coincide with geographical fact. Largely because no mention of de Fuca's voyage has been found in the Spanish archives, it has been generally dismissed as apocryphal. There are those, however, who maintain that Juan de Fuca did indeed sail into these waters. Captain Walbran, who knew this coast as thoroughly as any man, was satisfied that de Fuca had indeed discovered the strait and that De Fuca's Pillar off Tatooche Island is his 'high Pinacle, or spired Rocke.'

JUAN PEREZ SOUND, E. of Moresby I. (E-4). Named by G.M. Dawson in 1878 in honour of Juan Perez, who in 1774 discovered the Queen Charlotte Islands. (See *Estevan Point*.)

JUBILEE MOUNTAIN, W. of upper Columbia R. (C-11). Apparently commemorates the diamond jubilee of Queen Victoria in 1897.

JUDGE HOWAY, MOUNT, N. of Stave L. (B-8). Named after Frederic William Howay (1867-1943), county court judge at New Westminster, a leading BC historian, and president of the Royal Society of Canada 1941-2. The Howay and Scholefield *British Columbia from the Earliest Times to the Present* is a major work in its area. At his death Judge Howay left his books to UBC. Here they were joined a few years later by those of his fellow student at Dalhousie Law School, lifelong friend and fellow enthusiast for BC history, Robie Reid. The books now constitute the Howay-Reid collection at UBC, a source that has been of major help in compiling this book.

JUSKATLA INLET, Masset Inlet, Graham I. (F-3) Over a century ago, G.M. Dawson gave 'belly of the rapid' as the meaning of Juskatla. Other meanings that have been given are 'strong tide inlet' and 'inlet on the inside of Djus Island.'

K

KACHOOK CREEK, flows N. into Jennings R. (L-4). To use the quaint phrasing of a toponymical file in Ottawa, 'part of a Tlinkit Indian expression imploring mosquitoes to leave.'

KAGAN BAY, Skidegate Inlet (F-3). *Kaagan* is Haida for 'mouse' (specifically for the white-footed mouse). It is also the Haida name for nearby Slatechuck Mountain.

KAIEN ISLAND, Prince Rupert (G-4). Takes its name from a Tsimshian Indian word meaning 'foam.' Combinations of tide rip and heavy rain produce quantities of foam that can extend for a mile or so to the south of the island.

KAINS LAKE, W. of Port Hardy (C-6). After Tom Kains, Surveyor-General of British Columbia (1891-8). He was ahead of his time in urging photo-topographic surveying in mountainous country.

KAKIDDI LAKE, E. of Edziza Peak (J-4). A corruption of the Sekani Indian term for 'narrow water.'

KALAMALKA LAKE, S. of Vernon (C-10). The Okanagan Indian word *cheloot-soos* (meaning 'long lake cut in the middle') was applied specifically to the narrow strip of land separating Kalamalka Lake from Wood Lake to the south. This strip was sometimes called the Railway because it resembled a railway embankment, while Kalamalka Lake was formerly known as Long Lake.

Kalamalka was a well-known old Indian who once lived at the head of this lake. Kay Cronin in her *Cross in the Wilderness* (pp. 132-3) tells the touching story of Chief Kalamalka and his wives. In his old age, Kalamalka was very anxious to become a Christian and repeatedly asked Father Le Jacq to baptize him. Each time the good father protested that he could not do so until Kalamalka gave up his heathen practice of having four wives. Loyal to his wives, Kalamalka produced reasons against putting aside any of them: one was the mother of his oldest son, another was lame from the terrible frostbite she had suffered once when saving him amid the winter snows, and so the story continued. At length Father Le Jacq was so moved by the old Indian's constancy to his wives, along with his tremendous desire to be a Christian, that he appealed on his behalf to the bishop, only to hear his own ruling repeated – Kalamalka must settle for a single wife.

Coming back sadly from New Westminster, Father Le Jacq received from Kalamalka the tidings that at last he had only one wife. She turned out to be none of the four, but a good-looking young woman! The four wives had held a conference, decided that a new young wife could take over a lot of the work, and had sent the chief to find a new wife while they went into retirement.

And so from that day on, old Kalamalka had one wife but supported all five women, was baptized, and, presumably, was happy.

The word *kalamalka* can be identified as an Okanagan Indian man's name, making very suspect a theory that it is a Hawaiian name brought into the country by one of the Kanakas employed by the HBC.

KALEDEN, S. of Penticton (B-10). When the townsite was laid out in 1909, one of the lots was offered as a prize for whoever suggested the best name for the new settlement. The winner was the Reverend Walter Russell, who suggested 'Kaleden,' a compound derived from the Greek *kalos,* meaning 'beautiful,' and 'Eden,' the biblical garden of paradise.

KAMLOOPS, confluence of N. Thompson R. and S. Thompson R. (C-9). John Tod, the veteran HBC man who in 1841 took over the fort here after the murder of Chief Factor Black, noted in his memoirs: 'The Shuswap Indians called the place "Kahm-o-loops", meaning "the meeting of the waters."'

The first white man to visit the area was David Stuart of the Pacific Fur Company, who came by way of the Okanagan in 1811-12. He was followed in May 1812 by his associate, Alexander Ross. For ten days Ross traded with some 2,000 Indians 'at a place called by the Indians Cumcloups.' Later that summer Stuart returned and built the first fort, apparently on the site of modern downtown Kamloops. In November he was followed by Joseph LaRocque, who built the North West Company's Fort Thompson on the other side of the river, in the wide angle formed by the confluence of the North and South Thompson. Soon after, when the NWC bought out the Pacific Fur Company, Stuart's fort was abandoned, and only Fort Thompson remained in business. Ross Cox, writing during the earliest days of Fort Thompson, observed: 'Messrs LaRocque and M'Donald, who wintered among them, state that the Kamloops are less friendly than any other tribe among whom we have posts established. They are addicted to thieving and quarreling, wear little clothing, and are extremely dirty in their persons.' In 1842 John Tod built a new fort on the site of North Kamloops. In 1862 it was abandoned and a fourth fort built close to where Stuart's original fort seems to have stood. In 1912 the HBC observed the centenary of Fort Thompson by opening its large department store.

The names Fort Kamloops, Fort Thompson, and Thompson River Post were used interchangeably during the fur-trading period. When a post office was established here in 1870, the name chosen was Kamloops. The City of Kamloops was incorporated in 1893.

KANAKA BAR, Fraser Canyon (C-9). The Kanakas were Hawaiian Islanders. Even before 1834, when the HBC established a post at Honolulu, its trading ships brought a number of Kanakas to British Columbia. After leaving the HBC's employment, some of them washed for gold at Kanaka Bar.

The Thompson Indian name for Kanaka Bar means 'crossing-place,' referring to the Fraser River.

KARLUKWEES, Turnour I. (C-6). A Kwakwala Indian word meaning 'curved or circular beach.'

KARMUTZEN RANGE, W. of Nimpkish L. (C-6). From the Kwakwala Indian word for 'waterfall.' Karmutzen is the Indian name for Nimpkish Lake and was probably given because of the numerous waterfalls on the steep west side of the lake.

KASHUTL INLET, Kyuquot Sound (C-6). From the Nootka Indian language (or maybe from the Chinook jargon, which borrowed many words from the Nootka language), meaning 'broken, destroyed, killed,' perhaps with reference to some disaster or battle.

KASKA CREEK, flows E. into Liard R. (L-5). *Kaska,* meaning 'old moccasins,' was a term of scorn that Tahltan Indians applied to the neighbouring Kaska Indians in the Dease River area.

KASLO, W. side of Kootenay L. (B-11). Originally the settlement here was known as Kane's Landing after George and David Kane, who built the first house here on land preempted by George in 1889. When a post office was opened in 1892, it took the name of Kaslo from Kaslo Creek, which enters the lake at this point. David Kane, who became the second mayor of Kaslo in 1894, used to explain that its name was an anglicized spelling of 'Kasleau,' the creek having been named after John Kasleau, a prospector who had arrived many years earlier as one of an HBC party that had come to get lead for bullets from the site of the future Bluebell mine. There is independent evidence for the existence of 'Johnny' Kasleau, and Kane's account has generally been accepted. On the other hand, we have the rival 'blackberry theory' set forth by I. William Cockle, an early postmaster at Kaslo, who wrote to the Chief Geographer at Ottawa on 17 August 1905:

> The name Kaslo according to the evidence of an old Kootenay Indian named Sebastian who claims that his grandfather told him the names of stopping camps on Kootenay Lake is derived from the word Cassoloe, a blackberry. The affix 'a' to this denotes a place where blackberries grow, this when modified by use results in the word 'Ah-Kas-loe,' by which name the place was known to the Indians – Kaslo, The place where blackberries grow.

KATEEN RIVER, flows SW into Khutzeymateen R. (G-5). Probably derived from the Nisgha Indian word that means 'see fish weirs or traps.' (Cf. *Kiteen River.*)

KATHLEEN, MOUNT, W. of Peachland (B-9). John Moore Robinson, the founder of Peachland, Summerland, and Naramata, named this mountain after his wife, Catherine. Of Irish descent, Robinson preferred 'Kathleen' to 'Catherine.'

KATHLYN LAKE, N. of Smithers (G-6). This beautiful lake was formerly called Chicken's Lake after an old Indian, nicknamed Chicken because he always called the grouse that he hunted and sold to the first settlers 'chickens.' When the Grand Trunk Pacific Railway was built, its photographers took pictures of this beauty spot for use in advertising pamphlets. Obviously the name Chicken's Lake was far too inelegant, so it was renamed after Kathlyn, daughter of W.P. Hinton, vice-president and general manager of the railway.

KATZ, W. of Hope (B-9). A shortened form of Katzie, which is derived from the Halkomelem name for a many-coloured moss that once covered the present site of the Katzie Indian Reserve.

KAWKAWA LAKE, E. of Hope (B-9). From the Halkomelem word meaning 'much crying [of loons],' from an Indian legend.

KEATS ISLAND, Howe Sound (B-8). After Admiral Sir Richard Goodwin Keats. When Keats and HMS *Superb* were under Nelson's command in 1803, the great admiral wrote, 'I esteem his person alone as equal to one French 74, and the *Superb* and her captain to two 74-gun ships.' Keats was Governor of Newfoundland in 1813. He died in 1834 and was carried to his grave by six full admirals.

KEDAHDA LAKE, S. of Jennings R. (L-4). From the Tahltan Indian word meaning 'moose antler.'

KEEFERS [KEEFER'S], Fraser Canyon (C-9). After George Alexander Keefer, CE (1836-1912), in charge of building the CPR between North Bend and Lytton. At one time he and his family resided here.

KEITHLEY CREEK, flows SE into Cariboo L. (E-9). It was here in July 1860 that 'Doc' Keithley struck it rich and set off the Cariboo gold rush.

KELLY LAKE, SW of Clinton (D-9). After Edward Kelly, who in 1863 preempted the nucleus of the large Kelly Ranch near here.

KELLY LAKE, S. of Dawson Cr. (H-9). A corrupted form of Callahoo, originally the name of an Iroquois family in the service of the NWC. Descendants settled here around 1900.

KELOWNA, Okanagan L. (B-10). The fur traders and trappers of the early nineteenth century called the place L'Anse au Sable (Sandy Cove).

The name Kelowna (originally pronounced so that the second syllable rhymed with 'allow') entails a curious story. In 1862 one August Gillard preempted here. For his abode he had a strange dwelling, half shanty and half underground Indian *keekwillee*. Gillard was a great hairy man, and one day, when he crawled out of his dugout, some passing Indians, seeing a resemblance to a bear coming out of its den, laughingly cried out the word *keṁxtús* (anglicized as *kimach touche* and meaning 'black bear's face'). This became the local name for Gillard and his residence. In 1892, when Bernard Lequime had John Coryell, CE, lay out the townsite, the question arose as to the name for the new settlement. The old story of 'Kimach Touche' was recalled, but this name seemed too uncouth. Then someone came up with the bright idea of substituting the word *kelowna*, Okanagan for 'female grizzly bear,' and Kelowna it became.

KELSALL LAKE, near Yukon border (L-1). After R. Kelsall of the BC-Yukon boundary survey party, 1908.

KELSEY BAY, S. side of Johnstone Strait (C-7). After Mr. and Mrs. W. Kelsey, who arrived in the area in 1914 and established themselves and their three daughters at Kelsey Bay in 1922.

KEMANO, SE of Kitimat (F-6). This, the site of the huge hydroelectric power project of the Aluminum Company of Canada, takes its name from a small Indian band, a branch of the Kitimats. The meaning of the word is not known.

KEMBALL, MOUNT, SW of Kaslo (B-11). After Colonel A.H.G. Kemball, CB, DSO, commanding the 54th (Kootenay) Battalion, CEF, when killed at Vimy in 1917. He lived in the area before the war.

KENDRICK INLET, Nootka Sound (B-6). After John Kendrick, who, with his fellow American Gray, traded off this coast in 1788-91. They are said in one instance to have received from the Indians sea otter skins worth $8,000 in return for trade goods worth $100. Kendrick was killed in 1794 when a British ship, accidentally using a loaded cannon to salute the Hawaiian king, sent the shot into Kendrick's ship, the *Washington,* killing him and several of his crew.

KENNEDY, SE of Mackenzie (H-8). After John A. Kennedy, manager of the PGE Railway (now BCR) from 1948 to 1952.

KENNEDY LAKE, N. of Ucluelet (B-7). After Arthur Edward Kennedy (1810-83), the third and last Governor of Vancouver Island (1864-6). Knighted in 1868, he was successively Governor of the West African settlements, Hong Kong, and Queensland. (Also KENNEDY ISLAND at the mouth of Skeena River.)

KENNEY DAM, head of Nechako R. (F-7). When the Aluminum Company of Canada built this dam in 1952 to create a head of water for its Kemano power project, it named it after Edward T. Kenney (1888-1974), the Minister of Lands.

KENNICOTT LAKE, NW of Telegraph Cr. (K-4). After Robert Kennicott, at one time curator of the Museum of Natural History at Northwestern University. Around 1860-1 he collected natural history specimens in the Arctic for the Smithsonian Institution. Persuaded to join the Collins Overland Telegraph project, and to take charge of exploration in Yukon and Alaska, he found the responsibilities extremely heavy and died of a heart attack in Alaska in 1866.

KENT, S. of Harrison L. (B-9). When this municipality was incorporated in 1895, the name chosen was Kent (the English county famous for hops) since hop growing was an important local industry.

KEOGH RIVER, W. of Port McNeill (C-6). This Kwakwala Indian name means 'steelhead river.'

KEREMEOS, SW of Penticton (B-10). In 1860 the HBC closed down Fort Okanogan in American territory and transferred its staff and stock to Keremeos. The HBC post then founded was closed in 1872. Keremeos post office was opened in 1887. The name is derived from the Okanagan Indian word meaning '[land] cut across in the middle,' or 'a flat cut through by water.'

KERRISDALE, Vancouver (B-8). In 1905, when the BC Electric Railway put in a tram stop at what is now 41st Avenue in Vancouver, R.H. Sperling, general manager of the company, called on the young Scottish couple, Mr. and Mrs. William MacKinnon, who had recently settled nearby, and invited Mrs. MacKinnon to name the new station. She named it after Kerrysdale, her family's home in Gairloch, Scotland. The name of the station has become the name of this Vancouver district.

KERSLEY, S. of Quesnel (E-8). After Charles Kersley, who preempted land here in 1867.

KETTLE RIVER, boundary district (B-10). The Ne-hoi-al-pit-qua of the Indians and the Colvile River of the early whites. Of the various theories about the name 'Kettle River,' the two most likely are (1) that it comes from the boiling, seething Kettle Falls, known as La Chaudière to early explorers, and (2) that it comes from the round holes, shaped like cauldrons, that water has hollowed out in the rocks. Governor Simpson gave the second explanation in 1847.

KEWQUODIE CREEK, flows N. into Quatsino Sound (C-6). Possibly from the Kwakwala Indian word meaning 'shelter, calm place.'

KHUTZEYMATEEN RIVER, NE of Prince Rupert (G-5). Khutzeymateen is a Coast Tsimshian word meaning 'in valley.'

KHYEX RIVER, flows S. into Skeena R. (G-5). From a Tsimshian Indian word related to 'hiding.'

KIBBEE LAKE, N. of Bowron L. (F-9). After Frank Kibbee, who lived in the area from 1901 to 1925. A trapper who lost his scalp in a fight with a grizzly bear, Kibbee became the first warden for the province's Bowron Lake Game Sanctuary.

KICKING HORSE PASS, Alberta-BC boundary (D-11). Commemorates the fact that in August 1858 Dr. (later Sir) James Hector, geologist with the Palliser Expedition, was here kicked in the chest by one of his packhorses and sustained a nasty injury.

KICKININEE PARK, N. of Penticton (B-10). An anglicization of the Okanagan Indian word *kekeńí*, meaning 'little salmon.' Kickininee, or kokanee, is the name of the landlocked sockeye salmon found in many Interior lakes.

KIDPRICE LAKE, E. of Morice L. (F-6). After 'Kid' Price, an early prospector. He died in Victoria in 1958, aged about 100.

KIKOMUN CREEK PARK, L. Koocanusa (B-12). From a Kootenay Indian word that may mean 'deer lick.'

KIKWILLI CREEK, flows W. into N. Thompson R. (D-9). *Kikwilli* is Chinook jargon meaning 'underneath' or 'below.' The Interior of British Columbia is dotted with kikwilli holes, shallow depressions where the Indians once had semisubterranean winter dwellings.

KILBELLA RIVER, N. side of Rivers Inlet (D-6). From the Oowekyala Indian word meaning 'long river' – as opposed to nearby Chuckwalla River, meaning 'short river.'

KILDALA ARM, E. of Kitimat Arm (F-5). From the Haisla Indian word meaning 'long way ahead.'

KILDIDT SOUND, SW side of Hunter I. (D-5). From the Heiltsuk Indian word meaning 'long way inland' or 'long inlet.'

KILDONAN, SW of Port Alberni (B-7). The first fish cannery on the west coast of Vancouver Island was erected here in 1903. After the Wallace Bros. Packing Co. acquired the plant in 1910, the company named it after Kildonan in the brothers' native Scotland.

KILGARD, E. of Abbotsford (B-8). Named in 1910 by the Maclure brothers in conjunction with their brickworks. The first syllable represents the slurred pronunciation of 'kiln.' As for the second syllable, J.C. Maclure could only say, 'We don't know of another "Kilgard" anywhere, and it has a good hard firebrick sound.' (See *Clayburn*.)

KILPALA RIVER, flows E. into Nimpkish L. (C-6). According to Chief James Sewid, this is the Kwakwala Indian word meaning 'long stretching point.'

KILPILL MOUNTAIN, W. of Murtle L. (E-10). In 1932-3, when the provincial government paid a bounty for dead wolves, settlers knowing that wolves abounded here put out plenty of bait spiked with 'kill pills' (strychnine).

KILTUISH INLET, S. from Gardner Canal (F-5). From the Haisla Indian word meaning 'long and narrow stretch of water leading outward.'

KIMBERLEY, NW of Cranbrook (B-12). Much uncertainty surrounds the origin of this name. It appears likely, however, that Colonel Ridpath, an American mining magnate, conferred the name in 1896 or 1897, presumably hoping that his property here would prove as rich as the diamond-mining centre of Kimberley in South Africa, which had been named after the first Earl of Kimberley, who in 1871, as Colonial Secretary, had placed the diamond mines under British protection.

The settlement was first known as Mark Creek Crossing and then as Clark City.

KIMPTON CREEK, E. of Radium Hot Springs (C-12). After an early settler, Rufus Kimpton, who once stole an Anglican church. When the CPR decided in 1899 to move its divisional headquarters from Donald to Revelstoke, this church was one of the buildings scheduled for removal. Kimpton and his wife were very attached to the little church, however, so clandestinely he and his friends took down the building, shipped the timbers to Windermere, and reassembled the church here. Somewhere along the way, the bell was stolen and is now in the church at Golden.

KIMSQUIT, E. side of Dean Channel (E-6). From the Heiltsuk Indian word meaning 'place of the canyon people' (perhaps referring to the nearby canyon of the Dean River). The Kimsquit are a subgroup of the Bella Coola Indians.

KINASKAN LAKE, E. of Mt. Edziza (J-4). From a Tahltan Indian word meaning 'raft crossing' or 'the lake that you have to raft.'

KINBASKET LAKE, upper Columbia R. (D- and E-10, D-11). This immense reservoir behind Mica Dam takes the name of a former small lake engulfed by the flooding of the Columbia River valley here. Walter Moberly has left an account of his naming of this lost lake in 1866:

> we crossed the [Columbia] river, and at a short distance came to a little camp of Shuswap Indians, where I met their headman, 'Kinbaskit.' I now negotiated with him for two little canoes made of the bark of the spruce, and for his assistance to take me down the river. Kinbaskit was a very good Indian, and I found him always reliable ... We ran many rapids and portaged others, then came to a Lake which I named 'Kinbaskit' Lake, much to the old chief's delight. (*Rocks and Rivers of British Columbia*, pp. 51-2)

From 1973 to 1980, Kinbasket Lake was McNaughton Lake, being originally named after General A.G.L. McNaughton, CMG, DSO, MSc, LLD, who commanded the Canadian forces in Britain 1939-43 and then became Minister of National Defence. He headed the Canadian team that negotiated the Columbia River Treaty with the United States. It is nice to see the old Indian's name on the map, but General McNaughton deserves an equivalent monument.

KINCOLITH, mouth of Nass R. (G-5). A Haida expedition had been raiding up Nass River and as it approached the open sea on its homeward journey, some of the prisoners struggled so hard to get free that the Haidas, fearing their canoes would upset, landed at this point and killed all who were resisting. The heads of the victims were left displayed on the bluff here, thus giving rise to the name Kincolith, 'the place of the skulls' (sometimes 'place of scalps').

KING ISLAND, Central Coast (E-6). Named by Captain Vancouver after his 'late highly-esteemed and much-lamented friend, Captain James King of the navy.'

KING, MOUNT, N. of Kicking Horse R. (D-11). Named in 1886 after Dr. W.F. King, successively chief inspector of surveys and chief astronomer of Canada.

KINGCOME INLET, NE of Queen Charlotte Strait (C-6). After Rear-Admiral John Kingcome, RN, Commander-in-Chief of the Pacific Station from 1862 to 1864.

KING GEORGE, MOUNT, N. of Palliser R. (C-12). This mountain, 11,226 feet or 3,422 metres high, the loftiest in the Royal Group of the Rocky Mountains, is named for King George V. Other mountains in the group are named for his consort, Queen Mary, and their children, the Princes Albert, Edward, George, Henry, and John, and their daughter, Princess Mary.

KINNAIRD, S. of Castlegar (B-11). About 1904 the CPR set up a 'box car station' here and named it Kinnaird after Lord Kinnaird, a shareholder in the company. Kinnaird was amalgamated with Castlegar in 1974.

KINNEY LAKE, Mt. Robson Park (F-10). After the Reverend George Kinney, who in 1909, with Curly Phillips, made the first ascent to Mount Robson's summit crest. Kinney made his climb in bad weather and found a dangerous snow comb at the top. Since it is uncertain whether he reached the highest point on the mountain or a snow dome a few hundred feet from it, alpine purists credit Foster, MacCarthy, and Kain in 1913 with the first ascent of the mountain.

KINSKUCH RIVER, flows SE into Nass R. (H-5). From the Nisgha Indian word meaning 'place of transporting things in stages [in difficult passage]' – that is, place of portage.

KINUSEO FALLS, midway between Prince George and Dawson Cr. (G-9). *Kinuseo* is the Cree Indian word for 'fish,' the pools both above and below these falls being full of trout. The falls themselves are 180 feet or fifty-five metres high, not counting the cascades above.

KISGEGAS, E. of confluence of Babine R. and Skeena R. (H-6). From the Gitksan Indian word meaning 'people [who live] where there are small white gulls.'

KISHINENA CREEK, E. of Sage Cr. (B-12). This Indian name appears to have some connection with balsam fir, a coniferous tree.

KISKATINAW RIVER, flows NE into Peace R. (H-9). From the Cree Indian word for 'cutbank.'

KISPIOX, N. of Hazelton (H 6). From the Nisgha Indian name for a Gitksan village. It means 'people of the hiding place.' A legend tells of the Nass Indians killing the inhabitants with only one woman, who hid, surviving the massacre.

KITCHENER, E. of Creston (B-11). After Horatio Herbert Kitchener, first Earl Kitchener (1850-1916). In 1896 he commenced the reconquest of the Sudan, avenging the death of General Gordon. He was Commander-in-Chief in South Africa during the last phase of the Boer War and later in India. He became Secretary of State for War in 1914 and died in 1916 when HMS *Hampshire* was sunk while en route to Russia.

KITCHI MOUNTAIN, N. of McGregor R. (F-9). A Cree Indian word meaning 'great' or 'mighty.'

KITEEN RIVER, flows N. into Cranberry R. (H-5). From the Nisgha Indian word that means 'see fish weirs or traps.' (Cf. *Kateen River*.)

KITELSE LAKE, E. of Kitimat (G-5). From the Tsimshian Indian word meaning 'people of the mussels.'

KITIMAT, S. of Terrace (G-5). A Coast Tsimshian word meaning 'people of the falling snow.' Many years ago some Tsimshian people, arriving in great canoes to take part in some winter ceremonies, were greeted by their hosts during a heavy fall of snow. The name that the Kitimat people used for their village was 'Dsemosa,' meaning 'the place of logs.' During high tides masses of logs would wash up on the beach and remain there.

KITKATLA, S. of Prince Rupert (F-4). A Coast Tsimshian Indian name meaning 'people of the salt' – that is, village by the sea.

KITKIATA INLET, W. side of Douglas Channel (F-5). From the Coast Tsimshian Indian word meaning 'people of the poles,' referring to the ceremonial staffs borne by chiefs.

KITLOPE RIVER, flows into head of Gardner Canal (F-6). From the Coast Tsimshian Indian word meaning 'people of the rock.'

KITSAULT RIVER, flows S. into Alice Arm (H-5). From a Nisgha Indian word meaning 'space behind.' This probably refers to the location of Kitsault Indian village at the head of Alice Arm in relation to Nass Valley and Observatory Inlet.

KITSEGUECLA, SW of Hazelton (H-6). A Tsimshian Indian name, somewhat corrupted, meaning 'the people of Jigyukwhla,' the latter being the nearby mountain.

KITSELAS, NE of Terrace (G-5). From the Tsimshian Indian word meaning 'people of the canyon.'

KITSILANO, Vancouver (B-8). Derived from the name of a Squamish Indian who came from a village on the Squamish River (near its confluence with Cheakamus River) and settled in Stanley Park, just east of Prospect Point, about 1860. When the CPR was opening up land around 1905 and wanted a name for the new subdivision, Professor Charles Hill-Tout suggested modifying this Squamish Indian's name so that it corresponded to Capilano across the inlet.

Before this renaming, the Kitsilano area was known as Greer's Beach, after Samuel Greer, a redoubtable Irishman who settled here with his family in 1884. Greer claimed to have bought the land from two Indians and sold lots for $80 each. The CPR replied with advertisements in the Vancouver press warning the public against buying from Greer. Amid all the excitement, he shot and wounded a deputy sheriff. Sir Matthew Begbie, the province's chief justice, when subsequently dealing with Sam's claims, described him as a liar and a forger and sent him to jail. From its beginning Kitsilano has been anti-Establishment.

KITSUMKALUM RIVER, flows into Skeena R. at Terrace (G-5). From the Tsimshian Indian word meaning 'people of the plateau.'

KITWANCOOL, W. of Hazelton (H-5). From a Gitksan Indian word meaning 'people of the narrow place [in the river].'

KITWANGA, SW of Hazelton (H-5). From the Gitksan Indian word meaning 'people of the place of rabbits.'

KIWETINOK RIVER, Yoho NP (D-11). From the Cree Indian word meaning 'on the north side.'

KLAHOWYA LAKE, Spatsizi Wilderness Park (J-5). *Klahowya* is the word of greeting in the Chinook jargon. When one shouts 'Klahowya' across this lake, the opposite bluffs return the greeting.

KLAKLAKAMA LAKES, W. of Nimpkish R. (C-6). Possibly from the Kwakwala Indian word meaning 'red surfaces.'

KLANAWA RIVER, W. of Nitinaht L. (A-7). From the Nitinaht Indian word meaning 'blubber coming down stream.'

KLAPPAN RIVER, flows NW into Stikine R. (J-5). From a Tahltan Indian word meaning 'lower meadow.'

KLASHWUN POINT, N. coast of Graham I. (G-3). G.M. Dawson gave this name, an anglicized version of a Haida word meaning 'point of little mountain.'

KLASKISH INLET, E. side of Brooks Bay (C-6). From a Kwakwala Indian word possibly meaning 'seaside beach.'

KLASTLINE RIVER, flows NW into Stikine R. (K-4). From the Tahltan Indian word meaning 'confluence' or 'junction of waters.'

KLATTASINE CREEK, flows NW into Homathko R. (D-7). In 1864 occurred the 'Chilcotin Massacre,' when fourteen men, building Waddington's road north from Bute Inlet, were murdered by Chilcotin Indians. After a bungled search for the guilty Indians, a search that lasted months and covered a very large area, the hunted men surrendered. Five of them, including Chief Klatsassin (or Klattasine), were tried and hanged. After the trial, Judge Begbie wrote, 'Klatsassin is the finest savage I have met with yet, I think.'

KLEENA KLEENE, on Klinaklini R. (D-7). From the Kwakwala Indian word for eulachon grease, the prized staple of the Indian diet. The grease was rendered from the eulachon, small oil-laden fish caught in large numbers when they ascended coastal rivers in the early spring. Coastal Indians found a ready market with the Interior bands for their excess eulachon grease, which was transported inland along the 'grease trails' from the heads of various coastal inlets.

There is no linguistic justification for the meaning 'river with many turns,' which is so commonly advanced by people living in the area.

KLEHINI RIVER, Alaska boundary (L-1). From a Tlingit Indian word said to mean 'queen salmon.'

KLEMTU, Swindle I., Central Coast (E-5). Formerly known as China Hat because of the appearance of nearby Cone Island when viewed from the south. This Coast Tsimshian Indian word may mean 'to anchor' (in the secluded bay here), or 'to tie something.'

KLIKSOATLI HARBOUR, N. Denny I. (E-5). This fine harbour near Bella Bella has a Heiltsuk Indian name meaning 'big thing [rock, etc.] inside on the water.'

KLITSA MOUNTAIN, SW of Sproat L. (B-7). From the Nootka Indian word meaning 'always white.'

KLO CREEK, SE of Kelowna (B-10). KLO stands for the Kelowna Land and Orchard Co., formed in 1904.

KLOIYA BAY, E. of Kaien I. (G-4). From the Tsimshian Indian word meaning 'place where people [when they were on the move] went to hide their valuables.' A letter received by Captain Walbran in 1909 from Mrs. Odile Morison explains: 'Kloiya being in an out-of-the-way place from the regular route to the Skeena river where the Tsimpseans usually went to procure and dry their salmon ... the people used to take all their belongings to Kloiya and hide or cache their treasures' before proceeding to the fishing grounds.

KLOOTCHMAN CANYON, Stikine R. (J-4). This is the Chinook jargon word for a woman or a wife. An early writer noted that in this canyon the Indian men laid down their paddles and let the women take complete charge of navigation.

KLUACHON LAKE, head of Iskut R. (J-4). From a Tahltan Indian word meaning 'bigger fish.'

KLUSKUS LAKES, S. of Euchiniko Lakes (F-7). From the Carrier Indian word meaning 'place of small whitefish.'

KNEWSTUBB LAKE, head of Nechako R. (F-7). After Frederick W. Knewstubb, who, as a hydraulic engineer with the provincial government's Water Rights Branch, spent about twenty years investigating the water resources of British Columbia. His was the original concept of diverting water from the Nechako headwaters westward through the Coast Mountains to generate massive blocks of electric power. Very properly, when the Kenney Dam was built, the lake behind it was given Knewstubb's name.

KNIGHT INLET, S. coast (C-6 and 7). Captain Vancouver noted that in July 1792: 'To this [arm] after Captain Knight of the navy, Mr. Broughton gave the name of KNIGHT'S CHANNEL. The shores of it, like most of those lately surveyed, are formed by high stupendous mountains rising almost perpendicularly from the water's edge.' During the American Revolution, Captain Knight (later Admiral Sir John Knight, 1748?-1831) was a fellow prisoner with Broughton, who commanded Vancouver's second ship, the tender *Chatham*.

KNOCKAN HILL, N. of Portage Inlet, Victoria (A-8). From the Straits Salish Indian word meaning 'rock(s) on top,' a place mentioned in a Songhees Indian legend.

KNOUFF CREEK, flows SW toward N. Thompson R. (C-9). After James Vincent Knouff, a former packer on the Cariboo Road who settled in the area in the 1860s. In 1871 the photographer Baltzly noted that Knouff's was the last farm as one travelled up the North Thompson River valley.

KNOX MOUNTAIN, N. of Kelowna (B-10). After Arthur B. Knox, who ranched at the foot of the mountain from 1883 to 1902.

KNUTSFORD, S. of Kamloops (C-9). Captain Robert B. Longridge, who began ranching in this district near Kamloops in 1912, named the local post office after Knutsford in Cheshire, where he had earlier lived.

KOCH CREEK, flows SE into Little Slocan R. (B-11). After William C.E. Koch, a 'well and favourably known' local settler and sawmill operator. When a post office was opened at the CPR's Koch Siding, he became postmaster.

KOEYE RANGE, E. of Fitz Hugh Sound (D-6). From a Heiltsuk Indian word possibly meaning 'sitting on the water.'

KOKANEE CREEK, NE of Nelson (B-11). From the Okanagan Indian word meaning 'little salmon,' applied to the landlocked sockeye salmon. (See also *Kickininee Park.*)

KOKISH RIVER, E. of Nimpkish L. (C-6). This name probably comes from the Kwakwala Indian word meaning 'notched beach' or 'something broken – the beach.'

KOKSILAH RIVER, SE of Duncan (A-8). Of the various meanings advanced for this name, the most interesting is that of Jack Fleetwood, who has known the Cowichan language since boyhood. According to Fleetwood, in the mid-1800s the HBC pastured cattle on the open area at the mouth of the river and built a fence to prevent the cattle from escaping. The Indians applied to this corral their word for a kind of container. *Koksilah* has also been translated as 'place having snags.'

KOOCANUSA, LAKE, Kootenay R. (B-12). A synthetic name compounded from the first syllables of Kootenay and Canada plus USA. The lake itself is artificial, having been created by Libby Dam in Montana.

KOOKIPI CREEK, flows N. into Nahatlatch R. (B-9). This name is derived from a Thompson Indian word meaning either 'leader' or 'chief.'

KOOTENAY LAKE, SE BC (B- and C-11). Named after the Kootenay (Kootenai, Kootanae, Coutonai, Kutenai) Indians. This name is probably derived from the Blackfoot Indian pronunciation of these Indians' name for themselves – Ktunaxa. The meaning of the word is not known – the often quoted 'water people' is based on a false etymology.

The Kootenays should not be confused with the Lakes (Okanagan-Colvile) Interior Salish Indians, who utilized the territory to the west of Kootenay Lake, while the Kootenays had that to the east. In the early days, the Kootenays were often referred to as the Flatbow Indians, and hence we find references to Flatbow Lake and Flatbow River. When David Thompson travelled down the Kootenay River in 1808, he named it McGillivray's River, after Duncan and William McGillivray, his superiors in the NWC.

KOOTENAY JOE CREEK, flows into Kootenay Lake S. of Argenta (C-11). Kootenay Joe, an Indian chief living near Creston, often canoed up this creek to camp, fish, and hunt.

KOSKIMO BAY, Quatsino Sound (C-6). Anthropologist Wilson Duff translated this name as 'people of Kosaa,' who took their name from a place on the north coast of Vancouver Island at the mouth of Stranby River, where they had lived in earlier times.

KOSTER, NE of Clinton (D-9). After Henry Koster, a part-Indian boy adopted in the late nineteenth century by the landlady of the Clinton Hotel. Hearing the local government agent say 'These half-breeds are no good!' he set out to prove him wrong and embarked upon a career that made him one of the cattle barons of the Cariboo.

KOTCHO RIVER, flows SE into Hay R. (K-9). From Slave Indian words meaning 'big fire.'

KRAJINA ECOLOGICAL RESERVE, W. coast of Graham I. (F-3). After Dr. Vladimir Krajina of the Department of Botany, UBC. During World War II, Dr. Krajina, a botanist and ecologist at Charles University in Prague, was a leader of Czech underground intelligence operations. After the Communist takeover, he escaped and came to British Columbia. He was largely responsible for the province becoming the first to create ecological reserves and provided the province with a classification of its biogeoclimatic zones.

KRESTOVA, NW of confluence of Slocan R. and Kootenay R. (B-11). When the Doukhobors, a Russian religious sect, settled here, they named their community Krestova, meaning 'the place of the cross.'

KRUGER, MOUNT, W. of Osoyoos L. (B-10). After Theodore Kruger (1829-99). A native of Hanover, Germany, he took part in the Fraser River and Cariboo gold rushes, later becoming the HBC manager and then an independent merchant at Osoyoos.

'KSAN, Hazelton (H-6). This is the Gitksan and Nisgha Indian name for the Skeena River – it has no other meaning.

KUALT HILL, W. of Salmon Arm (C-10). This could be from a Shuswap Indian word meaning 'east' that the Indians used to refer to Shuswap people in the dialect area extending from Monte Creek to Salmon Arm. Another similar word, *Kwā'ut*, was given by Teit as the name of the principal village, located near here, of the Shuswap Lake band.

KULDO, upper Skeena R. (H-6). From the Gitksan Indian word meaning '[in the] wilderness,' the implication being that the people here are uncivilized.

KULLEET BAY, N. of Ladysmith (B-8). From an Island Halkomelem word that may mean 'shelter.'

KUNECHIN POINT, junction of Salmon and Sechelt Inlets (B-8). The Hydrographic Survey named this point, which is far removed from the original Kunechin, a former Sechelt Indian village at the head of Jervis Inlet. *Kunechin* may be translated as 'going as far as you can go.'

KUNGHIT ISLAND, S. of Moresby I. (E-4). From the Haida Indian name used to denote the Cape St. James area. Its exact meaning is uncertain, though one source suggests that it may mean 'to the south.' Was formerly known as Prevost Island.

KUPER ISLAND, NW of Saltspring I. (A-8). After Captain Augustus Leopold Kuper, RN (1809-85), commanding HMS *Thetis* on the Pacific Station from 1851 to 1853. In 1864 Kuper, now a rear-admiral, led a combined force of British, American, French, and Dutch ships in a successful action against the Japanese.

KUSKANAX CREEK, flows SW into Upper Arrow L. (C-11). From an Okanagan-Colvile Indian word meaning 'a point of land sticking out,' descriptive of the creek's delta. The Lakes Indians, who speak a dialect of Okanagan-Colvile, have not lived in the Arrow Lakes area since the early 1900s; they now live on the Colvile Indian Reservation in Washington State.

KUSKONOOK, S. end of Kootenay L. (B-11). Comes from the Kootenay Indian word meaning 'edge or end of lake.'

KWADACHA RIVER, flows SW into Finlay R. (J-7). Descriptive, *kwadacha* being a Sekani Indian word for 'white water.' (Also KWADACHA WILDERNESS PARK.)

KWALATE POINT, W. side of Knight Inlet (C-7). From the Kwakwala Indian word meaning 'place of the [edible] salmon-berry sprout.'

KWINAMUCK LAKE, N. of Aiyansh (H-5). From the Nisgha Indian word meaning 'place of cow parsnip.' The name refers to the use of the stems as drinking straws.

KWINATAHL RIVER, flows E. into Nass R. (H-5). From the Nisgha Indian word meaning 'place of strips of cedar bark.'

KWOIEK CREEK, flows SE into Fraser R. S. of Lytton (C-9). From the Thompson Indian word meaning 'gouged out,' referring to a large chunk missing from the canyon wall.

KWOMAIS POINT, SE side of Boundary Bay (B-8). From the Straits Salish or Halkomelem Indian word meaning 'dog face.'

KYE BAY, NE of Comox (B-7). *Kye* is said to be derived from an Indian word of the extinct Pentlatch language, meaning unknown. (It has been claimed that an early settler used the Scottish word for 'cattle' as the name for this bay.)

KYUQUOT SOUND, SE of Checleset Bay (C-6). Named after the local Indians. In his account of his sufferings after the Nootka Indians captured him in 1803, John Jewitt says: 'A considerable way further to the northward are the *Cayuquets*; these are a much more numerous tribe than that of Nootka, but thought by the latter to be deficient in courage and martial spirit, Maquinna having frequently told me that their hearts were little, like those of birds.'

L

LABOUCHERE CHANNEL, W. of Bella Coola (E-6). After the HBC paddle-wheel steamer *Labouchere*, which made her maiden voyage to British Columbia in 1858-9. She was named after the Rt. Hon. Henry Labouchere, Britain's Colonial Secretary from 1855 to 1858.

LAC DES ROCHES, E. of Bridge L. (D-9). 'Lake of the Rocks,' on account of the many rocky islets in it. On the original HBC brigade trail (via Little Fort) from Alexandria to Kamloops, this lake has preserved the name given to it by the French-Canadian voyageurs.

LACH GOO ALAMS, Port Simpson (G-4). This is the Tsimshian Indian phrase meaning 'place of the wild rose with the small hips' and is now another name for Port Simpson. Before Fort Simpson was established here in 1834 by the HBC, the small island at Port Simpson was a camping ground of the Tsimshian people. Once the fort was built, hundreds of Indians arrived and built a village in the immediate vicinity.

LAC LA HACHE, NW of 100 Mile House (D-9). Means 'Axe Lake.' Interviewing in 1946 a great-granddaughter of Peter Skene Ogden, the famous HBC Chief Factor, we were told that, after a mule with a load of hatchets fell into the lake (possibly through the ice), it became known as Lac la Hache. Confirmation is supplied by an 1871 entry in the diary of G.A. Sargison in the Provincial Archives: 'Derives its name from the Hudson's Bay Company having lost a lot of axes loaded on an animal.'

LAC LE JEUNE, S. of Kamloops (C-9). Originally known as Fish Lake. Renamed in honour of Father Jean-Marie Raphael Le Jeune, OMI (1855-1930), who arrived from France as a missionary priest in 1879. He served first in the East Kootenay, then at Williams Lake, and finally at Kamloops. Using the Chinook jargon printed in the French Duployan shorthand, he published for the Indians a number of books and one newspaper, the *Kamloops Wawa* (*wawa* is the Chinook jargon word for 'talk').

LADNER, Fraser R. delta (B-8). After the first white settlers, William H. Ladner and Thomas E. Ladner, who took up land here in 1868. Born in Cornwall, England, the Ladner brothers came to British Columbia in 1858. They prospected and mined in the Cariboo and engaged in store keeping and running pack trains before marrying in 1865 two sisters, Mary and Edney Booth, and turning to agriculture.

LADY FRANKLIN ROCK, Yale (B-9). When Lady Franklin, the widow of Sir John Franklin, the Arctic explorer, was on this coast in 1861, she travelled up the Fraser River to Yale, where a banner proclaimed the entrance to the Fraser Canyon to be 'Lady Franklin's Pass.' Her name now remains only in connection with the great rock in the river here.

The Indians believed that the chief of a band of water monsters was changed into this rock by The Transformer.

LADYSMITH, S. of Nanaimo (A-8). Originally the anchorage here was known as Oyster Harbour because of the abundant oyster beds on the flats.

James Dunsmuir, owner of the nearby coal mines, was having a townsite laid out here at the time of the Boer War. When word was received on 1 March 1900 that the British forces had finally relieved their besieged countrymen in Ladysmith, Natal, Dunsmuir decided to name his new town Ladysmith in commemoration of the event.

Ladysmith, Natal (founded in 1851), was named after the wife of the colonial Governor, Sir Harry Smith. Born Juana Maria de los Dolores de Leon, she was of noble rank and a descendant of Ponce de Leon. She was only thirteen when the British Army sacked Badajos in 1812. Their home wrecked by looters, their ears bleeding where their earrings had been torn away, her elder sister and she fled to the British camp to seek some officer who would give protection. Sir Harry was captivated by the girl and married her later the same year, just before the Battle of Salamanca, and a very successful and happy marriage it was.

LA FORME CREEK, flows W. into Columbia R. (D-10). After George La Forme, who was a packer in the area. About 1895 his pack train was snowed in around here, and the animals had to be shot.

LAIDLAW, SW of Hope (B-9). Originally St. Elmo, after a novel, but when the Canadian Northern Railway laid two tracks across the farm of W.F. Laidlaw, he insisted that its station be given his name.

LAKELSE LAKE, S. of Terrace (G-5). Properly pronounced 'La-kélse,' it is the Tsimshian Indian word for 'fresh water mussel.'

LAMA PASSAGE, off Bella Bella (E-5). A corrupted version of *Llama,* the name of the brig bought from her American owners by the HBC in 1832 and used in the founding of Fort McLoughlin (on the site of old Bella Bella) the following year. (See *Port McNeill.*)

LAMALCHI BAY, W. side of Kuper I. (A-8). Named after the band of Cowichan Indians that lived here. Their village on this bay was bombarded by HM gunboat *Forward* in 1863 during the hunt for the murderers of Frederick Marks and his daughter on Saturna Island.

Lamalchi is an Island Halkomelem word that may contain the meaning of 'slumped' and 'dirty water.' Alternatively it may mean 'large tidal flat.'

LAMARQUE PASS, E. of Tucho L. (K-6). This pass through the Cassiar Mountains is named for the well-known surveyor E.C.W. Lamarque, who discovered it in 1934. In his typescript memoirs, he described it as 'about 5 miles long, of a wide U shape and, with the exception of a few scattered balsam firs and willows, quite open and treeless.'

LAMBERT CHANNEL, between Denman I. and Hornby I. (B-7). After Lionel Lambert, RN, flag lieutenant to Rear-Admiral Baynes on his flagship HMS *Ganges* 1857-60.

LAMBLY CREEK, flows SE into Okanagan L. (B-10). Formerly Bear River or Creek. The name was changed in 1922 in honour of Charles A.R. Lambly, member of a pioneer Okanagan family, who had been government agent at Rock Creek, Camp McKinney, Osoyoos, and Fairview. He died in 1907.

LANEZI LAKE, Bowron L. Park (F-9). This is the Carrier Indian word for 'ten.' The lake is ten miles long.

LANGARA ISLAND, North QCI (G-3). Named in 1792 by the Spanish navigator Caamaño after the Spanish admiral Don Juan de Langara. For a considerable period, this was known as North Island, the Spanish name being in abeyance.

LANGDALE, W. side of Howe Sound (B-8). This ferry terminal on Howe Sound is named after Robinson Henry Langdale (1835-1908), a Yorkshireman who preempted land on Langdale Creek in 1892.

LANGFORD LAKE, W. of Victoria (A-8). After Captain Edward E. Langford (1809-95), once of the 73rd Regiment and a Sussex landowner. Captain Langford, his wife, and five good-looking daughters arrived in Victoria in May 1851, being the first English family to emigrate to the colony. Captain Langford had been hired as manager of the Esquimalt farm operated by the HBC's subsidiary, the Puget Sound Agricultural Company. Although Captain Langford did not entirely neglect his duties around the farm, he was much more interested in being a genial country squire and keeping open house for the young officers from the Royal Navy who would make eligible husbands for his girls. He returned to England in 1861.

LANGLEY, W. of Abbotsford (B-8). Takes its name from nearby Fort Langley (q.v.).

LANTZVILLE, N. of Nanaimo (B-7). After Harry Lantz, an American who invested in a coal mine at Nanoose.

LARCOM ISLAND, Observatory Inlet (H-5). After Lieutenant-Commander T.H. Larcom, who commanded HM gunboat *Forward*, 1868-9, at the end of her service on this coast.

LARDEAU, N. end of Kootenay L. (C-11). After an early prospector. Lardo post office, opened in 1899, became Lardeau in 1947.

LAREDO SOUND, S. of Aristazabal I. (E-5). Named in 1792 by Jacinto Caamaño, commanding the Spanish frigate *Aranzazu*. Probably Caamaño had in mind the Spanish port of Laredo on the Bay of Biscay.

LASKEEK BAY, E. coast of Moresby I. (E-4). Said to be an adaptation of the Tsimshian word (meaning 'on the eagle') applied to a large Indian village formerly here, known to the Haidas as Tanu.

LASLUI LAKE, expansion of Stikine R. (J-6). From the Sekani Indian word meaning 'fish lake.'

LASQUETI ISLAND, S. of Texada I. (B-7). Named by Narvaez in 1791 after Juan Maria Lasqueti, a prominent Spanish naval officer.

LASTMAN LAKE, W. of Taseko L. (D-8). Chilcotin Indians raided the camp here of a party of Shuswaps from Soda Creek, killing all but one man. The survivor escaped and got home. Subsequently a revenge party from Soda Creek caught up with the Chilcotins at Kleena Kleene and inflicted casualties before returning to Soda Creek.

LAUREL POINT, Victoria Harbour (A-8). Recalling Victoria in 1850, J.R. Anderson wrote that the 'laurels' were really the arbutus trees that made this Indian burial ground a beautiful spot.

LAURIER PASS, E. of Finlay Reach, Williston L. (I-8). Named after Sir Wilfrid Laurier, Prime Minister of Canada, by Inspector J.D. Moodie of the NWMP. Moodie discovered this pass through the Rockies in 1897 while leading a patrol of four constables on a notable trail-finding expedition from the Peace River country to the Yukon. Immediately to the south of the pass is MOUNT LAURIER. A slightly higher mountain to the southwest is named after Lady Laurier.

LAUSSEDAT, MOUNT, N. of Golden (D-11). After Colonel Aimé Laussedat (1819-1907), the Frenchman who in 1849 first applied photography to surveying.

LAVINGTON, E. of Vernon (C-10). After Lavington Park, the Sussex home of James Buchanan, the wealthy Scottish distiller who bought land here in the early 1900s. In 1921 he bought the Coldstream Ranch. He twice won the Derby, gave King George V the money to restore the nave of the Chapel Royal at Windsor, donated very generously to charities, and in 1922 was created Baron Woolavington of Lavington.

LAWYERS PASS [LAWYER'S PASS], NW of Thutade L. (J-6). The lawyer was Stuart Henderson, who, after he had secured the acquittal of Gunanoot at his trial for murder (see *Gunanoot Lake*), accompanied him on a prospecting trip from Bulkley House to Toodoggone River. The expedition has been described as 'abortive.'

LAZO, CAPE, E. of Comox (B-7). A shortened form of Punta de Lazo de la Vega, the name given by Narvaez in 1791. In Spanish *lazo* means 'snare' and *vega* means 'an open plain,' but Narvaez was apparently using the name of somebody he wished to honour.

LEANCHOIL, E. of Golden (D-11). The mother of Lord Strathcona, of CPR fame, was Barbara Stuart of the manor of Leth-na-Coyle (Leanchoil), Inverness-shire, Scotland.

LEBAHDO, Slocan R. (B-11). This community takes its name from the Chinook jargon word for 'shingle' (for roofing). From the French *le bardeau*.

LE BOURDAIS LAKE, NW of Quesnel Forks (E-9). After Louis Le Bourdais, who grew up to be a cowpuncher, then turned telegrapher and served at Lac la Hache, Golden, Vernon, and Quesnel. From 1937 until his death in 1947, he was MLA for the Cariboo. To promote the Cariboo, he presented his fellow members with succulent beefsteaks wrapped in cellophane with small sacks of edible alfalfa. He was a marvellous teller of Cariboo stories.

LECTURE CUTTERS, THE, N. of Mt. Sir Richard, Garibaldi Park (B-8). Named in 1965 at the suggestion of Professor Roy Hooley of UBC, who was aware how students managed to spend so much time in the mountains.

LEE CREEK, Shuswap L. (C-10). Apparently after William Lee, a prospector in the area. Lee was severely injured in 1886 in a fight with his partner.

LEECH RIVER, N. of Sooke (A-8). After Peter John Leech, who arrived in British Columbia as a corporal in the Royal Engineers in 1858 but remained here after his detachment was disbanded. He was 'astronomer' with the Vancouver Island Exploration Expedition of 1864, which found gold on this stream. The resulting gold rush soon petered out. Leech later became city engineer of Victoria. He died in 1899.

LEFROY, MOUNT, Yoho NP (D-11). Named by James Hector after Major-General Sir John Henry Lefroy (1817-90). Between 1842 and 1844, Lefroy travelled over 5,500 miles in the Canadian northwest making magnetic and meteorological surveys. He headed the Toronto Observatory from 1842 to 1853. During the Crimean War, he investigated hospital conditions in Constantinople and became a friend of Florence Nightingale. He was Governor of the Bahamas from 1871 to 1877 and of Tasmania from 1880 to 1882.

LEHMAN, MOUNT. See *Mount Lehman*.

LEIGHTON LAKE, S. of Savona (C-9). After James Buie Leighton (1851-1946). He came to British Columbia in 1863 and settled at Savona in the 1870s, ranching, running the ferry, and holding the mail contract until the coming of the CPR. In 1909 the Leighton family created this lake by building a dam to improve their summer range.

LEJAC, S. shore of Fraser L. (G-7). After Father Jean-Marie Le Jacq, perhaps the most saintly of all the priests of the Congregation of Missionary Oblates of Mary Immaculate. Father Le Jacq came to Williams Lake in April 1867. In 1873 he founded the Mission of Our Lady of Good Hope on Stuart Lake, by Fort St. James. From here he covered an enormous territory.

Kay Cronin recalls interviewing an old Indian who had known Father Le Jacq: 'When I see Père Le Jacq come, my heart cry,' said Louis Billy. 'He come from Babine. He walk; his blanket, his portable altar, his grub, on his back. And when he come, he cover his feet with his cassock because he has no stockings. But I see there is blood in his shoes.' Not surprisingly Cronin concludes: 'Of all the Oblate Fathers who had, or have since, worked among the northern tribes, none seems to have made such a deep impression on them or commanded more respect than did Father Le Jacq. Even today there are many among the Indians who are firmly convinced that the man was a living saint. And the tales still told about him are near-legendary' (*Cross in the Wilderness,* p. 134).

Le Jacq died in New Westminster in 1899.

LELU ISLAND, SE of Prince Rupert (G-4). According to Captain Walbran, this island has the Chinook jargon name meaning 'wolf' because it was 'infested' with wolves for many years.

LEMMENS INLET, N. of Tofino (B-7). After John Nicholas Lemmens (1850-97), a Dutch priest who, after a number of years as a missionary on the west coast of Vancouver Island, became the fourth Bishop of Victoria (RC).

LEMON CREEK, S. of Slocan (B-11). After Robert E. Lemon, pioneer merchant who later became warden of Nelson jail.

LEMPRIERE, N. Thompson R. (E-10). Apparently after a civil engineer engaged in construction here of the Canadian Northern (now Canadian National) Railway. The railway was completed in 1915, and Lempriere first appears on a map of 1917.

LENNARD ISLAND, S. of Tofino (B-7). After Captain C.E. Barrett-Lennard of the Thames Yacht Club, who, with his friend Captain N. Fitzstubbs, circumnavigated Vancouver Island in 1860 in his yacht *Templar.* His *Travels in British Columbia* was published in London in 1862.

LEO CREEK, flows SW into Takla L. (H-7). Formerly Leon Creek after Leon Prince, a member of a notable Indian family at Fort St. James.

LIARD RIVER, Yukon-BC border (L-5 and 6). Earlier known as the Rivière aux Liards. The liard is a cottonwood tree.

LIBERATED GROUP, W. of Chilko L. (D-7). So named after the women's liberation movement. Mountains in this group are named after notable women such as Agnes Macphail, Nellie McClung, and Charlotte Whitton.

LIGHTNING CREEK, W. of Barkerville (F-8). 'Lightning' was old Yankee slang for very hard work. In 1861 Bill Cunningham and two fellow prospectors discovered this creek. They had great trouble descending the steep slopes to the stream, and Cunningham exclaimed at one point, 'Boys, this is lightning!' His companions then made this the name of the stream.

LIGHTNING LAKES, Manning Park (B-9). C.P. (Chess) Lyons, making his survey before the setting up of the park, found Thunder Lake already named and the other three lakes collectively named the Lightning Lakes. He sorted things out by naming the four lakes Lightning, Strike, Flash, and Thunder.

LIKELY, W. end of Quesnel L. (E-9). Originally known as Quesnel Dam, renamed after 'Plato John' Likely (1842-1929). He took a great deal of gold out of Likely Gulch, Cedar Point, and other workings in this area. A native of New Brunswick, he was a great admirer of Plato and Socrates and would lecture on them to miners who came to his 'Philosophers' Grove' of giant cedars by Quesnel Lake. Occasionally he would take the more ardent of his disciples to a retreat on a little island in the lake. Some of the miners, appreciative of his teaching, gave him tips about good places to hunt for gold and thus contributed to his prosperity. He lies in an unmarked grave in Kamloops cemetery.

LILLIPUT MOUNTAIN, Yoho NP (D-11). Seen at a distance, rock pillars here resemble a crowd of little people.

LILLOOET, near confluence of Bridge R. and Fraser R. (C-9). Dr. Jan van Eijk, a Dutch linguist specializing in the Lillooet Indian language, disagrees with the meanings one usually finds for Lillooet: 'wild onion' or, less frequently, 'end of trail' (referring to trail from the coast). Van Eijk explains that *lillooet* is the anglicized form of the Indian word, origin unknown, applied to an undefined area near the present Mount Currie Indian Reserve on the northeast side of Lillooet Lake (hence the name of the lake).

Lillooet townsite was originally known as Cayoosh Flat (see *Cayoosh Creek*) but received its present name about 1860 because it was here that the trail from Lillooet Lake reached the Fraser River. The HBC's short-lived Fort Berens was on the opposite bank of the Fraser River from Lillooet. In 1862 Parsonville, on the site of Fort Berens, became mile zero for the numbering of the miles along the Cariboo Wagon Road.

Early mentions of the Lillooet Indian band give a fascinating variety of spellings, ranging from Governor Simpson's Lilowit to Dr. John McLoughlin's Lille-what and Littlewhite.

LIMA POINT, Digby I., S. of Prince Rupert (G-4). After Frederick Lima, paymaster, HMS *Malacca*, on the Pacific Station in 1866-7.

LINDBERGH ISLAND, off Lasqueti I. (B-7). So named, at the request of its American owner, shortly after Colonel Lindbergh's famous trans-Atlantic flight. At the time it was noted by Captain H.D. Parizeau, the West Coast hydrographer, that this is a very tiny island to name after so famous a man.

LINDELL BEACH, Cultus L. (B-8). When a post office was opened here in 1953, it was named after A.F. Lindell, who settled beside Cultus Lake before World War I.

LINDEMAN LAKE, S. of Bennett L. (L-2). Named in 1883 after the secretary of the Bremen Geographical Society, the German sponsors of the expedition (headed by Arthur and Aurel Krause) that went into the Tlingit country in 1881-2.

LINDQUIST LAKE, S. of Whitesail L. (F-6). After Charles Lindquist, the Swedish trapper and expert canoeman who was the first to penetrate to here.

LINKLATER CREEK, flows SE into L. Koocanusa (B-12). After John (Scotty) Linklater, manager in the 1850s of the HBC's Fort Kootenay, located close to this stream.

LIONS' GATE, Vancouver (B-8). The landmark mountains to the north now called the Lions were originally known as either the Sisters or Sheba's Paps. Sometime around 1890 Judge Gray, noticing their resemblance to couchant lions, suggested that their name be changed and that the entrance to Vancouver harbour be called 'The Lions' Gate.' The name found acceptance perhaps because of the suggestion that here was a British equivalent to San Francisco's Golden Gate.

According to the notes of Major Matthews in the Vancouver City Archives, the Indian name for the Lions, 'Chee-chee-yoh-hee,' means 'the twins.'

LISTER, SE of Creston (B-11). After Colonel Fred Lister, administrator of the soldier settlement established here after World War I. He became MLA for the Nelson-Creston riding.

LITTLE FORT, N. Thompson R. (D-9). A.R.C. Selwyn, reporting on his geological surveys of 1871-2, notes: 'we camped on a fine flat above The Little Fort, an old and now deserted Hudson Bay Company's trading post.' This little outpost of Fort Kamloops was established by Paul Fraser in 1850 and abandoned in 1852. Before 1935 Little Fort post office was named Mount Olie.

LIUMCHEN CREEK, flows N. into Chilliwack R. (B-9). From the Halkomelem word meaning 'water gushing out' (from the outlet of an underground stream).

LIZA CREEK, N. of Carpenter L. (C-8). Immortalizes an Indian woman who reputedly was so fond of a certain horse that whenever it was sold the new owner acquired Liza too. She is said to have been a very kindly person who, during the great flu epidemic of 1919, wore herself out visiting the sick before she herself caught the disease and died.

LLOYD GEORGE, MOUNT, Kwadacha Wilderness Park (J-7). R.M. Patterson has a sufficiently dry comment upon P.L. Haworth's successful attempt to get this fine mountain named after David Lloyd George, Prime Minister of Great Britain in the final years of World War I:

> It has been said, and with some truth, that the Rockies are the worst named mountain system in the world ... Haworth, in a fit of wartime enthusiasm, decided to suggest that one further alien name be added to the ill-assorted register: as soon as he got out he would propose to Ottawa that the high mountain he had seen that day, holding the Great Glacier in its lap, should be called Mount Lloyd George. With regrettable haste his suggestion was adopted. Time and the verdict of history have not added to the stature of the little Welsh politician.

LOCARNO PARK, Vancouver (B-8). Commemorates the signing in 1925 at Locarno, Switzerland, of a pact that many believed would usher in an era of 'no more wars.'

LODESTONE MOUNTAIN, W. of Princeton (B-9). From the magnetite deposits on this mountain.

LOGAN INLET, E. coast of Moresby I. (E-4). Named by G.M. Dawson after Sir William Logan (1798-1875), for many years director of the Geological Survey of Canada.

LOGAN LAKE, SW of Kamloops (C-9). Logan is a corruption of Tslakan, a Savona Indian who traded in furs and raised horses in the 1860s. He took great pride in being an Indian and insisted that whites employ the prefix 'Mr.' when using his name.

LOIS LAKE, E. of Powell River (B-7). After Captain Babbington of the tug *Lois* had obligingly transported some surveyors from Texada Island to Stillwater, they offered to name some geographic feature after him. Babbington vehemently refusing to let them do so, they named this lake after his tugboat.

LOLO, MOUNT, NE of Kamloops (C-9). After Jean Baptiste Lolo (1798-1868), possibly of mixed Iroquois and French-Canadian descent. He had a great admiration for St. Paul and hence, while at Fort Fraser, picked up the nickname of St. Paul, which stayed with him for the rest of his life. Later he moved to Kamloops, where he achieved such influence among the Shuswaps as to be accounted a chief. In *Four Years in British Columbia and Vancouver Island*, Commander Mayne has left the following account of his meeting with him at Kamloops in 1859: 'In the centre room, lying at length upon a mattrass stretched upon the floor, was the chief of the Shuswap Indians. His face was a very fine one, although sickness and pain had worn it away terribly. His eyes were black, piercing, and restless; his cheek-bones high, and the lips, naturally thin and close, had that white, compressed look which tells so surely of constant suffering' (p. 119). To Mayne's amazement the ailing Lolo insisted on rising from his bed of sickness, mounting his horse, and accompanying him and an HBC man on their ride to the top of the nearby mountain, 'which we christened Mount St. Paul [today's Paul Peak], in honour of the old chief.' (See *Paul Lake*.)

LONGSTAFF, MOUNT, Mt. Robson Park (F-10). After Dr. T.G. Longstaff, who mountaineered in the Himalayas as well as in British Columbia.

LONSDALE, North Vancouver (B-8). After Arthur Pemberton Heywood-Lonsdale, of Shavington Hall, Shropshire. With his kinsman J.P. Fell, he once owned a block of waterfront land extending from Moodyville to the Capilano River. The land was opened to settlers in 1903.

LOOP BROOK, Rogers Pass (D-11). Named for the great double loop in the CPR track here before the construction of Connaught Tunnel.

LOQUILLILLA COVE, S. side of Nigei I. (C-6). Probably derived from the Kwakwala Indian word meaning 'to fish for halibut.'

LORD RIVER, flows N. into Taseko L. (D-8). Lord was the maiden name of Mrs. J.D. Mackenzie, who accompanied her husband into this area in 1920.

LORIN LAKE, S. of Canim L. (D-9). After Dr. Lorin O. Lind, West Vancouver dentist and outdoorsman, who drowned in 1970 when his plane crashed into this lake.

LOST LAGOON, Vancouver (B-8). Named by Pauline Johnson, the poet of Mohawk and English descent, who was fond of canoeing here. It was cut off from the rest of Vancouver harbour by the building of the Stanley Park causeway.

LOUDOUN CHANNEL, Barkley Sound (A-7). The *Imperial Eagle,* the ship in which Captain Barkley sailed into Barkley Sound in 1787, was originally named the *Loudoun.*

LOUGHBOROUGH INLET, between Knight Inlet and Bute Inlet (C-7). Named by Vancouver after Alexander Wedderburn, first Baron of Loughborough

(1733-1805), a diligent Scot who took elocution lessons from Sheridan to rid himself of his accent and rose to be Lord Chancellor of England and Earl of Rosslyn.

LOUIS CREEK, flows N. into N. Thompson R. (D-9). According to the CNR, the creek is named after Chief Louis of the Kamloops Indians, who used to come here to fish for salmon. On the other hand, Mary Balf, formerly of the Kamloops Museum, says that the name comes from Louis Barrie, a prospector who found some gold in the area in 1860.

LOUISE ISLAND, E. of Moresby I. (E-4). After Princess Louise, fourth daughter of Queen Victoria and wife of the Marquess of Lorne, Governor-General of Canada from 1878 to 1883.

LOUSCOONE INLET, S. end of Moresby I. (E-4). Name transferred from Louscoone Point. This Haida word means 'good point' or 'nice point.'

LOVELY WATER, LAKE, W. of Brackendale (B-8). Not the invention of a real estate agent, this name was given in 1914 by two mountaineers, Basil Darling and Alan Morkill. (See *Tantalus Range*.)

LOWER POST, Liard R. (L-5). Lower Post, where Dease River enters the Liard, got its name with reference to an earlier 'upper post' established near McDame Creek, farther up Dease River.

LOWHEE CREEK, W. of Barkerville (F-9). Named after a secret society at Yale University by Dick Willoughby, who struck it rich here in 1861.

LUCERNE, W. of Yellowhead Pass (E-10). Named in 1912 by the Canadian Northern Railway because the site resembles the mountainous area surrounding Lucerne, Switzerland.

LUCKAKUCK CREEK, flows N. into Chilliwack Cr. (B-9). Originally a small springwater stream, this creek has a Halkomelem name meaning 'to straddle a log.'

LULU ISLAND, Fraser delta (B-8). Named by Colonel R.C. Moody in 1862 after Miss Lulu Sweet, a young actress in the first theatrical company to visit British Columbia.

LUMBY, E. of Vernon (C-10). After Moses Lumby (1842-93). He came to British Columbia from England in 1862, mined in the Cariboo, farmed near Kamloops, carried mail up to the Big Bend gold camps, settled in Spallumcheen in 1870, and became a justice of the peace and vice-president of the Shuswap and Okanagan Railway. In 1891 he became government agent at Vernon. He has been described as 'a portly gentleman with a usually smiling face.' After his death White Valley was renamed Lumby in his honour.

LUMPY LAKE, SW of Prince George (F-8). It is not the lake but the surrounding terrain that is lumpy – with eroded dunes, eskers, and a rough logging road.

LUND, NW of Powell River (B-7). When the brothers Frederick and Charles Thulin settled here in December 1889, they named their new home after Lund, the city in Sweden from which they had come. For lack of other transportation, they sometimes had to row to Vancouver and back.

LUSSIER RIVER, flows into Kootenay R. near Skookumchuck (B-12). Named by David Thompson in 1808 after one of his men, who had lost his kit in the Moyie River.

LUXANA BAY, Kunghit I., QCI (E-4). A note on Captain Duncan's 1788 chart of Luxana Bay, in the Dalrymple atlas, reads, 'this bay was named by Captain Duncan "Lux Aena" which, in the Indian language, signifies "Handsome Women."'

LUXOR CREEK, flows into Columbia R. N. of Radium (C-11). After the Shriners' Temple at St. John, New Brunswick.

LYALL HARBOUR, Saturna I. (A-8). After Dr. David Lyall, surgeon on HM survey ship *Plumper*, 1857-9.

LYELL ISLAND, E. of Moresby I. (E-4). Named by G.M. Dawson in 1878 after Sir Charles Lyell (1797-1875), the famous geologist.

LYNN CREEK, North Vancouver (B-8). Also LYNN VALLEY. After John Linn (note the correct spelling), one of the contingent of Royal Engineers who came to British Columbia in 1859. He remained in the colony after his unit was disbanded in 1863 and, with his family, lived at the mouth of this creek. He died in 1876.

LYTTON, confluence of Fraser R. and Thompson R. (C-9). Site of the Indian village of Camchin. *Camchin*, a Thompson Indian word, means either 'cross mouth' (referring to crossing the mouth of the Thompson River) or 'shelf that crosses over,' there being flat areas on both sides of the Fraser River. The short-lived HBC's Fort Dallas was also located here.

On 11 November 1858, Governor Douglas wrote that, 'as a merited compliment and mark of respect,' he had named the settlement after Sir Edward Bulwer-Lytton, Secretary of State for the Colonies. Colonel Moody observed somewhat pessimistically, 'It will require much perseverance and determination on our parts to prevent "Lytton" becoming fixed as "Lyttonville" or "Lytton City". The latter is not bad, if it was not so intensely American.'

Sir Edward Bulwer-Lytton (1803-73) is remembered today chiefly as a novelist and dramatist. Among his many works are *The Last Days of Pompeii* and

Richelieu. As Colonial Secretary he took a real interest in the infant colony of British Columbia. When the first detachment of Royal Engineers sailed from Southampton in 1858, he travelled down to Cowes and boarded the ship for a visit, during which he spoke to the men, emphasizing his concern for their welfare and his feeling that the success of the new colony would largely depend upon their exertions.

MABEL LAKE, NE of Vernon (C-10). According to A.G. Harvey, Charles E. Perry, a CPR survey engineer, named the lake in the 1870s after Mabel Hope Charles (b. 1860), the eldest child of William Charles, successively the HBC manager in Hope, Kamloops, and Victoria.

MACALISTER, N. of Williams L. (E-8). This station on the BCR is named after James M. Macalister, at one time postmaster here.

MACAULAY POINT, Esquimalt (A-8). After Donald Macaulay, bailiff for the Viewfield Farm near Esquimalt, established in 1850 by the Puget Sound Agricultural Company (HBC). This farm, some 600 acres in extent, included Macaulay Point.

MACDONALD, MOUNT, Rogers Pass (D-11). After Sir John A. Macdonald (1815-91), a Father of Confederation and the first Prime Minister of Canada. Familiar to Canadians is Macdonald's famous quip that the country would 'rather have John A. drunk than George Brown sober.' Perhaps more worthy of recollection is another comment, which Macdonald passed on his Liberal competitor: 'The great reason why I have always been able to beat Brown is that I have been able to look a little ahead, while he could on no occasion forgo the temptation of a temporary triumph.'

MACHMELL RIVER, flows W. into Owikeno L. (D-6). From an Oowekyala Indian word perhaps meaning 'with fog patches.'

MACHRAY, MOUNT, Mt. Robson Park (F-10). After Robert Machray (1831-1904), Archbishop of Rupert's Land.

MACJACK RIVER, N. of Quatsino Sound (C-5). A name derived from the Kwakwala Indian word meaning 'near behind.' The river approaches the sea behind a long, fairly high point of beautiful sand that has been formed parallel to the seashore.

MACKENZIE, S. end of Williston L. (H-8). Named after Sir Alexander Mackenzie (1764?-1820), the first white man to cross Canada to the Pacific Ocean. A.M. Sheppard, the first mayor, wrote: 'In our search for a name for our instant town, it became our knowledge that while there was a river and a district named after this very famous explorer of our country, there had not been a town or city named in his honour. Once this was known the choice became simple.'

The first three families arrived in July 1966, and two years later a town of 1,200 was inaugurated. The new town came into being as a result of the development of a major forest industry in the area.

MACKENZIE, MOUNT, Tweedsmuir Park (E-6). This and nearby Mackenzie Valley and Mackenzie Pass all remind us that Sir Alexander Mackenzie passed this way in 1793. Also SIR ALEXANDER MACKENZIE PARK (including Mackenzie Rock) on Dean Channel.

MACKENZIE, MOUNT, SE of Revelstoke (C-10). After Alexander Mackenzie (1822-92), Prime Minister of Canada from 1873 to 1878.

MACKTUSH CREEK, flows E. into Sproat Narrows (B-6). From a Nootka Indian word possibly meaning 'cross over to the other side.'

MacMILLAN PARK, Cameron L. (B-7). These 136 hectares surrounding the magnificent tall trees of Cathedral Grove were given to the province by MacMillan Bloedel Ltd. The park is named after H.R. MacMillan, an outstanding British Columbian. A university-trained forester, he came to the province in 1910 and in 1912 became its first Chief Forester. He resigned in 1916 to enter the lumber business, in which he prospered thanks to outstanding intelligence and hard work, becoming the titan of British Columbia's lumber industry.

MACOUN, MOUNT, S. of Rogers Pass (D-11). After John Macoun (1831-1920), botanist and naturalist. After British Columbia joined Canada, he visited the new province in the service of the Geological Survey of Canada.

MAD RIVER, flows SW into N. Thompson R. (D-10). The Reverend George Grant described it in 1872 as 'a violent mountain affluent.'

MADEIRA PARK, Sechelt Pen. (B-7). Named by Joseph Gonsalos, a native of the Madeira Islands, who settled here in the early 1900s.

MAGGIE LAKE, NE of Ucluelet (B-7). After a male Indian, whose claim to have discovered it was regarded with suspicion by the locals.

MAGNA BAY, N. shore of Shuswap L. (C-10). Called Magna (Latin for 'great') since it is one of the largest bays on Shuswap Lake. It was formerly known as Steamboat Bay but, because this name was not acceptable to the postal authorities, the name had to be changed.

MAGUIRE, MOUNT, SE of Sooke (A-8). After Lieutenant Rochfort Maguire, HMS *Herald,* in local waters in 1846-7.

MAHATTA RIVER, S. side of Quatsino Sound (C-6). From the Kwakwala Indian word meaning 'having sockeye salmon,' the name of a former Koskimo summer village here.

MAHOOD LAKE, Wells Gray Park (D-9). After James Adams Mahood, land surveyor, who died in 1901. Mahood conducted a CPR survey party along the shore of the lake in 1872.

MAIDEN CREEK, flows E. into Bonaparte R. (C-9). Walter Moberly heard the following Indian legend from Sir James Douglas. Once an Indian maiden was betrothed to a young Indian chief, a great warrior and hunter. During the winter he went away for a very long time on a hunting trip. His maiden waited anxiously for him since they were to be married that winter. Spring came and the maiden sat, still watching for her lover, near the junction of this creek with the Bonaparte River. Finally she saw him approaching, but with him was a wife whom he had married when visiting a distant band. The maiden died of grief, and her band buried her where she had kept her watch. From her breasts grew two little mounts, and they are the twin knolls seen today near the mouth of Maiden Creek.

MAILLARDVILLE, NE of New Westminster (B-8). After the Reverend Edmond Maillard, OMI, of St. Malo in Brittany. He became the first priest of the Roman Catholic parish of Our Lady of Lourdes, founded in 1909 to minister to the spiritual needs of French-Canadian millworkers recently brought from Quebec to work in the Fraser Mills. Maillard died in France in 1966, aged eighty-six.

MAITLAND ISLAND, SW of Kitimat (F-5). After H. Maitland Kersey, managing director of the Canadian Development Company, who, with Louis Coste (see *Coste Island*), examined Kitimat harbour in 1898.

MAJERUS FALLS, Murtle R., Wells Gray Park (E-9). After Michael Majerus, trapper and homesteader, who lived in the Clearwater area from 1913 until his death in 1958.

MAKAI POINT, W. entrance to Juskatla Inlet (F-3). Named by the Hydrographic Service after Makai, a Tlingit Indian adopted into the Haida band at Masset. He was a violent and dangerous man, and when he was converted to Christianity by the Anglican missionary W.H. Collison, the Haidas were much impressed. (See W.H. Collison, *In the Wake of the War Canoe,* pp. 198-202.)

MALAHAT RIDGE, W. of Saanich Inlet (A-8). Opinion is divided over the meaning of this Saanich Indian word. One school champions 'infested with caterpillars,' with reference to a year when the tent caterpillars stripped the trees of their leaves. The other school favours 'place where one gets bait.'

An earlier Indian name for the mountain was Yaas, the home of a legendary rainmaker – the Indians believed that if one pointed at the mountain it would rain.

MALAKWA, NE of Sicamous (C-10). The Chinook jargon word for 'mosquito.'

MALASPINA STRAIT, E. of Texada I. (B-7). Also MALASPINA PENINSULA. After Captain Alexandro Malaspina, an Italian who, with two ships of the Spanish

Navy, explored widely in the Pacific from 1789 to 1794. He sailed along the BC coast in 1791 but never entered the waters east of Vancouver Island. For political reasons the Spaniards imprisoned Malaspina not long after his voyage. He was liberated and banished to Italy in 1803, where he died in 1809.

MALCOLM ISLAND, N. of Port McNeill (C-6). After Admiral Sir Pulteney Malcolm (1758-1838). In 1816, as Commander-in-Chief of the St. Helena station, he had charge of Napoleon during his final captivity. Named in 1846 by Commander George T. Gordon of HMS *Cormorant.*

MALIBU, Princess Louisa Inlet (C-8). Takes its name from a summer resort, the Malibu Club of Canada, opened here in 1946 by T.F. Hamilton of Seattle, a manufacturer of airplane propellors.

MAMALILICULLA, Village I. (C-6). This beautifully situated Indian village, deserted since its people moved to Alert Bay, bears the euphonious name of the Indian band that held second place among the Kwakiutl. Only the Fort Rupert band outranked them. One possible translation for *mamaliliculla* is 'seem to be swimming.'

MAMIT LAKE, N. of Merritt (C-9). From an Okanagan Indian word applied to the whitefish found in the area.

MAMQUAM RIVER, flows W. into Squamish R. (B-8). The meaning of this Squamish Indian name is uncertain. One source says that it is onomatopoeic, echoing the murmur of smooth-running stretches of the river.

MANNING PARK, Hope-Princeton highway (B-9). After Ernest C. Manning (1890-1941), Chief Forester of British Columbia at the time of his death in a plane crash. Manning was one of the first to warn the people and politicians of British Columbia that our timber resources are finite, and he recommended that steps be taken immediately to put the forests on a sustained-yield basis. He was an able exponent of the value of forests for recreation and was instrumental in starting the system of provincial parks, which for years were administered by the BC Forest Service.

MANSON CREEK, Omineca (H-7). Presumably after Donald Manson, who from 1844 to 1857 administered the HBC's New Caledonia district from Fort St. James. (See also under *Manson Ridge.)*

MANSON RIDGE, formerly Manson Mtn., E. of Hope (B-9). After Donald Manson, who entered the employ of the HBC in 1817. After service from 1831 to 1834 at Fort Simpson, he served at Fort McLoughlin, at the Thompson River Post (Kamloops), and at Fort Stikine. In 1844 he took over the command of New Caledonia from Chief Factor Ogden. Manson was a bully, and his reputation as such kept the company from making him a Chief Factor, the rank that normally went with his post. However, the company left him in his command until 1857. In 1855 Manson and Paul Fraser, who had an even worse reputation as a bully, were seated in their tent on Manson Mountain when one of their men, felling a tree, accidentally(?) let it fall on their tent. Fraser was killed, but Manson escaped injury.

MANSONS LANDING [MANSON'S LANDING], Cortes I. (C-7). After Michael Manson, a Shetland Islander who, with his brother John, maintained a trading post here from 1887 to 1895. Later he represented the Comox riding in Victoria.

MANZO NAGANO, MOUNT, E. of Rivers Inlet (D-6). Commemorates the first Japanese settler in Canada. Nagano arrived in New Westminster in 1877 and earned his living as a salmon fisherman.

MAQUINNA POINT, Nootka Sound (B-6). Maquinna was the hereditary name of a succession of Nootka chiefs. About 1791 the Spaniards named this point after the most famous of the line, that Chief Maquinna who in 1788 sold a small patch of land on Nootka Island to Captain Meares (see *Meares Island*). Captain Vancouver met Maquinna in 1792. In 1803 the next Chief Maquinna led an attack on the American ship *Boston,* massacring all but two of the crew. The last Maquinna to maintain authority over the Nootka Indians died in 1901.

MARA LAKE, S. of Sicamous (C-10). After John Andrew Mara, one of the Overlanders of 1862. After his epic journey west, he became a merchant, blossomed as a capitalist entrepreneur, presided as Speaker over the BC legislature, and went on to become an MP in Ottawa.

MARCEL HILLS, N. of West Road R. (F-8). Named by the veteran surveyor Forin Campbell, who recalled, 'These hills were very wavy hills and there was an Englishman with me who said they looked like a woman's marcel.'

MARGUERITE, S. of Quesnel (E-8). After an Indian woman who lived near here until her death at a great age around World War I. She has been described as 'a wonderful old person.'

MARIA SLOUGH, NE of Agassiz (B-9). Jeremiah Gilbert Bristol (Bristol Island) had a sulky Indian wife who let her sister, Maria, do all the work. Bristol, in his will, left Maria land in this area. The sternwheeler *Maria* plied the nearby Fraser during the gold rush. The Halkomelem name of this slough means 'turn at the head.'

MARINA ISLAND, W. of Cortes I. (C-7). After the Indian woman who was interpreter, adviser, and mistress to Hernando Cortes in his conquest of Mexico. From 1849 to 1906, it was Mary Island, but then the Spanish name was restored.

MARINERS REST [MARINERS' REST], W. of Gambier I. (B-8). A place where the remains of seafarers are committed to the deep.

MARK, MOUNT, N. of Horne L. (B-7). Named in 1860 after Mark Bate, who came to Nanaimo from England in 1857, was employed by the HBC, and subsequently became manager of the Vancouver Coal Mining and Land Company. In 1875 he became the first mayor of Nanaimo and was repeatedly reelected, often by acclamation. In 1887, when he was believed to be unwilling to stand again, 90 per cent of Nanaimo's voters signed a 'requisition' calling upon him to do so.

MARKTOSIS, Flores I. (B-6). From the Nootka Indian word meaning 'camp on either side of bay.'

MARPOLE, Vancouver (B-8). Formerly Eburne (q.v.), but renamed Marpole in 1916. After Richard Marpole (1850-1920), who, after early experience with English railways, came to Canada, where he became a CPR contractor in 1881. Subsequently he entered the employment of the company and in 1886 became superintendent of construction and operation for the CPR's Pacific Division.

In *The Valley of Youth*, C.W. Holliday has an interesting anecdote about Marpole:

> A difficulty at first in some places was the shortage of women, especially in Vernon during the first year when numbers of bachelors had collected there. A bunch of these bachelors were, one evening, sitting around the lounge of the new Kalamalka Hotel, the dining room of which had a fine large floor. Mr. Marpole, the superintendent of the CPR, from Vancouver, happened to be there, and regarding this floor, he said, 'Why don't you young fellows give a dance?' ... 'Oh, no chance,' said someone, 'there are not enough girls.' 'Suppose you haul us in a carload, Marpole!' said Hankey. 'By Jove!' said Marpole, slapping his thigh, 'you fellows will have the dance. That's just what I will do.' And that is exactly what he did, for he invited a bevy of Vancouver girls and brought them, complete with chaperon, in his private car all the way up to Vernon. (pp. 327-8)

MARRON RIVER, S. of Penticton (B-10). *Marron* is French for a tamed animal that has reverted to the wild state. The Marron River probably got its name from the wild horses that were once numerous here.

MARSHALL CREEK, flows SE into Carpenter L. (C-8). After John Marshall, who established a flour mill near Soda Creek in 1868. Later he had a mill at Lillooet. He was also a prospector.

MARSHALL POINT, Texada I. (B-7). Without any indication as to whom Marshall might be, Captain Vancouver, in late June 1792, gave the name of Point Marshall to 'the NW point of the island of Fevada' (Texada). Later his Point Marshall became first Cohoe Point and then, in 1945, Kiddie Point after

Thomas Kiddie, a Texada mining engineer. Meanwhile a second promontory, a mile to the south, became Marshall Point. The name had, in Kaye Lamb's phrase, 'migrated.'

MARTEL, N. of Spences Bridge (C-9). After Eli (Joe) Martel, third owner of the ranch here. 'Small of stature but big in endeavour,' Martel established an eleven-acre orchard and, loading his fruit in a wagon, peddled it along the Cariboo Road.

MARTLEY, MOUNT, SE of Pavilion (C-9). After Captain John Martley of the 9th Regiment of Foot, who, selling his commission in Britain, secured at Pavilion one of those military land grants intended to make settlers of ex-officers. He has been described as 'a great landed proprietor who had little ready cash.' He died, aged sixty-eight, in 1896.

MARY HILL, NW of junction of Pitt R. and Fraser R. (B-8). Named after Colonel Moody's wife, Mary. The Royal Engineers thought it would be a good idea to build a citadel here to defend New Westminster against any American invasion.

MARYSVILLE, S. of Kimberley (B-12). After the Virgin Mary, the town being built on St. Mary's River, which had been named by one of the OMI fathers at nearby St. Eugene's Mission.

MARY TOD ISLAND, Oak Bay (A-8). After the second daughter of John Tod, a veteran HBC officer who began his retirement at Oak Bay in 1850. (For John Tod see *Tod Inlet*.)

MASSET INLET, Graham I. (F-3). The Haida name 'Maast' was originally limited to a small island in Masset Sound. The meaning of the word *maast* is unknown.

MATHER CREEK, flows SE into Kootenay R. (B-12). After Robert D. Mather. A merchant at Wild Horse during the gold excitement of the 1860s, he subsequently preempted land at the mouth of this creek.

MATILPI, E. side of Havannah Channel (C-6). Derived from the Kwakwala Indian word meaning the 'point of Matagila.' Matagila was the legendary hero of the Matilpi clan.

MATSQUI, S. of Mission (B-8). From the Halkomelem word meaning 'easy portage' or 'easy travelling,' apparently referring to the ease with which the Indians could ascend creeks from the Fraser and drag their canoes over the height of land to the old Sumas Lake or to tributaries of the Nooksack River.

MAURELLE ISLAND, NE of Quadra I. (C-7). After Francisco Antonio Maurelle, second in command aboard the *Sonora* when Quadra took her up the coast in 1775, probably as far north as the Nass River.

MAXWELL, MOUNT. See *Mount Maxwell Park.*

MAYNE ISLAND, Gulf Islands (A-8). After Richard Charles Mayne, RN, who arrived on this coast in 1857 as lieutenant on the survey ship *Plumper*. He served here until 1861, when he was promoted to commander and returned to England. In 1862 he published *Four Years in British Columbia and Vancouver Island,* a book that should be in any library of British Columbiana. Rear-Admiral R.C. Mayne, MP, died in 1892 while attending a banquet at the Mansion House, London.

MAYOOK, E. of Cranbrook (B-12). In 1906 the local postmaster, R.B. Benedict, wrote to Ottawa that the settlement was named after a local Kootenay Indian family of some means. He added, 'They are respectable and speak English fairly well.'

McARTHUR, LAKE, Yoho NP (D-11). After James Joseph McArthur, DLS (1856-1925), of the Dominion Topographical Survey, who discovered both this lake and nearby Lake O'Hara.

McBRIDE, SE of Prince George (F-9). Began as one of the stations of the newly constructed Grand Trunk Pacific Railway (now part of the CNR). Named after Sir Richard McBride (1870-1917), who in 1903, at the age of thirty-three, became the youngest premier in the history of British Columbia. Knighted in 1912, McBride resigned as premier in 1915 and became the province's agent-general in London, where he died two years later. 'Dicky' McBride was a courtly, warm, outgoing person full of energy and ambitions for his province.
 Also MOUNT MCBRIDE (on VI, W. of Buttle L.).

McCONACHIE PEAK, W. of Duti R. (J-6). After Grant McConachie, a noted bush pilot long before he became president of Canadian Pacific Air Lines. Back in those earlier days, he flew a government survey party, headed by Frank Swannell, into the Thutade Lake area. McConachie, by way of a joke, asked Swannell what he would have to pay Swannell to get a mountain, a lake, or a stream named after him. Swannell replied, 'A bottle of rum would be a fair price.' McConachie took along a bottle of rum on the next trip. His fellow pilots Gil McLaren, Oakes, and Kubicek also got features named after them.

McCONNELL HILL, NE of Lac Le Jeune (C-9). After Archibald McConnell, local rancher, who – at the age of sixty-five – entered a horse in the 1892 BC Derby, rode it himself, and won.

McCRAE, MOUNT, N. of Kwadacha R. (J-7). After Lieutenant-Colonel John McCrae, author of 'In Flanders Fields.'

McCREIGHT LAKE, NW of Campbell R. (C-7). After John Foster McCreight (1827-1913), first Premier of British Columbia (1871-2). From 1880 to 1897, he was a justice of the Supreme Court of British Columbia. A contemporary described McCreight as 'a nervous fidgety queer-tempered man.' On the other hand, Patricia Johnson, his biographer, has praised him, declaring: 'He stood for integrity in a place and age where success was often judged more important. He stood for discipline in an era of self-expression, and for principle rather than personality.'

McCULLOCH, SE of Kelowna (B-10). After Andrew McCulloch (1864-1945), general superintendent from 1918 to 1933 of the Kettle Valley Railway (part of the CPR). As chief engineer, he located and constructed the line between Hope and Midway, a great achievement.

McDAME CREEK, flows SE into Dease R. (L-5). After Harry McDame, who discovered gold on this creek in 1874. He was the longtime partner of another black West Indian, J.R. Giscome (see *Giscome*).

McDOUGALL LAKE, W. of Murtle L. (E-10). After Peter McDougal (*sic*), who came from the United States with Mike Majerus (see *Majerus Falls*) in 1913.

McDOUGALL RIVER, flows E. into McLeod R. (G-8). After James McDougall, clerk in the service of the NWC. After Simon Fraser built a trading post on McLeod Lake in 1805, possibly the first permanent white settlement in British Columbia, he left McDougall in charge of it.

McGILL, MOUNT, Glacier NP (D-11). After McGill University.

McGILLIVRAY, Crowsnest Pass (B-12). Originally Bullhead, but renamed after D. McGillivray, a contractor engaged in the construction of the CPR's Crowsnest line.

McGILLIVRAY RIDGE, E. side of Athabasca Pass (E-10). Some time before 1814, this massive mountain was given its original name, McGillivray's Rock, after the great William McGillivray of the NWC. (See Marjorie Wilkins Campbell, *McGillivray Lord of the Northwest*.)

McGREGOR RIVER, flows W. into Fraser R. NE of Prince George (G-9). This – along with its tributaries Captain Creek, James Creek, and Herrick Creek – commemorates Captain James Herrick McGregor, first president of the Corporation of BC Land Surveyors, president of Victoria's Union Club in 1914, and an occasional poet – see *The Wisdom of Waloopi,* which he had printed for private circulation only. He was killed in World War I at Ypres in April 1915.

McHARG, MOUNT, Alberta-BC boundary (C-12). After Lieutenant-Colonel William Hart-McHarg. A dedicated militia officer and a champion marksman, he practised law first in Rossland and then, after service in the Boer War, in Vancouver. On 23 April 1915, commanding the 7th Battalion CEF ('The British Columbians'), he was mortally wounded while reconnoitring between the lines. His friend Major J.S. Matthews later wrote of him: 'His massive face suggested nothing of the frailty of his body, which, through chronic ill-health (indigestion) weighed about 140 pounds. Frequently his diet was merely biscuits and milk. He was a cool quiet man of commanding personality and a bachelor.'

Vancouver's Georgia Viaduct, completed shortly after his death, was originally McHarg Viaduct.

McINTYRE BLUFF, S. of Vaseux L. (B-10). After Peter McIntyre, one of the Overlanders of 1862. Earlier he was an 'Indian fighter' and a guard on the Pony Express in the American West. Late in 1886 he received a Crown grant to the land beside this great precipitous cliff.

McINTYRE LAKE, N. of McLeod L. (H-8). This McIntyre, after working for the HBC, set up his own store at Fort McLeod. His backer is said to have been a wealthy American woman whose scruples kept her from taking money made from the trapping of animals. This suited McIntyre fine, and he pocketed the profits that the store made from buying furs.

McKAY PEAK, W. of Ladysmith (B-7). After Joseph William McKay (1829-1900), Chief Trader with the HBC, who played a major part in the company's development of coal deposits. In 1855 he became one of the six members of the first Legislative Assembly of Vancouver Island. In 1883, having retired earlier from the service of the HBC, he took a post with the federal Indian Department.

McKELVIE, MOUNT, N. of Tahsis, VI (B-6). After Bruce A. McKelvie (1889-1960), journalist and BC historian. At one time president of the BC Historical Association, McKelvie wrote a number of books, including *The Early History of British Columbia, The Pageant of British Columbia,* and *Maquinna the Magnificent.*

McKINLEY LAKE, S. of Horsefly L. (E-9). After Archibald McKinley, a clerk in the service of the HBC, who in 1840 married Sarah Julia, daughter of Chief Factor Peter Skene Ogden. A son ran a store and farmed at Lac la Hache.

McLEAN CREEK, flows into Skaha L. (B-10). After Roderick McLean, who in 1860 built the HBC post at Keremeos and later farmed there.

McLEAN LAKE, W. of Cache Cr. (C-9). After Chief Trader Donald McLean, one of the very few HBC officers with a deserved reputation for brutality. He is said to have come out of retirement at his Bonaparte River ranch to aid the government forces in the Chilcotin War because he saw a chance to raise his record of nineteen Indians slain to an even score. Before he could get his twentieth, an Indian in ambush shot him through the heart. He was the father of the notorious 'McLean boys' hanged many years later for murdering a police officer.

McLEAN FRASER POINT, W. coast of Moresby I. (E-4). After Dr. C. McLean Fraser (1872-1946), at one time director of the biological research station at Departure Bay and for many years head of UBC's Department of Zoology.

McLENNAN RIVER, joins Fraser R. at Tête Jaune Cache (E-10). After Roderick McLennan, who camped on the river with his CPR survey party in 1871.

McLEOD LAKE, S. of Williston L. (G-8). This is the Trout Lake where Simon Fraser established a post in 1805. Soon both the lake and the fort were renamed in honour of Archibald Norman McLeod, one of the most energetic officers of the NWC.

McLEOD, MOUNT, W. of Dease L. (K-4). Named by G.M. Dawson in honour of Chief Trader John M. McLeod of the HBC, the discoverer of Dease Lake, which it overlooks.

McLOUGHLIN BAY, E. side of Campbell I., Central Coast (E-5). In 1833 the HBC established Fort McLoughlin, named after Dr. John McLoughlin, Chief Factor in command of its Columbia Department. When the fort was abandoned in 1843, McLoughlin's name remained attached to the bay.

McLURE, N. Thompson R. (D-9). Named after John McLure. Arriving in 1906, he farmed here until his death, aged eighty-four, in 1933.

McMILLAN ISLAND, opposite Fort Langley (B-8). After James McMillan of the HBC, who made a reconnaissance in this area in 1824 and then, on a second visit in 1827, founded Fort Langley. He commanded the fort during its first year.

McMURDO, SE of Golden (D-11). After Archie McMurdo, prospector and first white settler in the district.

McMURPHY, N. Thompson R. (D-10). After an engineer who, during construction of the Canadian Northern Railway, lost his footing on a slippery rock and fell to his death near here.

McNEILL BAY, Victoria (A-8). Commonly referred to by the unofficial designation of Shoal Bay. It is named after Captain William H. McNeill (1803-75), of the HBC's maritime service, who married one of the Macaulay daughters (see *Macaulay Point*) and built a comfortable house on 200 acres of land on this bay. (See also *Port McNeill*.)

McQUEEN LAKE, NW of Kamloops (C-9). After Isaac Brock McQueen, an Overlander of 1862. He came to the Kamloops area in 1865, logged and ranched here, and died in 1894.

MEADE CREEK, flows S. into Cowichan L. (A-7). After Robert Meade, an English 'remittance man' who arrived in the district about 1895. He retained all the graces of a highly educated Englishman and owned the only dress suit in the area.

MEAGER CREEK, flows NE into upper Lillooet R. (C-8). Centre of British Columbia's most promising area for thermal energy development. Named after J.B. Meager, who owned timber licences on the creek.

MEARES ISLAND, Clayoquot Sound (B-7). After John Meares (1756-1809). Meares, having learned of the enormous profits that Captain Cook's sailors had realized in selling their sea otter skins to the Chinese, made a trading voyage to the Pacific coast in 1786 and a second one in 1788 with the ships *Felice* and *Iphigenia*. Putting in at Nootka on this latter expedition, he purchased some land from Chief Maquinna. This he used as a base for the building of the *North West America*, launched in September 1788.

Some months later Don Estevan José Martinez arrived at Nootka and declared Meares's purchase null and void since Spain owned Nootka by right of prior discovery. When the *North West America* next put in at Nootka, she and several other British ships were seized by the Spaniards. In May 1790 Meares presented a memorial to the British House of Commons, asking for reparations from Spain. National indignation swept Britain, and war with Spain seemed imminent, but Spain abandoned her exclusive claims to Nootka and agreed to pay Meares an indemnity. In 1790, capitalizing on the Nootka excitement, Meares published his *Voyages Made in the Years 1788 and 1789*. He was an unreliable person, and his account of his expeditions contains various misrepresentations.

MEEK LAKE, NE of Dease L. (K-5). After R.J. Meek, in charge of the McDame detachment of the BC Provincial Police in the 1930s. He served in the RCAF and RAF in World War II, winning the Conspicuous Gallantry Medal and the Distinguished Flying Cross.

MELDRUM CREEK, W. of Williams L. (E-8). After Thomas Meldrum, reputedly the first white man to have the courage to live in the Chilcotin Indians' country following the massacre of the Waddington road-builders in the Chilcotin Uprising of 1864.

MEMALOOSE CREEK, flows E. into Similkameen R. (B-9). A Chinook jargon word meaning variously 'to kill,' 'to die,' 'to decay,' or 'to become rotten.'

MEMORY ISLAND PARK, Shawnigan L. (A-8). A memorial to two Victoria men, Allan Mayhew and Kenneth Scharff, who lost their lives in World War II.

MENZIES BAY, N. of Campbell River (C-7). Named by 1847, probably after Archibald Menzies (1754-1842), botanist and surgeon with Vancouver on HMS *Discovery*. Menzies had earlier visited this coast as a surgeon on the trading vessel *Prince of Wales*. Among his many discoveries while on Vancouver's expedition was the *Arbutus menziesii*, known in the United States as the madrona, whose orange-red trunk is so commonly seen near our shoreline. Menzies had a tiny greenhouse on the quarterdeck of the *Discovery* in which he kept specimens he was taking back alive to the Royal Botanical Gardens at Kew.

MERRITT, SW of Nicola L. (C-9). Earlier names of Forksdale and Diamond Vale were laid aside in 1906 when the settlement was named after William Hamilton Merritt, one of the promoters of the railway, through the Nicola Valley to Spences Bridge, that opened up the country for coal mining.

MERVILLE, N. of Courtenay (B-7). When this area was developed by returned soldiers after World War I, they named the settlement Merville after the place in France where the Canadians had had their first field headquarters.

MESACHIE LAKE, S. of Cowichan L. (A-7). The Chinook jargon word meaning 'bad' or 'evil.' The Cowichan Indians believed that an old man (called the Mesachie Man in legends) lived in the lake and that anyone trespassing on his territory would be drowned in the lake's dark and forbidding waters.

MESILINKA RIVER, flows SE into Omineca R. (I-7). From a Sekani Indian word meaning 'stranger.'

METCHOSIN, SW of Victoria (A-8). The future Sir James Douglas, when he visited the area in 1842 to choose the site for Fort Victoria, mentioned Metcho-sin. A number of meanings are given for this Straits Salish word. Many years ago a dead whale reportedly washed up on the beach here, and several meanings given for the word, such as the 'smelling of oil' and the 'place of oil,' could possibly relate to this occurrence.

METLAKATLA, W. of Prince Rupert (G-4). In October 1857 there arrived among the Tsimshian Indians a lay missionary, William Duncan, despatched from England by the Church Missionary Society. In 1862, determined to isolate his converts from heathenism, Duncan resolved to found a new, Christian settlement. Thus, on May 27th he arrived by canoe with some fifty Indians at Metlakatla harbour – the name Metlakatla being a Tsimshian Indian word meaning 'a passage connecting two bodies of salt water.'

Duncan was a good but eccentric and very strong-willed man. Through his personal ascendancy over the Indians, he created a model community at Metlakatla. On the other hand, he quarrelled with his fellow missionaries and with his bishop. He refused to let his Indian converts receive Holy Communion – fearing that they would confuse partaking of the body of

Christ with the cannibal dancer's biting of spectators. In 1887, having received an ultimatum from the Church Missionary Society, he defied both it and Bishop Ridley and, moving with over 800 of his Indians, founded New Metlakatla in Alaska.

METSANTON LAKE, E. of Caribou Hide, Cassiar (J-6). This is a Sekani Indian word meaning 'full belly' and refers to good fishing at this lake.

MEXICANA POINT, Hope I. off N. VI (C-6). Named in 1792 by Galiano and Valdes after their little schooner *Mexicana* in which, with the *Sutil,* they circumnavigated Vancouver Island that year.

MEZIADIN LAKE, E. of Stewart (I-5). Although this name is derived from the Tsetsaut Indian language (which became extinct about 1940), the name is believed locally to mean 'beautiful,' referring not so much to the scenery as to the excellent fishing provided by the lake.

MICHEL CREEK, S. of Crowsnest Pass (B-12). In 1906 Frank Harmer, the postmaster at Elk Prairie, wrote to James White, the Chief Geographer at Ottawa, that the settlement of Michel took its name from 'an old French trapper and hunter called Michel' who once had a cabin on nearby Michel Prairie. The person in question would be the half-Indian Pierre Michel, who accompanied the Flathead Indians on several of their campaigns and was admired by them for his bravery. At times he served the HBC as an interpreter.

Michel has also been identified, less convincingly, as Michael Insula, or Red Feather, 'The Little Chief' of the Flatheads in the 1850s, and as Michael Phillipps, the HBC clerk at Fort Shepherd in 1864.

MICHEL PEAK, Dawson Range, Glacier NP (D-11). After Friedrich Michel, one of the CPR's Swiss guides, based at Glacier House (opened in 1886), who made various first ascents around 1900.

MIDDLE RIVER (G-7). Refers to its position. Driftwood River runs south into Takla Lake; Tachie River runs out of Trembleur Lake and southeast into Stuart Lake; and in the middle of this waterway is Middle River linking Takla Lake with Trembleur Lake.

MIDGE PEAK, W. of Canal Flats (q.v.) (C-11). After the steam launch *Midge*, which W.A. Baillie-Grohman brought in from the United States. He managed to get her admitted duty free as an agricultural implement after jocularly assuring the customs officer that he would need it to pull a plough across the flooded part of his land.

MIDWAY, W. of Grand Forks (B-10). When the townsite here was laid out in 1893, it was given the name of Boundary City. This name proved too similar to that of the nearby Boundary Creek post office, so Captain R.C. Adams of

Montreal, one of the owners of the townsite, changed the name the next year to Midway. He is said to have taken the name from the Midway Pleasance at the Chicago's World Fair of 1893. However, there are a number of reasons that made Midway a particularly suitable name for this new settlement: (1) Midway was located approximately midway between Penticton and Marcus, Washington, then its nearest railway point; (2) Midway stands near midpoint on the old Dewdney Trail, from its beginning at Hope to its terminus at Wild Horse Creek, near Fort Steele; and (3) Midway stands approximately midway between the Rockies and the Pacific.

MIETTE PASS, NW of Yellowhead Pass (F-10). Named after Baptiste Miette, a voyageur in the service of the NWC. He is remembered chiefly for his ascent of Roche Miette near the eastern entry of Jasper Park in Alberta. Walter Moberly recounted the story late in the nineteenth century in *The Rocks and Rivers of British Columbia*: 'One day, seized with a desire to get to the top of the rock, he, after a most difficult and dangerous climb, succeeded. The venturesome Miette then sat on the edge of the cliff, dangling his legs over it and smoked a pipe, enjoying the fine view his elevated station afforded; and from that day it has been known as "La Roche à Miette" by the Indian and half-breed hunters' (p. 88).

MILBANKE SOUND, Central Coast (E-5). Named in 1788, by Captain Duncan of the *Princess Royal,* after Vice-Admiral Mark Milbanke.

MILL BAY, W. side of Saanich Inlet (A-8). Takes its name from the sawmill built here by Henry S. Shepherd in 1861 and soon purchased by William Sayward.

MILL STREAM, Esquimalt (A-8). Site of the first machine-operated sawmill on Vancouver Island. In 1848 a millwright was brought from England to build it.

MILLER LAKE, Mt. Revelstoke NP (D-10). After A.E. Miller, school inspector at Revelstoke, who explored the area and published articles extolling its beauty.

MILLIGAN CREEK, flows SW into Beatton R. (I-9). After Lieutenant George M. Milligan (1888-1918), MC, FRGS, BCLS, who surveyed in the Peace River country in 1911-15 and was killed in World War I.

MILNER, NE of Langley (B-8). After Alfred, first Viscount Milner (1854-1925), British statesman and colonial administrator.

MILNES LANDING [MILNE'S LANDING], Sooke Inlet (A-8). Takes its name from Edward Milne, who opened a grocery store here in 1893.

MILTON, MOUNT, W. of Albreda R. (E-10). One of the most unusual expeditions in the history of British Columbia occurred in the summer of 1863, when the epileptic English nobleman William Wentworth Fitzwilliam, Viscount Milton (1839-77), his personal physician Dr. Walter Cheadle, an utterly useless classical scholar named O'Beirne, and a part Indian guide and the latter's Indian wife and son, entered the province by the Yellowhead Pass and found their way down the trackless wilderness of the Albreda and North Thompson Rivers until, emaciated and starving, they arrived at Kamloops. The expurgated version of their travels, *The North-West Passage by Land*, was published in 1865 and, being a highly readable book, went into a number of editions. In it one reads how, as they travelled down the Albreda, they saw 'a magnificent mountain, covered with glaciers, and apparently blocking up the valley before us. To this Cheadle gave the name of Mount Milton.'

MINERS BAY [MINERS' BAY], Active Pass (A-8). During the Fraser River gold rush of 1858-9, this was the favourite overnight camping spot for prospectors making, by open boat, the two-day crossing from Fort Victoria to the mouth of the Fraser River.

MINETTE BAY, Kitimat (G-5). When Louis Coste, chief engineer, Department of Public Works, Ottawa, was in the area in 1898, he named this bay after his wife. (See *Coste Island*.)

MINISKIRT, NW of Victoria (A-8). Nearby Skirt Mountain (originally Skirt Hill) was named as early as 1858. When in 1976 a name was wanted for a minor eminence to the northeast of it, Miniskirt seemed a logical choice. This is the complete official name.

MINNEKHADA PARK, Coquitlam (B-8). This Sioux Indian name meaning 'water rattling by' was given by an American farming here before World War I. In 1934 E.W. Hamber, Lieutenant-Governor of British Columbia from 1936 to 1940, built a 'Scottish hunting lodge' here. Park and lodge are now owned by the Greater Vancouver Regional District.

MINNIE LAKE, S. of Nicola L. (C-9). After Minnie (b. 1873), daughter of Byron Earnshaw, first settler here.

MINSTREL ISLAND, entrance to Knight Inlet (C-6). Local tradition has it that a 'minstrel boat' arrived here when a survey crew was working in the area. That boat was almost certainly HMS *Amethyst*, which in 1876 took the Governor-General (Lord Dufferin) and his lady on a cruise upcoast to Metlakatla. A member of her crew (Patrick Riley, *Memories of a Bluejacket*, p. 87) mentions that *Amethyst* had an amateur troupe of 'nigger minstrels' – these entertainers were generally white and made up with black faces – who provided entertainment for their shipmates and, presumably, any visitors, so it seems highly probable that Minstrel Island commemorates a performance in these waters. Nearby are Bones Bay and Sambo Point. Both 'Mr. Bones' and 'Sambo' were stock characters in the minstrel shows that were so popular in the late nineteenth century.

MINTO, N. of Chilliwack (B-9). In 1900 the little sternwheel ferry *Minto* was built at Harrison for C.R. Menten, who used it for transporting passengers and freight between Chilliwack and the CPR's station on the north side of the Fraser River. Originally known as Menten's Landing, it gradually became Minto Landing and finally Minto.

MINTON CREEK, flows NE into Chilcotin R. (D-8). After Michael Minton, who settled in the area about 1870.

MIOCENE, NE of Williams L. (E-9). After the old Miocene Mine on the Horsefly River. Miocene is part of the Tertiary period in geological time.

MIRACLE BEACH PARK, SE of Campbell River (B-7). The story goes that after the Cape Mudge Indians had befriended a supernatural stranger, giving him food and shelter, he rewarded them with wealth and prosperity but warned them not to become proud in consequence. When they became overweening, however, the Cowichans attacked them. The stranger came a second time and saved them. As a reminder of the lesson taught to them, he worked a great miracle, turning an Indian princess into Mitlenatch Island.

MISSION, N. of Abbotsford (B-8). Takes its name from St. Mary's Indian mission, founded here by Father Fouquet, OMI, in 1861. The school, convent, and church were built in 1885 when the original site became part of the CPR right of way. Mission, the junction point, narrowly escaped being renamed either New Seattle or Gladstone when in 1899 the CPR built its spur line to Sumas, Washington. Until 1973 its official name was Mission City.

MISSION CREEK, Kelowna (B-10). Takes its name from the mission founded north of here in 1859 by Father Pandosy, Father Richard, and Brother Surel of the Oblates of Mary Immaculate. In 1860 the mission was moved to Mission Creek.

MISSION RIDGE, N. of Seton L. (C-8). Named after the mission on Seton Lake. The trail linking the mission with the Bridge River area went through MISSION PASS.

MISSUSJAY CREEK, flows E. into Stikine R. (J-4). Flowing into Missusjay Creek is Misterjay Creek. Mr. and Mrs. Jay kept a roadhouse on the Stikine River.

MISTAYA MOUNTAIN, Alberta-BC boundary (D-11). The Stoney Indian word for 'grizzly bear.'

MITCHELL BAY, Malcolm I. (C-6). After Captain William Mitchell (1802-76), skipper of various ships in the service of the HBC. A Royal Navy lieutenant who travelled with Mitchell in 1852 described him thus: 'as kindly a Scotsman as ever followed the sea.' MITCHELL INLET in the Queen Charlotte Islands is also named after Captain Mitchell.

MITCHELL ISLAND, N. of Lulu I., Fraser delta (B-8). After Alexander Mitchell, who in 1882 became the first man to farm on this island.

MITCHELL RANGE, E. of Kootenay NP (C-12). After William Roland Mitchell, travelling companion of Captain Arthur Brisco when the latter participated in Palliser's explorations during 1858-9. Mitchell was disappointed to find Palliser's program left little time for shooting expeditions.

MITLENATCH ISLAND, SE of Quadra I. (B-7). The name of this well-known nature reserve comes from the Island Comox Indian language. It means 'calm at the end,' in the sense of affording shelter from the wind.

MOBERLY, N. of Golden (D-11). This is the oldest place of continual white residence along this stretch of the Upper Columbia, a cabin having been built here in 1871 by Walter Moberly, CE (1832-1915). Born in Steeple Ashton, Oxfordshire, he worked on surveys in northern Ontario before he came to British Columbia by way of Cape Horn in 1858. One of the more important explorers of this province, he ranged far and wide over British Columbia and discovered the Eagle Pass (q.v.) between Shuswap Lake and Revelstoke. In 1885 he published *The Rocks and Rivers of British Columbia,* a little book of reminiscences.

MOBERLY LAKE, N. of Chetwynd (H-9). Named after Henry Moberly, brother of Walter (see preceding entry). Henry spent many years in the Peace River country as an HBC employee, free trader, trapper, and prospector. In 1865 he built himself 'a comfortable shack' on the shore of Moberly Lake. Like his brother, he wrote a book of reminiscences, his being entitled *When Fur Was King.*

MOFFAT CREEK, flows N. into Horsefly R. (E-9). After Tom Moffitt (note the spelling), who came into the Cariboo with Peter Dunlevy's party in 1859. Despite the warnings of his partners that this creek showed none of the signs of a gold-bearing stream, he insisted upon working it – and proved that they were right.

MONASHEE MOUNTAINS, between Shuswap L. and Columbia R. (C- and D-10). A Gaelic, not an Indian, name. The story goes that around 1880 Donald McIntyre, the highlander who staked the Monashee Mines, was prospecting in this area. The day had been one of mixed snow and rain and strong winds. Toward evening the wind fell and the clouds cleared, while the setting sun cast a beautiful peaceful light on a nearby mountain. 'Monashee!' McIntyre exclaimed. This word, correctly spelled in Gaelic, is *monadh*, meaning 'mountain,' and *sith* (pronounced 'shee'), meaning 'peace.'

MONCK PARK, Nicola L. (C-9). When Major C.S. Goldman, owner of the Nicola Lake Stock Farm, retired in 1951 at the age of eighty-one, he gave land for this park, which is named after one of his sons, who distinguished himself in World War II.

MONCKTON, MOUNT, SW of Lower Post (L-5). After Philip Marmaduke Monckton, BCLS (1892-1965). He carried out extensive surveys in northwestern British Columbia, where his hardiness and ability to live off the country won him the name of 'The Wolverine' from the Tahltan Indians. Had he lived he would have inherited another title, Viscount Galway, which passed to his son.

MONEY MAKERS ROCK, Ganges Harbour (A-8). Boaters put holes in their hulls so often here that the local repair yard takes in plenty of money.

MONIAS, on BC Railway, SW of Fort St. John (I-9). This station's name is from the Cree Indian word for 'white man.' (See also *Moonias Mountain*.)

MONKMAN PASS, NE of Prince George (G-9). This pass, providing access between central British Columbia and the Peace River country, is 162 feet lower than the Yellowhead Pass. Alexander Monkman, a Peace River trader and trapper, discovered it by chance during a 1922 hunting trip.

MONTAGU CHANNEL, Howe Sound (B-8). After HMS *Montagu*, which served under Lord Howe when he won his great victory 'The Glorious First of June.' Captain James Montagu, who commanded her, was killed in that battle. He had asked for the command of HMS *Montagu* since she had been named after one of his ancestors.

MONTE CREEK, flows N. into S. Thompson R. (C-10). The map in A.C. Anderson's *Handbook and Map to the Gold Regions of Frazer's and Thompson's Rivers* (c. 1858) has 'Monteé' where Monte Creek enters the South Thompson. Presumably this is a printer's error for 'Montée' (French *la montée*, meaning 'height of land'). The height of land in question is that between the Okanagan and South Thompson valleys, which had to be crossed by the early fur brigades.

Monte Creek post office was formerly known as Duck and Pringle's. Duck Range and Ducks Meadow in this area take their names from the pioneering Duck family. Jacob Duck settled here in 1863.

MONTIGNY CREEK, flows E. into N. Thompson R. (D-9). After Edouard Montigny, one of the HBC's engagés at Kamloops in the 1840s, when the New Caledonia brigades used the Little Fort route. He helped Peers to lay out the Hope-Tulameen brigade trail.

MONTNEY, N. of Fort St. John (I-9). After Montaigné, a Beaver Indian chief who perished with many of his followers in the flu epidemic of 1919.

MONTROSE, E. of Trail (B-11). Named by A.G. Cameron, a Trail lawyer, after his hometown in Scotland.

MOONIAS MOUNTAIN, S. of Narraway R. (G-9). The Cree Indian word for 'white man.' Monias, on the BCR south of Peace River, is the same word.

MOORE CHANNEL, Englefield Bay, Moresby I. (E-3). After George Moore, RN, master of HMS *Thetis,* who surveyed the channel in 1852 during the rush of American gold-seekers to the area.

MORAN, N. of Pavilion (C-9). After J. Moran, son-in-law of Patrick Welch, one of the original contractors for the building of the PGE Railway (now the BCR).

MORESBY ISLAND, S. of Skidegate Channel, QCI (E-3 and 4, F-3 and 4). After Rear-Admiral (later Admiral of the Fleet) Sir Fairfax Moresby, Commander-in-Chief of the Pacific Station from 1850 to 1853. The island was named by Moresby's son-in-law, Commander Prevost, HMS *Virago,* in 1853.

During the years 1819-23, Moresby notably distinguished himself in suppressing the African slave trade.

MORFEE LAKES, E. of Mackenzie (H-8). A.L. Morfee, an RCAF officer, flew photographic planes in 1929 during a PGE Railway survey.

MORIARTY, MOUNT, SE of Port Alberni (B-7). After William Moriarty, RN, first lieutenant on HMS *Plumper,* on this coast 1857-61.

MORICE RIVER, flows N. into Bulkley R. (G-6). Named after Father Adrien Gabriel Morice, OMI (1859-1938). For twenty years, 1883-1904, amid his missionary endeavours, Father Morice systematically explored north-central British Columbia. Working with only a watch, a telemeter, a compass, a mountain barometer, and a sounding line, he produced the first real map of the area.

Controversy exists as to whether Victoria deliberately suppressed Morice's names for the lakes, rivers, and mountains first discovered by him, or if he supplied his information only after government surveys had established a different set of names. Certainly the relationship between the peppery French priest and the stolid civil servants in Victoria was not a happy one. Among Father Morice's books, mention must be made of his *History of the Northern Interior of British Columbia.*

MORICE LAKE and MORICETOWN are also named after Father Morice.

MORIGEAU, MOUNT, W. of Columbia L. (C-11). After the first white settler in this area, François Morigeau, who arrived before 1820 and founded a family that is now widely represented in this district. Father De Smet has left the following account of his meeting with Morigeau in 1845:

> The monarch who rules at the source of the Columbia is an honest emigrant from St. Martin, in the district of Montreal, who has resided for twenty-six years

in this desert. The skins of the rein and moose deer are the materials of which his portable palace is composed; and to use his own expressions, he *embarks on horseback* with his wife [a daughter of Chief Peter Kinbaskit] and seven children, and *lands* wherever he pleases ... Many years had Morigeau ardently desired to see a priest; and when he learned that I was about to visit the source of the Columbia, he repaired thither in all haste to procure for his wife and children the signal grace of baptism. (*Life and Letters,* pp. 498-9)

MORKILL RIVER, flows SW into Fraser R. (F-9). After Dalby Brooks Morkill, BCLS (1880-1955), who worked in this area in 1912-13. He is remembered by the Corporation of BC Land Surveyors as the man who introduced the 'Ceremony of the Spats.' In 1929, when president of the corporation, Morkill came to the annual meeting wearing spats. Derisive comments were made about these very 'unsurveyor-like' articles of apparel. Unmoved, Morkill kept them on until the moment when he handed over to the incoming president. Then he took them off and presented them, autographed, to his successor. Since that time spats have been the regalia of the president of the Corporation of BC Land Surveyors, though the original much-autographed spats have long since been replaced.

MORRISSEY, S. of Fernie (B-12). After James Morrissey, who, with John Ridgway, around 1878 cut out a trail from Elko to Crowsnest Lake. They are said to have found the first samples of coal in this area while blazing this trail.

MORSE BASIN, SE of Prince Rupert (G-4). After F.W. Morse, vice-president of the GTPR when it established Prince Rupert as its western terminus in 1906. Also MOUNT MORSE.

MOSLEY CREEK, flows S. into Homathko R. (D-7). After Edwin Mosley or Mosely, one of the three survivors of the Chilcotin Massacre of 1864, in which fourteen white road builders perished on the banks of the Homathko. Other creeks in the area are named for Tellot and Klattasine, leaders of the Indians.

MOUNT CURRIE, settlement and mountain E. of Pemberton (C-8). In 1851 John Currie, who had come to Quebec with his Scottish parents, ran away from home and headed for the California goldfields to make his fortune. Unsuccessful as a gold-seeker both there and in the Cariboo, he turned rancher and around 1885 finally settled near Pemberton with his Lillooet Indian wife. In 1898 Currie had disastrous losses when he tried to drive cattle from Pemberton to Howe Sound for selling on the Vancouver market. He died in 1910.

MOUNT LEHMAN, NW of Abbotsford (B-8). After Isaac Lehman, who preempted land here around 1875. Later he moved to the Cariboo, subsequently was a blacksmith in New Westminster, and ended up as an undertaker in Ashcroft.

MOUNT MAXWELL PARK, Saltspring I. (A-8). For over a century, the locals have refused to apply to their Maxwell's Mountain its official name of Baynes Peak, after Rear-Admiral Baynes, RN. Accordingly, in 1938, they insisted on this name for their park, memorializing the Maxwell family, which had lived continuously on the island since the mid-1860s.

MOWICH CREEK, Manning Park (B-9). *Mowich* is the word for 'deer' in the Chinook jargon.

MOYIE RIVER, flows SW into Kootenay R. (B-12). David Thompson named it McDonald's River after his clerk, Finan McDonald, while Governor Simpson called it the Grand Quête in honour of an Indian chief. The name Moyie or Mooyie is very old, however, and is a corruption of the French *mouillé,* meaning 'wet.' Lees and Clutterbuck, who were this way in 1887, had some bitter things to say about 'this water-logged Mooyie valley.'

Ms. MOUNTAIN, S. of Mt. Donner, Strathcona Park (B-7). Named in 1980 for the female leaders of the first party to climb this peak.

MUCHALAT INLET, E. of Nootka Sound (B-6). According to the early missionary Father Brabant, *muchalat* is a Nootka Indian word meaning 'deerstalkers.'

MUDGE, CAPE, S. end of Quadra I. (B-7). Named by Vancouver in 1792 after Zachary Mudge (1770-1852), his first lieutenant on HMS *Discovery.* Mudge was again on this coast in 1796 as first lieutenant on HMS *Providence.* He commanded the frigate *Blanche* when it surrendered to a French squadron in 1805, but a subsequent court martial acquitted him and congratulated him on his 'very able and gallant' conduct. He subsequently rose to the rank of full admiral in the Royal Navy.

MUIR POINT, E. of Sooke Inlet (A-8). After John Muir, who arrived on Vancouver Island with his wife and four sons in 1849. He worked for the HBC in the coalfields at Fort Rupert and Nanaimo, then moved to Sooke in 1853, acquiring the estate of Captain W.C. Grant, the first independent settler on Vancouver Island.

MUMMERY, MOUNT, N. of Golden (D-11). After A.F. Mummery of the Alpine Club (Great Britain), killed in 1895 while climbing Mount Nanga Parbat in the Himalayas.

MUNCHO LAKE, S. of Liard R. (K-7). This lake, skirted by the Alaska Highway and surrounded by a provincial park, takes its name from the Tagish Indian word meaning 'big lake.'

MUNDAY, MOUNT, SE of Mt. Waddington (D-7). After W.A.D. 'Don' Munday, who, accompanied by his wife, Phyllis, made numerous expeditions into this area between 1926 and 1936. (See *Waddington, Mount.*)

MURCHISON, MOUNT, W. of Squamish (B-8). After Sir Roderick I. Murchison (1792-1871), one of the more formidable of the Victorians. Aided by a commanding presence, great energy, a highly supportive wife, and a private fortune, he cut out for himself a notable career, becoming president of the Geological Society, president for many years of the Royal Geographical Society, and director-general of Britain's Geological Survey. He was a co-author of a book on the geology of Russia and published many papers on the geology of Britain and Switzerland, but a lack of imagination kept him from matching the achievements of Lyell and Sedgwick, with whom he worked.

MURDERER CREEK, flows NW into Cariboo R. (E-9). The story goes that during the gold rush days Boone Helme, a notorious Montana bandit, robbed and killed three men between Keithley Creek and Quesnel Forks. He is supposed to have buried his loot under a cedar tree on this creek.

MURPHY LAKE, E. of Williams L. (E-9). After Patrick Murphy, a homesteader whose family left the area after his death in the great flu epidemic at the end of World War I.

MURRAY RIVER, flows N. into Pine R. (H-9). After N.F. Murray, CE, who surveyed the region for the British Columbia Forest Service. In World War I, he enlisted in the 67th Battalion and was killed in action in France.

MURTLE LAKE, Wells Gray Park (E-10). Named in 1874 by Joseph Hunter, a CPR surveyor, after his birthplace in Scotland.

MUSKWA RIVER, flows E. into Fort Nelson R. (K-8). This is the Cree Indian word for 'black bear.'

MUSQUEAM, N. mouth of Fraser R. (B-8). This Indian village, now within the boundaries of Vancouver, marks the farthest point reached by Simon Fraser when he made his descent in 1808 of the river bearing his name. Because of the hostility of the Indians here, Fraser decided against any exploration of the coast. Returning to an unidentified village above New Westminster, he found the natives there curious to know 'how we had the good fortune to escape the cruelty of the *Masquiamme.*' *Musqueam* is Halkomelem and means 'place always to get [the root of] irislike plant.' Unfortunately the plant is now extinct here.

MYNCASTER, SE of Bridesville (B-10). After Thomas and William McMynn, farmers hereabouts, who came to British Columbia in 1887.

MYRA FALLS, Strathcona Park (B-7). These falls and the creek on which they occur are named after vivacious young Myra Ellison, who in 1910 took part with her father, the Hon. Price Ellison, in the government exploratory survey that preceded the setting up of Strathcona Provincial Park.

N

NABESCHE RIVER, flows S. into Peace Reach, Williston L. (I-8). This is the Sekani Indian word for 'ottertail.'

NADEN HARBOUR, N. coast of Graham I. (F-3). A Haida name indicating that there were numerous settlements around the harbour.

NADINA RIVER, flows E. into François L. (F-6). From the Carrier Indian word meaning 'a log thrown across a creek to serve as a bridge.'

NADSILNICH LAKE, S. of Prince George (F-8). Based on a Carrier Indian word possibly meaning 'swampy willow.'

NAHATLATCH RIVER, flows E. into Fraser R. (B-9). From the Thompson Indian word meaning 'deep down from both sides.' The name has also been connected with the Thompson word for 'icy.'

NAHMINT RIVER, W. side of Alberni Inlet (B-7). Named by the explorer John Buttle after the Nahmint Indians.

NAHUN, N. of Westbank (C-10). This name is derived from an Okanagan Indian word referring to the cliffs in the area. One legend tells how the little rock island on the east side of Okanagan Lake was once part of the big rock near the wharf at Nahun.

NAHWITTI RIVER, W. of Port Hardy (C-6). Early nineteenth-century references to the Neu-wit-ties, New Whitty, and Newettees suggest the original pronunciation of the first syllable. The Kwakwala Indian name possibly refers to the Nahwitti Indians as 'being knowledgeable about tribal history.'

NAIKOON PARK, NE part of Graham I. (F-4). G.M. Dawson in 1878 noted that Naikoon was the Haida name for Rose Point (q.v.) and meant 'long nose.'

NAKA CREEK, flows into Johnstone Strait E. of Tsitika R. (C-6). From the Kwakwala Indian word meaning 'to drink.'

NAKIMU CAVES, Glacier NP (D-11). Said to be from an Indian word meaning either 'grumbling' or 'spirit sounds.' It is descriptive of the noise made by subterranean waters.

NAKUSP, Upper Arrow L. (C-11). From an Okanagan Indian word meaning 'closed in' or 'come together.' Illustrative is an anecdote preserved by Kate Johnson in her *Pioneer Days of Nakusp*:

> The late Frederick W. Jordan oft told the story of asking Chief Indian Louie how Nakusp got its name. Chief Louie was holding a Big [Old?] Chum tobacco bag in his hands at the time and this is the answer he gave:

'Indians come down lake in canoes, storm very bad, canoes nearly lost at Kuskanax Creek, but on entering big bay at the point (at this point Chief Louie pulled the string of the tobacco bag tight) Nequ'sp – "safe."' (p. 9)

Randy Bouchard and Dorothy Kennedy, in their typescript *Lakes Indian Ethnography and History* (1985), are inclined to agree with one of their Indian informants, who maintained that the name indicated a place where the lake closes in or narrows.

NAMU, Fitz Hugh Sound (D-6). A Heiltsuk Indian word probably meaning 'closely alongside,' referring to the nearby lake. The Indian name for the village site means 'ferrying place' or 'water transportation place.'

NANAIMO, VI (B-8). In July 1791 José Maria Narvaez, commanding the schooner *Saturnina,* explored here and named the waterways in the area Boca de Winthuysen, in honour of Francisco Xavier de Winthuysen, a Spanish naval officer. When the HBC established a settlement in 1852 to work the coal deposits, it was named Colvile Town, after Andrew Colvile, then Governor of the HBC.

The Indians in this area were known as the 'Sne-ny-mo,' meaning 'people of many names,' referring to the confederation that various villages in the area had formed for their better protection. From this Island Halkomelem word comes Nanaimo. Winthuysen Inlet disappeared as a name in the 1850s, and by 1858 Colvile Town had become simply Nanaimo.

The Indian name for Nanaimo River was 'Quamquamqua,' meaning 'strong, strong water' or 'swift, swift water,' referring to the current of the river.

NANCY GREENE LAKE, SW of Castlegar (B-11). After the Rossland girl who in 1967 became the world's champion woman skier.

NANOOSE HARBOUR, NW of Nanaimo (B-7). This is the name of a small band of Indians who are the farthest northern subgroup of the Island Halkomelem and are closely related to the Nanaimo Indians. Nanoose may mean 'a collection of families at one place.' On the other hand, there is a theory locally that it means 'pushing inward,' with reference to the shape of Nanoose Bay.

NARAMATA, N. of Penticton (B-10). John Moore Robinson, who founded Naramata in 1907, tells us that he got the name

> from the denizens of the spirit world through the mediumship of Mrs. J. M. Gillespie, one of the most prominent spiritualistic lecturers and mediums of the American Spiritualistic Church. Big Moose was a Sioux Indian Chief, and he dearly loved his wife of whom he spoke in the most endearing terms, and he gave us her name as Narramattah, and he said she was the 'Smile of Manitou.' It struck me that this would be a good name for our village which I thought of calling Brighton Beach. We therefore cut out the unnecessary letters and called the town Naramata.

Mrs. Gillespie's first husband had lived in Australia, and sceptics who do not believe in Big Moose suspect that Naramata is based on some Australian aboriginal name such as that of Paramatta, near Sydney.

NARAO PEAK, Yoho NP (D-11). From a Stoney Indian word meaning 'to spatter or fly out,' like grease from a fire. Another interpretation is 'hit in the stomach.'

NARRAWAY RIVER, Alberta-BC boundary (G-9). After A.M. Narraway, controller of surveys, Ottawa, who visited the river in 1922.

NARVAEZ BAY, Saturna I. (A-8). After José Maria Narvaez, commander of the *Saturnina,* who explored these waters in 1791.

NASPARTI INLET, E. of Brooks Pen. (C-6). Captain G.H. Richards, RN, gave this name, an adaptation of that of the local Indians, in 1862.

NASS RIVER, flows SW into Portland Inlet (H-5). The Tlingit Indians of Alaska made periodic visits here because of the abundance of eulachon and salmon. They gave to the river's mouth the name of Nass, which is interpreted as 'food depot' or 'satisfier of the belly.'

NATADESLEEN LAKE, S. of Kinaskan L. (J-4). The name is derived from a Tahltan Indian word meaning 'spread rapid.'

NATAL RIDGE, SE of Sparwood (B-12). This preserves the name of Natal, which with its sister village of Michel, has been absorbed into the District Municipality of Sparwood. Natal was apparently named at the time of the Boer War in honour of the British Crown Colony of Natal. Natal and the Cape Colony were engaged in hostilities with the Boer republics of Transvaal and Orange Free State.

NATALKUZ LAKE, S. of Cheslatta L. (F-7). From a Carrier Indian word possibly meaning 'narrow crossing between two lakes.'

NATION RIVER, flows NE into Parsnip Reach, Williston L. (H-8). Simon Fraser, recording his ascent of the Parsnip River, noted on 1 June 1806, 'We came to an encampment about 2 miles below River au Nation.' The next day he noted that the river was so named because the 'Big Men' (Sekanis) who live up it belonged to a different nation from those living at Trout (McLeod) Lake.

NAVER CREEK, flows W. into Fraser R. (F-8). A shortened form of Strathnaver (q.v.).

NAZKO RIVER, flows N. into West Road R. (F-8). This is the name that the Carrier Indians applied to the whole of the West Road River but that the whites apply only to the major tributary coming from the south. Father Morice wrote that there are two possible translations for Nazko: 'the river that flows "across" the land into the Fraser,' and 'Cross River, the river one has to cross in order to go north or south.'

NEAL, MOUNT, Garibaldi Park (C-8). After Dr. Neal M. Carter, biologist and mountaineer, d. 1978. Also CARTER GLACIER. He climbed in Europe, New Zealand, and Japan but is chiefly remembered for his ascents in British Columbia in the Rockies, the Selkirks, and, above all, the Coast Mountains, where he had a number of first ascents to his credit. Friend and associate of Don Munday. In 1922 and 1923, he made surveys of Garibaldi Park that resulted in the first useful map of the area. He is remembered for 'his quiet good humour and unfailing generosity' (*Canadian Alpine Journal* [1978]:60).

NECHAKO RIVER, joins Fraser R. at Prince George (F-8). Almost certainly from the Carrier Indian word meaning 'big river.'

NECOSLIE RIVER, flows into S. end of Stuart L. (G-7). A Carrier Indian word meaning 'arrows floating by,' from an incident in an Indian legend.

NEDS CREEK [NED'S CREEK], SW of Pritchard (C-10). After Edouard de Champ ('French Ned'), incinerated when his cabin burned down in 1873.

NEEDLES, W. side of Lower Arrow L. (B-10). Formerly a little village stood where the 'needles,' long thin sand spits, ran into the lake. Now both village and needles have disappeared in consequence of the building of the dam at the foot of the lake, and only a ferry terminal preserves the old name.

NELSON, Kootenay L. (B-11). Started as a mining camp after the staking in 1886 of the Silver King claim on Toad Mountain. Known briefly as Salisbury after the Marquess of Salisbury, Prime Minister of Britain, and as Stanley, after Lord Stanley, the Governor-General of Canada. Finally, in 1888, the town was named Nelson after Hugh Nelson (1830-93), then Lieutenant-Governor of British Columbia.

Nelson was a leading businessman back in the Crown Colony period. He was associated with Wells Fargo during the Cariboo gold rush, and from 1866 to 1882 he was vice-president and general manager of the Moodyville sawmill in present-day North Vancouver. Nelson was an ardent champion of confederation with Canada, and when British Columbia entered the dominion, he became the first MP for New Westminster, being elected by acclamation. He entered the Senate in 1879 and was appointed Lieutenant-Governor of British Columbia in 1887. He retired because of ill health in 1892 and died in England in 1893, having gone there in search of a cure.

NELSON, MOUNT, W. of Windermere L. (C-11). After Horatio, Viscount Nelson, the victor of Trafalgar. In his narrative for March 1809, David Thompson mentions 'the rude pyramid of Mount Nelson (for so I named it).' Also NELSON ISLAND, Jervis Inlet.

NELSON KENNY LAKE, W. of Strathnaver (F-8). After Nelson Clarke Kenny, BCLS, killed in World War I.

NEMAIA VALLEY, E. of Chilko L. (D-8). This fine valley is named after a Chilcotin Indian, possibly a chief who left Redstone village and, with a small band, settled in this valley.

NE-PARLE-PAS POINT, Peace Reach, Williston L. (I-8). Preserves the memory of the Ne-Parle-Pas Rapids, obliterated when the waters of the Peace River rose behind the W.A.C. Bennett Dam. *Ne-parle-pas* means 'does not speak,' and these rapids received their name because they made remarkably little noise. Voyageurs coming down the Peace River, not hearing the rapids, could be into them with practically no warning.

NEROUTSOS INLET, Quatsino Sound (C-6). Named about 1926 after Captain C.D. Neroutsos, assistant manager (later manager) of the CPR's British Columbia Coast Steamship Service.

NESAKWATCH CREEK, flows NW into Chilliwack R. (B-9). From the Halkomelem word for a reed that once grew abundantly along the banks of this stream and was used for making fishnets.

NESSELRODE, MOUNT, Alaska-BC boundary (K-2). After Count Karl Robert Nesselrode (1780-1862). In 1816 he became Russia's Foreign Minister and in 1844 imperial chancellor. For forty years he directed Russia's foreign policy.

NETALZUL MOUNTAIN, headwaters of Suskwa R. (H-6). G.M. Dawson described this as 'a great rugged mass of mountain' and reported that its Carrier Indian name meant 'watery mountain,' and indeed many streams do flow down from it.

NEW CALEDONIA. The department of the HBC extending from the Coast Range to the Rockies and north from Alexandria to approximately the fifty-seventh degree of latitude was usually termed New Caledonia.

According to family tradition, Simon Fraser gave the name because, after he had crossed the Rockies, he found that the country reminded him of his mother's descriptions of the Scottish Highlands. Earliest recorded use of the name comes in 1808.

For a while it seemed that New Caledonia might become the name of British Columbia, but this idea was discarded since the name New Caledonia had become attached to some French islands in the South Pacific. The name does not appear in the modern gazetteer of British Columbia. It survives, however, in clipped form in the Anglican Diocese of Caledonia, whose bishop has his cathedral in Prince Rupert.

NEW DENVER, Slocan L. (B-11). When the surveyor, Perry, needed a name for the settlement in 1891, he chose El Dorado, presumably hoping that the mines would make it as rich as the legendary city of gold. The next year, when it had become clear that not gold but silver and lead constituted the mineral wealth, a public meeting was called to decide on a more suitable name. At this meeting Thomas Latheen, who had come from Denver, Colorado, stoutly maintained that this little town would grow to be even greater than Denver, and he persuaded the citizens to adopt the name of New Denver for their settlement.

NEW WESTMINSTER, SE of Vancouver (B-8). Governor Douglas had intended that Derby (site of the original Fort Langley) should be the capital of the mainland Crown Colony of British Columbia. Colonel Moody of the Royal Engineers inspected this site in December 1858 and declared that military considerations ruled it out. (Being on the south side of the Fraser, Derby was vulnerable to attack in the event of an American invasion.) Moody had no doubt about where the new capital should be:

> In steaming up one fine reach at a spot 20 miles f^r the entrance of the Channel to the Frazer, my attention was at once arrested to it's [sic] fitness, in all probability, for the site of the first, if not the Chief Town in the Country. Further study of that ground as well as other sites has now convinced me that it is the right place in all respects. Commercially for the good of the whole community, politically for imperial interests & military for the protection of & to hold the country against our neighbours at some future day, also for all purposes of convenience to the local Government in connection with Vancouver's Island at the same time as with the back country. It is a most important spot. It is positively marvellous how singularly it is formed for the site of a large town (not a small one) to be defended against any foreign aggression. It is not adapted for a small military position, such as w^d be required for a mere military or naval post or depot. The features are too extensive for that ... Viewed f^r the Gulf of Georgia across the meadows on entering the Frazer, the far distant giant mountains forming a dark background – the City w^d appear throned Queen-like & shining in the glory of the midday sun. The comparison is so obvious that afterwards all hands on board the Plumper & indeed everyone joins in thinking the appropriate name w^d be 'Queenborough.'

The colonel's rhetoric did not impress everybody. Governor Douglas, for some quirky reason, would accept 'Queensborough' but not 'Queenborough.' The Colonial Secretary, W.A.G. Young, expressing the feeling of the citizens of Victoria, maintained that the latter, named for the sovereign, was already the Queen's borough. To end the discord, Governor Douglas wrote to Sir Edward Bulwer-Lytton:

> It will be received and esteemed as an especial mark of royal favour were her Majesty to name the capital of British Columbia either indirectly after her Royal Self, or directly after His Royal Highness, the Prince Consort, His Royal

Highness, the Prince of Wales, or some member of the Royal Family, so that the colonists of British Columbia, separated from friends and kindred in this their far distant home, may be ever gratefully reminded in the designation of their capital of the power that protects their hearths, of the watchful interest that guards their liberties, and of the gentle sway by which they are governed.

Back came a crisp reply: 'I am commanded to acquaint you that Her Majesty has been graciously pleased to decide that the capital of British Columbia shall be called "New Westminster."'

Alas, New Westminster's days as a capital city were to be few, for a year and a half after the union in 1866 of the colonies of British Columbia and Vancouver Island, the capital of the united colony was transferred to Victoria.

Queen Victoria's personal choice of the name of New Westminster accounts for the unofficial title of 'The Royal City,' which has been used so widely and proudly. As for Queensborough, many years later it became the name of a suburb.

The Halkomelem Indians gave the white men's town the name of 'Skwiy-ee-mihth' ('where many people died' in a fire c. 1860).

NEWCASTLE ISLAND, Nanaimo (B-8). In 1853, very aware of the importance of the coal discovered in this area, the HBC officers surveying Nanaimo harbour named this island after Newcastle-upon-Tyne, proverbial for its exports of coal.

NEWCOMBE INLET, Tasu Sound, Moresby I. (E-3). After Dr. Charles F. Newcombe (1851-1924), a Victoria physician who made various trips to the Queen Charlottes to study the anthropological, botanical, geological, and ethnographical aspects of the islands. He was a very active collector of Indian artifacts, acting as an agent for various museums.

NEWTON, SE of New Westminster (B-8). After Elias John Newton, saddler and harness-maker of New Westminster, who took up land here. His real name was Villeneuve, but surrounded by anglophone neighbours he changed it to Newton (which, like Villeneuve, means 'new town').

NEWTON, MOUNT, W. side of Saanich Pen. (A-8). After W.E. Newton, a farmer who arrived in Victoria in 1851 and married Emmeline, daughter of John Tod (see *Tod Inlet*).

NEY, MOUNT, N. of Tahtsa L. (F-6). After Charles S. (Charlie) Ney, an outstanding field geologist who worked in the area between 1964 and 1971. President of the Vancouver Natural History Society at the time of his death in 1975, he will long be remembered by its members. A cherubic smile would light up his features when he led its geological hikes, his long loose prospector's stride looking deceptively leisurely.

NICKEL PLATE MOUNTAIN, NE of Hedley (B-9). After the Nickel Plate claim, staked by C.H. Arundel and F. Wollaston in 1898, which became one of British Columbia's major gold mines.

NICOAMEN RIVER, flows N. into Thompson R. E. of Lytton (C-9). Some say that Nicoamen is from the Thompson Indian word meaning 'wolf' and that the river was so named because the water comes from a lake named 'wolf lake or water' or 'wolf's den.' On the other hand, an Indian informant told Dr. L.C. Thompson, a linguist from the University of Hawaii, that Nicoamen is derived from a word meaning 'means of carving out a valley' (probably referring to the stream).

It was just downstream from the confluence of the Nicoamen and Thompson Rivers that gold was discovered in 1856, an event that resulted in the Fraser River gold rush of 1858.

NICOLA LAKE, NE of Merritt (C-9). After a famous Thompson Indian head chief, Hwistesmexe'quen ('Walking Grizzly Bear'), 1785?-1865, who was given the name of Nicolas by early fur traders. The Indians pronounced this as 'Nkwala' (see *Nkwala, Mount*).

The fur traders recognized Nicolas as the most powerful and influential chief in the southern Interior of British Columbia. The Anderson map of 1849 shows both Lac de Nicholas and R. Nicholas. John Tod, the old HBC trader at Kamloops, drily notes in his memoirs that Nicholas or Nicola was 'a very great chieftain and a bold man, for he had 17 wives.'

NICOLUM CREEK, flows NW into Coquihalla R. (B-9). From the Thompson Indian word meaning 'thirsty creek.'

NICOMEKL RIVER, NE of Semiahmoo Bay (B-8). Carries the name of an Indian band wiped out by smallpox late in the eighteenth century.

NICOMEN ISLAND, lower Fraser R. (B-8). This name (pronounced 'Nick-cóhm-men') is derived from a Halkomelem word meaning 'level part' or, possibly, 'part [people] travel to.'

NIEUMIAMUS CREEK, flows S. into N. Bentinck Arm (E-6). The name comes from a Bella Coola Indian word meaning 'place of blackflies.'

NIGEI ISLAND, off N. coast of Vancouver I. (C-6). Until 1900 this island was known as Galiano Island, and then the name was changed to avoid confusion with Galiano in the Gulf Islands. Captain Walbran notes that Nigei is the hereditary name of the principal chief of the Nahwitti band. Nigei could also come from the Kwakwala Indian word for 'mountain.' The name is pronounced 'nee-gee,' with a hard 'g.'

NIMPKISH RIVER, SE of Port McNeill (C-6). From the Kwakwala Indian name for a mythical monster that looked like a halibut, was of immense size, and had been known to draw canoes under water. This monster was also believed to cause the powerful tide rip off the mouth of the Nimpkish River.

NINGUNSAW RIVER, flows NW into Iskut R. (I-4). From a Tahltan Indian word meaning 'rock under ground.'

NINSTINTS, Anthony I., QCI (E-4). This was the name of an extremely important Haida chief of the now-abandoned village on Anthony Island. Ninstints can be translated as 'one who is two.' Chief Ninstints made a fortune outfitting the last sea otter hunts in this area.

NIP AND TUCK PEAKS, SE of Snowcap L., Garibaldi Park (B-8). So named since they appear to be of much the same height.

NISKONLITH LAKE, SW of Chase (C-10). Named after a well-known Shuswap chief in the mid-1800s.

NITINAT LAKE, SW of Cowichan L. (A-7). Takes its name from the Nitinaht Indians, whom early explorers described as a fierce, warlike people. The Nitinaht people originally lived at Jordan River, and their name is derived from the Indian name for that area, meaning unknown. Linguist John Thomas says that the Nitinaht people do not have a special name for Nitinaht Lake and simply refer to it by a word meaning a 'large body of water.'

NIUT RANGE, W. of Tatlayoko L. (D-7). According to a Chilcotin Indian legend, Niut Mountain was the wife of Mount Tsoloss (which unfortunately has been renamed Mount Tatlow after a BC politician).

NKWALA, MOUNT, W. of Penticton (B-10). Formerly Niggertoe Mountain but renamed Mount Nkwala, Nkwala being the Interior Salish pronunciation of Nicolas. (See *Nicola Lake*.)

NOEL CREEK, flows N. into Cadwallader Cr., S. of Carpenter L. (C-8). After Arthur F. Noel, prospector and gold miner, and his wife, Delina. He developed the Golden Cache mine and the Little Joe, Pioneer, and Lorne properties before selling out to Bralorne. Delina, educated at a convent school in Quebec, came home to Lillooet and, at the age of nineteen, married Noel. She immediately took to going into the field with him. Leading the rugged life of a prospector, she had six miscarriages and prided herself that she had never carried a baby full term. In her house hung the skins of three grizzly bears. Built into her fireplace was a piece of ore from every producing mine in British Columbia.

NOOMST CREEK, flows NE into Bella Coola R. (E-6). The name of a Bella Coola village, which can be translated as 'place of crushed objects [berries].'

NOON BREAKFAST POINT, SE of Point Grey (B-8). Captain Vancouver's lieutenant, Peter Puget, uses this name in his journal. It was adopted officially in 1981.

NOOSKULLA CREEK, flows N. into Dean R. (E-6). From a Bella Coola Indian word meaning 'place of red huckleberries.'

NOOTKA SOUND, W. coast, VI (B-6). Nootka Sound was discovered in 1774 by Juan Perez, who was prevented by poor weather from landing but named the harbour here San Lorenzo. He was followed by Captain Cook in 1778, who bestowed the name King George's Sound. Cook erroneously reported that Nootka was the Indian name for the inlet.

The most likely explanation of the word *nootka* is that supplied by the missionary priest Father Brabant, who lived for many years among these Indians. His suggestion is that what the white men heard was *noot-ka-eh*, the imperative of the Indian verb meaning 'go around!' According to Dr. T. Hess of the University of Victoria, the Indians were probably saying *nu-tka-pičim*, meaning 'go around the harbour.' The Spanish subsequently changed their name to San Lorenzo de Nutka and later to Santa Cruz de Nutka, but the name was soon reduced to plain Nootka.

NORALEE, François L. (F-6). When Lee Newgaard and his wife, Nora, opened a post office here in 1937, they combined their first names to secure a name for it.

NORBURY LAKE, E. of Cranbrook (B-12). After the Hon. F. Paget Norbury, an Englishman who ranched in the district in the 1890s.

NORMAN LAKE, W. of Prince George (F-8). This and nearby Dahl Lake are named after two men who came to the area around 1908 over the old Telegraph Trail from Ashcroft. Besides homesteading they trapped to eke out a living. An old-timer remembered them and the brides they brought with them into the wilderness – 'the women were frightened and clung to each other.'

NORNS RANGE, W. of Slocan R. (B-11). The Norns were the Fates in Scandinavian mythology.

NORRISH CREEK, flows S. into Nicomen Slough (B-8). After William Henry Norrish, pioneer Scottish-Canadian farmer here.

NORTH BEND, Fraser Canyon (B-9). Originally known as Yankee Flats or Yankee Town. Much of the CPR mainline across British Columbia was built from west to east. Having travelled eastward through the Fraser Valley, the line turned at Hope and travelled north through the Fraser Canyon before resuming its eastward route. Because of this major deflection, the CPR's stretch north of Hope was referred to as the railway's 'north bend,' and when Yankee Flats became the divisional point for this part of the line, North Bend became its name.

NORTHUMBERLAND CHANNEL, E. of Nanaimo (B-8). After Algernon Percy, fourth Duke of Northumberland (1792-1865). Named by HBC officers in 1852 when the Duke was First Lord of the Admiralty.

NORTH VANCOUVER, Burrard Inlet (B-8). The history of this suburb of Vancouver begins about 1860 when Philip Hicks started a sawmill here. Hicks got so deeply in debt to Sewell Prescott ('Sue') Moody, who supplied him with logs, that he had to let Moody have his mill. A small settlement named Moodyville grew up around Moody's mill. Moody perished in the wreck of the steamer *Pacific* off Cape Flattery in 1875. Some days after the ship was lost, a piece of wood washed up near Victoria bearing the inscription 'All lost, S.P. Moody.' After Moody's death the mill was managed by Hugh Nelson, one of Moody's partners (see *Nelson*). In 1902 the name of Moodyville was changed to North Vancouver. The Moody mill stood a little east of the foot of Lonsdale Avenue. (See also *Lonsdale*.)

NOSEBAG CREEK, flows NE into Carpenter L. (C-8). In his book *Beyond the Rockies* (p. 152), Lukin Johnston mentions meeting 'a delightful old gentleman named Jones – "Jonesey" for short, or sometimes "Nosebag Jonesey". From him Nosebag Creek takes its name.' Presumably Nosebag Jones was fond of his food. (A nosebag, partly filled with oats, would be slipped over the head of a horse to let it eat during a halt.)

NOTCH HILL, S. of Shuswap L. (C-10). So named by G.M. Dawson because of the gap in the hills here through which the CPR mainline runs.

NOWITKA MOUNTAIN, NW of Kimberley (B-11). After the *Nowitka*, one of the steamers that ran on the upper Columbia in the early days. Owned by the Columbia River Lumber Company, the steamer has been described as 'a makeshift thing.'

NUCHATLITZ INLET, W. side of Nootka I. (B-6). From the Nootka Indian word meaning 'people of place having mountain behind village.'

NULKI HILLS, S. of Vanderhoof (F-7). From the Carrier Indian word meaning 'big grey hills.'

NUMA MOUNTAIN, Kootenay NP (D-11). This is the Kootenay Indian word for both 'thunder' and 'lightning.'

NUMUKAMIS BAY, E. side of Trevor Channel, Barkley Sound (A-7). From the Nootka Indian word meaning 'something private, personally owned.'

NUSATSUM RIVER, flows NW into Bella Coola R. (E-6). From the Bella Coola Indian word meaning 'place of old or large spring salmon.'

NUTTLUDE LAKE, W. of Kinaskan L. (J-4). From a Tahltan Indian word meaning 'dome mountain.'

O

OAK BAY, E. of Victoria (A-8). Before it became a suburb of Victoria, it was, in Sir Charles Piers's phrase, 'a veritable bay of oaks.' The name appears on Captain Kellett's chart of 1847.

OATES, MOUNT, Hamber Park (E-10). After 'that very gallant gentleman' Captain L.E.G. Oates, who perished in 1912 during the Scott expedition to the South Pole. (Crippled with frostbite, he deliberately went to his death in an Antarctic storm rather than be an encumbrance as his comrades attempted, vainly, to get back to their base.)

OATSOALIS CREEK, E. side of Calvert I. (D-6). This name (the Indian name for Safety Cove, into which the creek empties) is derived from a Heiltsuk word meaning 'fairly deep bay.'

O'BEIRNE, MOUNT, Mt. Robson Park (E-10). After the eccentric Eugene Francis O'Beirne. (See *Milton, Mount.*)

OBSERVATORY HILL, Saanich (A-8). Formerly Little Saanich Mountain but renamed in 1917 following the building of the Dominion Astrophysical Observatory here. The Saanich Indians, impressed by the appearance of the dome when it was opened, applied to it their word for 'grimace' or 'snarl.'

OBSERVATORY INLET, N. coast (H-4 and 5). So named by Captain Vancouver in 1793 since he set up an observatory here to check the rate of his chronometers and correct his fixing of latitude and longitude.

OCEAN FALLS, Central Coast (E-6). From the impressive waterfall here, where Link Lake empties into Cousins Inlet.

OCEAN PARK, Boundary Bay (B-8). This name was chosen about 1910 by the Reverend W.P. Goard for his Ocean Park Syndicate, made up of prominent Methodists in the Lower Mainland who wished to secure land for their church's future recreational and educational purposes. A church camp was founded here.

ODARAY MOUNTAIN, Yoho NP (D-11). Derived from two Stoney Indian words that together mean 'many waterfalls.'

O'DELL, N. of Prince George (G-8). This BCR junction bears the name of a trapper active in the Summit Lake area in the 1920s.

ODIN, MOUNT, W. of Upper Arrow L. (C-10). Named by G.M. Dawson after the chief of the Scandinavian gods. Similarly Dawson named a neighbouring mountain after Thor, the god of thunder.

O'DONNEL RIVER, flows W. into Atlin L. (L-3). After a Major O'Donnel, an Irishman who staked a claim here in 1898.

OEANDA RIVER, NE Graham I., flows into Hecate Strait (F-4). Derived from the Haida name meaning 'raven river.'

OESA, LAKE, Yoho NP (D-11). Stoney Indian word meaning 'corner.' Lake Oesa lies in a corner where Abbot Pass meets the valley leading down to Lake O'Hara.

OGDEN, MOUNT, Yoho NP (D-11). After I.G. Ogden, first financial vice-president of the CPR.

OGDEN POINT, Victoria (A-8). Named in the first days of Fort Victoria after Peter Skene Ogden (1794-1854), one of the most important figures during HBC days early in the nineteenth century. His trapping expeditions for the company into the Snake River country between 1824 and 1830, intended so to deplete the area as to make it unattractive to American trappers, demanded incredible stamina and courage. These expeditions covered most of the American northwest and took him as far as California and Utah. (Ogden, Utah, is also named after him.) In 1835, now one of the company's Chief Factors, he took charge of the New Caledonia Department, with his headquarters at Fort St. James. After the resignation of Dr. John McLoughlin in 1846, Ogden and Douglas jointly administered the HBC's Columbia Department. In 1847, in what had become American territory, he rescued the survivors of the Whitman Massacre.

Always a heavily built man, Ogden became extremely fat in his later years. Throughout his life he seems to have been notable for his high spirits. Back in 1817, when Ogden was a young Nor'Wester, Ross Cox wrote appreciatively of 'the humorous, honest, eccentric, law-defying Peter Ogden, the terror of the Indians, and the delight of all gay fellows.' In 1841 Lieutenant Charles Wilkes, USN, reported of the Chief Factor, 'Mr. Ogden is a general favourite; and there is so much hilarity, and such a fund of amusement about him that one is extremely fortunate to fall into his company.'

Descendants of Peter Skene Ogden are living to this day in the Lac la Hache area.

OGILVIE PEAK, NE of Hope (B-9). After John Drummond Buchanan Ogilvie, in charge of Fort Hope late in the 1850s. He subsequently entered the colonial service and was the customs officer at Bella Coola when he was murdered in April 1865. Colonel Moody of the Royal Engineers thought very highly of Ogilvie: 'A splendid fellow ... He stands erect in a superb attitude, his voice is measured and sonorous and his words are most telling ... His reputation for judgment and discretion combined with a steady firmness has gained him a position of trust as the HB agent at Fort Hope. I hope to get him made a J.P.'

The Halkomelem Indian name for Ogilvie Peak means 'many breasts.'

O'HARA, LAKE, Yoho NP (D-11). After Lieutenant-Colonel Robert O'Hara of the Royal Artillery. Although he did not discover this beautiful lake, he was probably the first tourist to visit it. He was a prickly character.

O.K. RANGE, S. of Kleanza Cr. (G-5). When the northern spur is partly covered with snow, the letters OK stand out clearly.

OKANAGAN LAKE, S. Interior (B- and C-10). At least forty-seven different spellings of this name have been found, beginning with Lewis and Clark's 'Otchenaukane' in 1805 and David Thompson's 'Ookanawgan' in 1811. Sometimes there seem to be just about as many theories as to the derivation of the name, but the majority agree that the compound word *okanagan* contains the word for 'head.' One likely translation is 'looking toward the upper end [head?]'; another is 'seeing the top or head,' possibly referring to the summit of Mt. Chopaka. The anthropologist Teit wrote, 'Okanagan is said to be derived from the name of a place on Okanagan River, somewhere near the Falls, so named because it was the "head" of the river, at least in so far as the ascent of salmon was concerned.'

The settlement of Okanagan Falls takes its name from the pretty twin falls that were here before the modern flood-control system eliminated them. The Okanagan Indians had extensive salmon-drying racks here, and the place was usually thought of as their headquarters. The white settlement at the falls used to be known as Dogtown (see *Skaha Lake*).

OKANAGAN MISSION, south of Kelowna (B-10). Was founded by Father Charles Pandosy, OMI, in 1859. It was the first permanent white settlement in the valley. Once one enters the United States, Okanagan becomes 'Okanogan.'

O'KEEFE, N. of Okanagan L. (C-10). Named after 'The O'Keefe of Okanagan,' Cornelius O'Keefe, who arrived at the head of Okanagan Lake around 1867 and began acquiring land for the ranch that made him wealthy. A Roman Catholic from Quebec, of Irish descent, he was a genial, kindly man.

OKISOLLO CHANNEL, N. of Quadra I. (C-7). From the Kwakwala Indian word meaning 'little passage or channel.'

OLALLA, N. of Keremeos (B-10). From a Chinook jargon word, *olallie*, meaning 'berries.' The name was chosen because of the abundance of Saskatoon (service) berries here.

OLDFIELD, MOUNT, Prince Rupert (G-4). After Captain R.B. Oldfield of HMS *Malacca,* on this coast in the 1860s.

OLD FORT, Babine L. (H-6). This is the site of Fort Kilmaurs, established in 1822, the original HBC post on the lake.

OLD FRIEND MOUNTAIN, E. of Mackenzie (H-8). This outstanding mountain is a landmark for those travelling through Pine Pass.

OLIVER, S. Okanagan (B-10). After 'Honest John' Oliver (1856-1927), Premier of British Columbia from 1918 to 1927. Under his Liberal administration, the province carried out an irrigation project here as part of its plan for soldier settlement at the end of World War I.

Oliver came from the best sort of English working-class background. He arrived in British Columbia after some years in Ontario and by diking and draining made his farm one of the more prosperous ones in the delta of the Fraser River.

Various stories are told of 'Honest John's' thrift while premier. When in Vancouver he would pass by the resplendent Hotel Vancouver and stay in a cheaper place on a side street. His secretary, Morton, remembers how he took him into the White Lunch to eat. Catching a train to an Ottawa conference, he would book a lower berth, not the private compartment or drawing room that his ministers regarded as their prerogative.

Once, when he got a letter claiming damages because the Public Works Department had let a drain get plugged and the fields had flooded, Premier Oliver took the ferry to the mainland, drove out to Dewdney, and took a look at the plugged drain. What the farmer thought when he found the premier looking around in person, nobody knows.

Premier Oliver was a quick man with an aphorism. Here are two: 'Think before you work. Don't work first and think afterwards.' 'The man on top of the stack has the widest view, but he gets all the wind and the flying ants.'

OLIVER LAKE, south of Prince Rupert, is also named after Premier John Oliver.

OMINECA RIVER, flows NE into Williston L. (H-7). Father Morice reported that this name is derived from a Sekani Indian word meaning 'lake-like or sluggish river.'

100 MILE HOUSE, N. of Clinton (D-9). Gets its name from the fact that the mile-house built here when the Cariboo Wagon Road was pushed through in the early 1860s was just 100 miles from Lillooet.

In 1912 the Marquess of Exeter purchased an extensive ranch here. It was later managed by his son, Lord Martin Cecil, who died in 1988, aged seventy-eight.

OONA RIVER, Porcher I. (F-4). From the Coast Tsimshian word for 'skunk cabbage,' a plant that flourishes here.

OOTSA LAKE, N. of Tweedsmuir Park (F-6). From a Carrier Indian word pronounced 'yoot-soo,' meaning 'very low down' or 'way down toward the water.' Father Morice named this lake after H.J. Cambie, who travelled this way in 1876 (see *Cambie*), but the name was not accepted in Victoria.

OPABIN PASS, near L. O'Hara (D-11). Traditionally this Stoney Indian word has been translated as 'rocky,' but 'impurities' is preferred by a linguist specializing in the Stoney language.

OPATCHO LAKE, SE of Prince George (F-8). After an Indian sleigh dog renowned for his strength. He could pull by himself a sleigh that would normally require a team.

OPENIT PENINSULA, SW of Sydney Inlet (B-6). The name of this peninsula, by Hotsprings Cove, comes from the Nootka Indian word meaning 'place of calm waters.'

OPITSAT, SW end of Meares I. (B-7). Possibly from the Nootka Indian word meaning 'place having an island in front.'

OREGON JACK CREEK, SW of Ashcroft (C-9). After John ('Oregon Jack') Dowling, a rancher who maintained a roadhouse here, popular with teamsters on the old Cariboo Road. Mary Balf reports that it was 'somewhat unruly.' The *British Columbian* of 5 August 1863 reports the arrest in Lytton of the 'notorious' Oregon Jack on a charge of burglary after the government had posted a reward of $500 for him dead or alive.

ORLEBAR POINT, Gabriola I. (B-8). After Lieutenant V.B. Orlebar, RN, who commanded the gunboat *Rocket* on this coast from 1875 to 1882.

ORMOND LAKE, N. of Fraser L. (G-7). After Frank Ormond Morris, Surveyor-General of British Columbia 1950-1.

OSBORN BAY, E. of Crofton (A-8). After Captain (later Rear-Admiral) Sherard Osborn, RN (1822-75), Arctic explorer and author, who served with distinction in the Crimean War.

OSOYOOS, S. Okanagan (B-10). Comes from the Okanagan Indian word meaning 'sand bar across,' referring to the long strip of land that almost cuts Osoyoos Lake in two. (Cf. *Tsuius Creek*).

Nobody knows for sure how the initial *o* was added, though there is a story that Peter O'Reilly, magistrate at Hope in 1858, jocularly suggested that adding an *o* would give dignity to the Indian name. Osoyoos appears on the maps as early as 1860. The village of Osoyoos was incorporated in 1946. Old-timers still call the place 'Soo-yoos.' An early name for Osoyoos Lake, Forks Lake, came from its proximity to the confluence of the Okanagan and Similkameen Rivers.

OTTER CHANNEL, N. of Campania I. (F-5). After the *Otter,* the second steamboat operated on this coast by the HBC (the *Beaver* was the first). She arrived at Fort Victoria in June 1853 and was burned for her metal in 1890.

OTTERTAIL RIVER, Yoho NP (D-11). Translation of the Indian name for the river.

OUTRAM, MOUNT, SE of Hope (B-9). Named by the veteran HBC officer A.C. Anderson after his uncle Sir James Outram, one of the relievers of Lucknow when it was besieged during the Indian Mutiny.

OWIKENO LAKE, at the head of Rivers Inlet (D-6). Takes its name from the Indian band who live nearby. Several meanings have been advanced for this word. An early one was 'portage makers' or 'those who carry on the back,' referring to the Indians making the portage between Rivers Inlet and Owikeno Lake. A more recent translation is 'right-minded people' or 'people talking right.'

OYAMA, S. of Vernon (C-10). Not an Indian name but that of Prince Iwao Oyama (1842-1916), Japanese field marshal, captor of Port Arthur in the First Sino-Japanese War and commander in Manchuria in the Russo-Japanese War. The post office at Oyama was opened in 1906.

PACHENA POINT, S. of Barkley Sound (A-7). This word is derived from the Nitinaht Indian name for the site of Port Renfrew, but by mistake the anglicized name Pachena was applied to a point farther up the coast that had a nearly identical configuration. Pachena in its original form meant either 'sea foam' or 'foam on the rocks.'

PACIFIC, NE of Terrace (G-5). The Pacific division of the GTPR (now the CNR) began here.

PACIFIC LAKE, NE of Prince George (G-9). So named because its waters run to the Pacific, whereas nearby Arctic Lake drains into the Arctic watershed.

PACIFIC SPIRIT PARK, Point Grey, Vancouver (B-8). In 1989 the provincial government transferred title of 763 hectares of undeveloped land near UBC to the Greater Vancouver Regional District for a park. A competition to name the new park was held, the winner being Sherry Sakamoto with 'Pacific Spirit Park,' signifying 'Gateway to the Pacific and spiritual ground to becoming one with nature.'

A minor trail in the park is named for Iva Mann, the heroine who spearheaded the long, hard fight against real estate interests and others who had grandiose development plans. Thanks in part to her the public now has this park almost twice the size of Stanley Park, with thirty-five kilometres of trails and fine beaches.

PACKER TOM CREEK, flows W. into Dease R. (K-5). Packer Tom was an Indian gifted with a remarkable memory. It has been claimed that he could remember what he had done each day over a number of past years.

PACOFI BAY, Selwyn Inlet, Moresby I. (E-4). A 'manufactured' name derived from the Pacific Fish Company, which operated here in the early 1910s.

PAGET PEAK, Yoho NP (D-11). After the Very Reverend Dean Paget of Calgary, who made the first recorded ascent.

PALDI, E. of Cowichan L. (A-8). After a village in Punjab, India, birthplace of the Mayo brothers, who built a sawmill here.

PALLISER RIVER, flows SW into Kootenay R. (C-12). After Captain John Palliser (1817-87), who, under the direction of the British government, spent the period 1857-9 exploring the country between the forty-ninth parallel and the North Saskatchewan River and between the Red River and the Rockies. He was also charged with seeking for passes through the Rockies, and this extension of his duties took him into British Columbia.

PALMER, MOUNT, N. of Puntzi L. (E-7). After Lieutenant Henry Spencer Palmer of the Royal Engineers, who passed this way in 1862 with a couple of sappers while surveying the route between Bella Coola and Fort Alexandria.

PALMERSTON, CAPE, S. of Cape Scott (C-5). After Viscount Palmerston (1784-1865), Prime Minister of Great Britain.

PANDAREUS, MOUNT, NW of Squamish (B-8). This peak in the Tantalus Range is named after Pandareus, whose golden mastiff, according to Greek myth, was stolen by Tantalus.

PANTHEON RANGE, N. of Mt. Waddington (D-7). The mountaineering party that undertook the exploration of the area in 1964 decided 'to name the peaks after deities from various mythologies.' The catholicity of their nomenclature is evidenced by names such as Mount Astarte, Mount Juno, Manitou Peak, Osiris Peak, and Mount Vishnu.

PARKSVILLE, NW of Nanaimo (B-7). After Nelson Parks, first settler in the district and first postmaster.

PARRY BAY, SW of Victoria (A-8). After Rear-Admiral Sir William Edward Parry (1790-1855), Arctic explorer and Hydrographer of the Navy. Also PARRY PASSAGE, Queen Charlotte Islands.

PARSNIP RIVER, flows NW into Williston L. (H-8). R.M. Patterson mentions the 'almost tropical growth of the giant cow parsnip from which the river gets its name.' He found this plant growing up to seven feet and says that 'the din of the rain on the huge leaves was like the rush of a tremendous wind' (*Finlay's River,* p. 38). The plant is sometimes called Indian Rhubarb since the Indians eat the petioles or leaf stalks.

PARSON, SE of Golden (D-11). After Henry George Parson, Golden merchant, MLA, 1909.

PARTON RIVER, E. of St. Elias Mtns. (L-1). After G.F. Parton, head packer with the BC-Yukon boundary survey party of 1908.

PASLEY ISLAND, mouth of Howe Sound (B-8). After R.G.S. Pasley, flag lieutenant to Rear-Admiral John Kingcome, Commander-in-Chief of the Royal Navy's Pacific Station 1862-4.

PASSAGE ISLAND, mouth of Howe Sound (B-8). Named by Captain Vancouver because of its position midway in the passage (Queen Charlotte Channel) between Point Atkinson and Bowen Island.

PATERSON LAKE, W. of Campbell R. (C-7). After T.W. Paterson, Lieutenant-Governor of British Columbia from 1909 to 1914.

PATRICIA BAY, Saanich Inlet (A-8). In September 1912, when the Duke of Connaught, then Governor-General of Canada, visited Victoria, he was accompanied by his daughter, the attractive young Princess Patricia or 'Princess Pat.' Shortly thereafter nearby Union Bay was renamed Patricia Bay in her honour. Today it is commonly called Pat Bay.

PATTULLO, MOUNT, NW of Meziadin L. (I-5). After Thomas Dufferin (Duff) Pattullo, successively alderman of Dawson City, YT, and Prince Rupert, mayor of Prince Rupert, provincial Minister of Lands, and finally Premier of British Columbia 1933-41. The Pattullo Bridge across the Fraser River at New Westminster is also named after him.

Pattullo is a Scottish, not an Italian, name. Duff Pattullo was a dapper man in his attire and addicted to rather highly coloured rhetoric. His associates found him hard to know, but he surprised them at times with little acts of kindness.

PAUL LAKE, NE of Kamloops (C-9). Also PAUL PEAK, Kamloops. After the Indian chief Jean Baptiste Lolo, commonly known as 'St. Paul.' The latter name was frequently shortened to Paul. (For details see *Lolo, Mount.*)

PAVILION, N. of Lillooet (C-9). From the French *pavillion*, for a flag or a tent. In his journal for 30 September 1826, Archibald McDonald mentions 'An Indian from the Pavillon.' Commander Mayne tells us that, in accordance with the Indians' custom, a large white flag once waved here over the grave of one of their chiefs.

The Shuswap Indian name for Pavilion Lake meant 'where someone broke wind over the water.' (For the legend behind this name, see James Teit, *The Shuswap*, pp. 752-3.)

PAY BAY, W. coast of Hibben I. (E-3). This name appears on the 1852 chart made by Captain Kuper, RN. He named it after two paymasters, Luxmore and Rogers.

PEACE RIVER, NE BC (I-9). The Sekani Indians of north-central British Columbia knew this as Tse-tai-e-ka, meaning 'the river that runs by the rocks,' referring to its canyon through the Rocky Mountains. The Crees called it A-mis-kwe-i-moo-si-pi, meaning 'the Beaver Indian River.' As for the Beaver Indians, they came to call it Unjigah, Unchaga, or Unjaja, variant forms of their word for 'peace.' They called this 'the Peace River' because of Peace Point near Lake Athabasca. Some time before 1790, the Knistenaux (Cree) and Beaver Indians ended a territorial war by setting their boundary at this point in the river.

PEACHLAND, W. side of Okanagan L. (B-10). Formerly known as Camp Hewitt after Gus Hewitt, who had a mining camp here around 1890. In 1897 John Moore Robinson, who later founded Summerland and Naramata, laid out a townsite here. He probably chose the name 'Peachland' because the Lambly family had earlier grown the first peaches in the Okanagan on the flat land here.

PEARDONVILLE, SW of Abbotsford (B-8). Named in 1894 after Richard Peardon, a local farmer who became the first postmaster.

PEARKES, MOUNT, Jervis Inlet (C-8). After Major-General George Randolph Pearkes, VC, PC (Canada), CC, CB, DSO, MC, Croix de Guerre (1888-1984). Born in England, he came to Canada in 1906, subsequently joining the RNWMP. In World War I, he was commissioned in the field and rose to the rank of lieutenant-colonel. Between the wars he served in Canada's permanent army. In Britain in World War II, he commanded the 1st Canadian Division from 1940 to 1942. In the latter year, Japan having entered the war, he was flown to British Columbia to take over the Pacific Command, remaining GOCPacific until his retirement in 1945. First elected to the House of Commons in 1945, he was Minister of National Defence from 1957 to 1960.

From 1960 to 1968, he was Lieutenant-Governor of British Columbia, winning in abundance the respect and affection of the province. Perhaps of all British Columbians General Pearkes best merits the famous phrase *Chevalier sans peur et sans reproche.*

PEARSE ISLAND, Portland Inlet (G-4). After Captain Pearse of the US Army, commanding the first American military post in Alaska, on Tongass Island, in 1868. This island was claimed by the United States but awarded to Canada in 1903.

PEARSE ISLANDS, E. of Port McNeill (C-6). After Commander W.A.R. Pearse, captain of HMS *Alert* (see *Alert Bay*).

PECULIAR LAKE, NW of Summit L. (G-8). What makes this lake peculiar is a very long spit extending down the centre of the lake that almost makes it look like two parallel lakes.

PEDDER BAY, S. tip of VI (A-8). Named in 1846 by Captain Kellett, RN, apparently after a friend, Lieutenant William Pedder.

PEERS CREEK, flows W. into Coquihalla R. (B-9). After Henry Newsham Peers, the HBC clerk who in 1848-9 laid out the Brigade Trail up the Coquihalla River and Peers Creek, through Fool's Pass, down Podunk Creek, and across the Tulameen River. During the same period, he was in charge of building Fort Hope.

PEMBERTON, midway between Squamish and Lillooet (C-8). After Joseph Despard Pemberton (1821-93). A graduate of Trinity College, Dublin, Pemberton was brought to this country in 1851 by the HBC as Surveyor-

General of Vancouver Island, a post that he held until 1864. He travelled widely throughout Vancouver Island and the mainland.

In 1857, arriving at Nitinaht on the west coast of Vancouver Island after crossing from Cowichan Harbour on the east, Pemberton was confronted by the Nitinaht chief and his band back from a victory, the bloody heads of their victims mounted on poles, the long black hair waving in the breeze. Stepping forward, Pemberton demanded, in Queen Victoria's name, food and canoes. These he paid for with vouchers scribbled on leaves torn from his notebook. Needless to say, they were scrupulously honoured by the HBC.

Pemberton was a civil engineer as well as a surveyor, designing bridges and public buildings. Later in life he founded the real estate firm of Pemberton and Son in Victoria, where Pemberton Road and Despard Avenue are named after him. His daughter Harriet Susan remembered him in her memoirs as 'cheery, bright, and sanguine ... affectionate without ostentation, of a most amiable nature' (*BCHQ* 8 [1944]:111-25).

PEND D'OREILLE RIVER, flows W. into Columbia R., S. of Trail (B-11). Named after the local Indians. They were given this name by French-Canadian voyageurs because of their practice of wearing dangling shell earrings.

PENDER HARBOUR, Sechelt Pen. (B-7). Also NORTH PENDER and SOUTH PENDER ISLANDS. After Staff Commander (later Captain) Daniel Pender, RN, engaged from 1857 to 1870 in surveys of the BC coast on the *Plumper, Hecate,* and *Beaver.* He carried on this laborious work with remarkable zeal and thoroughness.

PENDLETON, MOUNT, NW of McDame (L-5). After Captain George Pendleton, who, after retiring from the sea (he was the captain of a sailing ship), bought a claim in this area and for many years washed for gold. He has been described as 'a small peppery man with a sense of command, who had an enormous fund of very interesting stories.'

PENELAKUT SPIT, NE shore of Kuper I. (A-8). Takes its name from an Island Halkomelem word referring to 'something buried.' In 1927 Jack Fleetwood was told by an old Kuper Island Indian that on this spit two large poles were once partly buried vertically in the sand and that between them a huge net was suspended at certain times of the year to catch migrating waterfowl.

PENNASK LAKE, SE of Douglas L. (B-9). There are two theories about this name. The first is that it comes from the personal name of an Indian, possibly meaning 'gathered to a point cloud.' The other is that it means 'plenty of lake trout at all times.' This lake was an important source of fish for both the Thompson and Okanagan Indians in the food-scarce winter months. Today it is still known as a prime fishing lake.

PENROSE, MOUNT, W. of Gun L. (C-8). After us Senator Boies Penrose (1860-1921) from Philadelphia. With W.G.C. Manson, a Bridge River big-game guide, he climbed this mountain.

PENTICTON, S. end of Okanagan L. (B-10). The Indian name used by Tom Ellis (see *Ellis Creek*) for his Penticton Ranch founded in the 1860s. Penticton became the name of the post office opened here in 1889 and of the townsite laid out in 1892. Penticton was incorporated as a city in 1948.

    The name 'Penticton' apparently comes from an Okanagan Indian word that means 'the always place,' probably in the sense of 'permanent abode.' While most elderly Indians who speak the Okanagan language accept the 'forever place' meaning, in the early 1950s a Penticton Indian named Gideon Eneas gave anthropologist Norm Lerman a different derivation. He said that the original name for Penticton was 'Snpnpiniyatn,' a Nicola-Similkameen word meaning 'place where deer net was used.' This Nicola-Similkameen language is now extinct. (Information supplied by Randy Bouchard.)

PERELESHIN, MOUNT, between Stikine R. and Scud R. (J-4). After Lieutenant Pereleshin, who led a Russian exploring party into this area in 1863.

PERRY CREEK, S. of Kimberley (B-11). After Frank Perry, or François Perrier, a French-Canadian Métis who discovered gold here in 1867 or 1868.

PERRY RIVER, flows S. into Eagle R. (C-10). After Albert Perry, Walter Moberly's assistant during his explorations in this area in 1865.

PERT PEAK, W. of Columbia L. (C-11). Commemorates one of the early river-boats on the Upper Columbia. Originally named the *Alert,* she was a fifty-foot bateau, rowed or poled by four men, and capable of carrying fourteen tons of freight. In 1890 she was fitted with a one-cylinder engine and side paddles and renamed the *Pert.*

PETER HOPE LAKE, SE of Stump L. (C-9). Named after a prospector who came to the Interior in 1863. A part-time guide, he claimed to know every inch of the Nicola country.

PETERSON CREEK, Kamloops (C-9). Originally Jacko Creek, it is named for John Peterson, who in 1868 preempted land that is part of the site of present-day Kamloops. In 1884 he sold this land on the creek to the New Township Syndicate.

PHILIP CREEK, flows NW into Nation R. (H-8). Named after Philip Nation, killed at Ypres in 1915. The Nation River (q.v.), however, was named much earlier and bears no reference to him.

PHILLIPPS CREEK, crosses international boundary by Roosville (B-12). After Michael Phillipps, son of the Reverend Thomas Phillipps of Densall,

Herefordshire. Arriving in British Columbia in 1863, he became an interesting transitional figure in the history of the East Kootenay district. Originally he was an HBC clerk at Fort Shepherd and married Rowena, daughter of Chief David of the Tobacco Plains Indians. Later he became the first homesteader in the district and the first Indian Agent in the area. He prospected along the Elk River in 1873 and was disgusted at finding nothing but coal. He may have discovered the Crowsnest Pass from the west in the same year, ignorant that it had already been found by the Palliser Expedition. He died in 1916.

PHOENIX, NW of Grand Forks (B-10). The mining town that grew up here around 1900 took its name from the Phoenix claim. The name has proved prophetic, for repeatedly when Phoenix appeared dead it has come back to life. After the mine's initial closure in 1920, it was reopened and worked in 1924, 1936-42, and 1959-78.

PHYLLIS'S ENGINE, Garibaldi Park (B-8). Back around 1914, when the BC Mountaineering Club had a camp in the vicinity of Castle Towers Mountain, one of the campers, Phyllis Dyke (Mrs. Beltz), exclaimed that a rocky eminence seen against the skyline looked like a steam locomotive. Jokingly her fellow campers took to referring to it as 'Phyllis's Engine.' After decades of unofficial use by mountaineers, the name was officially adopted by the government in 1979.

PIEBITER CREEK, flows W. into Cadwallader Cr. (C-8). After 'Piebiter' Smith, an early prospector with protruding teeth and a special fondness for pies. He went insane while working his claim here.

PIERCE RANGE, Muchalat Inlet, Nootka (B-6). After Lieutenant Pierce of the Royal Marines, who hoisted the British flag at Friendly Cove, Nootka Island, after the withdrawal of the Spaniards in 1795.

PIERS ISLAND, N. of Saanich Peninsula (A-8). Named after Henry Piers, RN, surgeon on HMS *Satellite,* on the Pacific Station 1857-60. The Saanich Indians knew this as 'Crow Island' because of the great number of crows here.

PILLAR LAKE, S. of Chase (C-10). After a column of rock and clay here, sixty feet high and capped with a large flat rock.

PILLCHUCK CREEK, flows SW into Squamish R. (B-8). This is the Chinook jargon for 'red water,' and the water here is a reddish colour.

PILLSBURY POINT, Prince Rupert (G-4). On 6 May 1906, the steamer *Tees* unloaded three civil engineers, two carpenters, and a load of supplies at Metlakatla. From there the leader, J.H. Pillsbury, conducted the men to Tuck Inlet, where they started clearing land for the future city of Prince Rupert.

PILOT BAY, E. side of Kootenay L. (B-11). A corruption of 'Pirate Bay,' its earlier name. The 'pirates' were a raiding party of Colvile Indians who made a night attack on some sleeping Kootenays and made off with their canoes, which they cached in this bay. The Kootenay Indians knew the bay as Yakhsoumah, their word for 'canoe.'

PIMAINUS CREEK, flows W. into Thompson R. (C-9). From the Thompson Indian word meaning 'the flat underneath or near a steep rise.' A flat stretch of land along the Thompson River here is hemmed in by hills.

PINANTAN LAKE, expansion of Paul Cr. (C-9). This name is derived from a Shuswap Indian term for which no meaning is known. (The lake was not named after Antonio Pene.) Pinantan Lake was an important root-digging area for the Shuswap people from both the Kamloops and Chase areas.

PINCHI LAKE, N. of Stuart L. (G-7). In 1811 Daniel Harmon noted the existence of 'Pinchy.' Father Morice wanted to call this Rey Lake after 'le premier chapelain de Montmartre,' but Victoria stayed with the Carrier Indian name, which means 'lake outlet.'

PINDER PEAK, S. of Nimpkish L. (C-6). After William George Pinder (1850-1936), a pioneer land surveyor who worked in almost every part of British Columbia.

PINGSTON CREEK, flows E. into Upper Arrow L. (C-11). After 'Al' Pingston, successively mate and captain of the *49*, the first steamboat on the BC stretch of the Columbia River. It crossed the international boundary, coming up from Marcus, Washington, in 1866.

PIONEER MINE, S. of Carpenter L. (C-8). Named after the Pioneer claim, staked in 1897 by Harry Atwood. He named his claim after the Pioneer Hotel in Lillooet, having been grubstaked by William Allen, the hotelkeeper here. By the 1980s Pioneer Mine had become a ghost town.

PITQUAH, E. of Lytton (C-9). From the Thompson Indian word meaning 'little spring.' A CNR file on station names gives a more graphic account – 'when it rains, the spring water up above increases, causing disaster.'

PITT ISLAND, N. coast (F-5). Apparently takes its name from the Pitt Archipelago (no longer in the gazetteer), of which Captain Vancouver wrote, 'I named it after the Right Hon. William Pitt, PITT'S ARCHIPELAGO.' For another naming after Pitt the Younger, see below.

PITT RIVER, flows SW into Fraser R. (B-8). Also PITT LAKE and PITT MEADOWS. The first mention of 'Pitt's River' comes in 1827 in the journal of James McMillan, the founder of Fort Langley. He apparently named it in honour of William Pitt the Younger (1759-1806), British Prime Minister during much of the Napoleonic Wars.

PLUMPER SOUND, W. of Saturna I. (A-8). After HMS *Plumper,* an auxiliary steam sloop of 484 tons, carrying twelve guns, which arrived at Esquimalt under the command of Captain G.H. Richards on 9 November 1857. Richards initiated a detailed survey of this coast that was not completed until 1870. In January 1861 the *Plumper* sailed for England, her officers having transferred to the newly arrived HMS *Hecate.* In 1863 Captain Richards returned to England with the *Hecate,* leaving Commander Daniel Pender to complete the survey, using the historic steamship *Beaver,* which had been leased from the HBC.

PODUNK CREEK, flows NE into Tulameen R. (B-9). After Willard Albert ('Podunk') Davis, a trapper and woodsman who died in 1943. Recalled the Reverend John Goodfellow: 'Podunk was a man to be reckoned with. Even in old age his frame suggested the rugged strength of younger days.'

POETT NOOK, E. of Imperial Eagle Channel (A-7). Named in 1861 after Dr. Poett, an English physician of means who had a practice in San Francisco and was interested in copper claims on Copper Island, Barkley Sound.

POILUS, MONT DES, N. boundary of Yoho NP (D-11). Named Mount Habel in 1898 after Jean Habel, a German mountaineer who had climbed in the area. Late in World War I, noting how the Canadians were naming mountain after mountain for French generals, *Les Annales* of Paris declared, 'We beg our allies ... to keep one mountain ... for the great hero of the age, the humble and fascinating Poilu.' The upshot was that Herr Habel's mountain was renamed Mont des Poilus.

POISON COVE, head of Mussel Inlet, Central Coast (E-5). Here in June 1793 Captain Vancouver lost one of his seamen, John Carter, who died from eating poisonous mussels.

POLETICA, MOUNT, Alaska-BC boundary (L-2). After Pierre de Poletica, Russian minister to the United States and a plenipotentiary in Russia's Alaska boundary negotiations with Great Britain in 1824-5. The latter resulted in the creation of the 'Alaska Panhandle.'

PONDOSY LAKE, S. of Eutsuk L. (F-6). A corrupted form of the name of Father Charles-Jean Pandosy (1824-91), one of the founders of the Oregon Mission of 1847. In 1859 he established Okanagan Mission near Kelowna.

POOLE CREEK, flows SW into Birkenhead R. (C-8). After Thomas Poole, who settled in the district before 1874. Poole and his two children were murdered here in 1885.

POOLE, MOUNT, S. of Skidegate Inlet (F-4). After Francis Poole, who visited the Queen Charlotte Islands in the mid-1860s. Of him G.M. Dawson noted:

'In 1863-64, Skincuttle Inlet was the scene of the exploits of a certain Mr. Francis Poole, calling himself a civil and mining engineer. He subsequently published a volume called *Queen Charlotte Islands*, which is chiefly remarkable for the exaggerated character of the accounts it contains.'

POOLEY ISLAND, W. of Mathieson Channel, Central Coast (E-5). After Charles E. Pooley (1845-1912), who came to British Columbia via Panama in 1862. For twenty-two years he was a member of the provincial legislature. A leading member of the bar, he grew rich attending to the legal business of the Dunsmuir interests.

POPKUM, E. of Rosedale (B-9). Linguists today translate this word as 'puffball,' regarding with suspicion Diamond Jenness's equating of the word with 'old spring salmon.'

PORCHER ISLAND, S. of Chatham Sound (G-4). After Commander E.A. Porcher, who was on this coast in 1865-8 commanding HMS *Sparrowhawk*.

PORLIER PASS, N. of Galiano I. (B-8). Named by Narvaez in 1791 after Antonio Porlier, an official in Madrid.

PORT ALBERNI, S.-central VI (B-7). Port Alberni post office was opened in 1900 as New Alberni and had its name changed to Port Alberni in 1910. For the story of Pedro de Alberni, see *Alberni Inlet*.

PORT ALICE, Neroutsos Inlet, Quatsino Sound (C-6). After Alice Whalen, mother of the brothers who founded the Whalen Pulp and Paper Company, which built the pulp mill here in 1917.

PORT CLEMENTS, Masset Inlet, QCI (F-3). Originally Queenstown, but this name was rejected by the postal authorities as duplicative. Accordingly the owner of the townsite renamed the settlement in 1913 after Herbert S. Clements, MP, who reciprocated by getting Ottawa to build a government wharf here.

PORT COQUITLAM. See *Coquitlam*.

PORT DOUGLAS, head of Harrison L. (B-8). Came into being in 1858 when Governor Douglas ordered work begun on the Harrison-Lillooet route to the Interior. In December of this year, Douglas was able to report that the townsite, named after him, had been laid out and that seventy lots were occupied. It is now practically deserted.

PORT ESSINGTON, mouth of Skeena R. (G-5). The site of this settlement was known to the Indians as 'Spokshute,' meaning 'the autumn camp ground.' Captain Vancouver applied the name Port Essington to the estuary of the Skeena, failing to realize that it was a river mouth. In giving this name, he was honouring his friend Captain (later Vice-Admiral) Sir William Essington (1753-1816).

Later the name Port Essington was limited to a trading post established by Robert Cunningham in 1871 near the mouth of the Ecstall River.

PORT GUICHON, Fraser R. delta (B-8). After Laurent Guichon (1836-1902), a native of France who took part in the Cariboo gold rush of 1861 and subsequently ranched with his brothers in the Nicola country (see *Guichon Creek*). In 1883 he left the Interior and bought and began farming a large tract of land here in the Fraser delta. The 'port,' once a halt for regular steamboat traffic, is now silted in.

PORT HAMMOND, SE of Pitt Meadows (B-8). After two young Englishmen, John Hammond, a farmer, and William Hammond, a civil engineer, who owned the townsite at the time of the building of the CPR. They had hopes of Port Hammond becoming a port of call for deep-sea shipping.

PORT HARDY, N. coast VI (C-6). After Vice-Admiral Sir Thomas Masterman Hardy, RN (1769-1839). As captain of HMS *Victory*, he held the dying Lord Nelson in his arms during the Battle of Trafalgar. After retiring from the sea, he became Governor of Greenwich Hospital, the home for old sailors maintained by the British government.

PORT HARVEY, E. of West Cracroft I. (C-6). After Captain Harvey of HMS *Havannah*. Naming this inlet, Captain Richards noted that Harvey 'never lost an opportunity of adding to our hydrographical knowledge.' Nearby is HAVANNAH CHANNEL (q.v.).

PORT KELLS, N. of Langley (B-8). After Henry Kells, who owned land here before the townsite was laid out in 1890.

PORT MANN, North Surrey (B-8). Named after Donald Mann, partner of William Mackenzie in the building of the Canadian Northern Railway. A native of Acton, Ontario, Mann was knighted in 1911. Port Mann was laid out as the Pacific terminus of the Canadian Northern, which was taken over by the dominion government in 1917 and became part of the CNR system in 1923.

PORT McNEILL, NE coast VI (C-6). After Captain William Henry McNeill (1801-75). A native of Boston, Massachusetts, he was first on this coast in 1825 with the trading vessel *Convoy*. In the summer of 1832, the HBC, desperately in need of an extra ship for its coastal trade, sent Chief Factor Duncan Finlayson to Honolulu, where he bought McNeill's brig *Llama* and hired McNeill and his two mates to navigate her. McNeill never became a British subject but served the HBC faithfully for the rest of his active life. He took command of the company's steamer *Beaver* in 1838. In 1849 he was placed in command of the company's new Fort Rupert, and in 1856 he was made a Chief Factor, at the top of the company's hierarchy. He retired in 1863. (See also *McNeill Bay*.)

PORT MELLON, Howe Sound (B-8). After Captain Henry Augustus Mellon. After some years with the Royal Navy, during which he saw service in the Indian Mutiny, Captain Mellon served with the Allan Line and Dominion Steamship Company. He came to Vancouver in 1886, where he served as a marine surveyor, marine insurance agent, examiner of mates and masters, and police magistrate.

When the British Columbia Wood, Pulp and Paper Company was founded in 1908, Mellon, the first vice-president, chose the millsite on West Howe Sound that now bears his name. Before the pulp mill was built, Port Mellon (then known as Seaside) was a popular destination for boat excursions from Vancouver.

PORT MOODY, Burrard Inlet (B-8). Named in 1860 by Captain Richards, RN, in honour of Colonel R.C. Moody (1813-87), commanding the Columbia Detachment of the Royal Engineers (not to be confused with those attached to the British Boundary Commission).

Born in Barbados, Moody was trained at the Royal Military Academy, Woolwich, where he subsequently became Professor of Fortification at the age of twenty-five. He was in command of the Royal Engineers in the north of Britain in 1858 when he was promoted brevet colonel, appointed commissioner of lands and works for British Columbia, given a dormant commission as Lieutenant-Governor, and sent off to the colony with his contingent of six officers and 158 NCOs and men. Colonel Moody returned to England after his unit was disbanded late in 1863 and subsequently rose to the rank of major-general.

Port Moody was originally intended to be the western terminus of the CPR. It lost much when the railroad was extended to Vancouver.

PORT NEVILLE, N. side of Johnstone Strait (C-6). This 'very snug and commodious port' was named by Captain Vancouver in 1792, without any indication as to Neville's identity. Captain Walbran thought that Vancouver probably intended Lieutenant John Neville of the Royal Marines, but this Neville had not yet achieved the distinction of being killed in the battle of 'The Glorious First of June,' 1794.

PORT RENFREW, SW coast, VI (A-7). Port Renfrew is on the inlet known as Port San Juan (named Puerto de San Juan by Quimper in 1790). As mail addressed to Port San Juan kept being misdirected to the San Juan Islands in the United States, in 1895 the settlers at Port San Juan decided to name their post office Port Renfrew in honour of Lord Renfrew, who at one time planned to settle Scottish crofters in the San Juan Valley.

PORT SIMPSON, N. of Prince Rupert (G-4). After Captain Aemilius Simpson, a native of Scotland and a distant relation of Governor George Simpson of the HBC. Aemilius Simpson entered the Royal Navy in 1806 at the age of thirteen as a midshipman. He retired as a half-pay lieutenant in 1816. He subsequently joined the HBC and became superintendent of its marine department on the Pacific coast. In 1831, commanding the *Dryad*, he founded a fort at the mouth of the Nass River. When he died there, in September of that year, the new establishment was named Fort Simpson in his honour.

The original site proving unsatisfactory, Fort Simpson was transferred in 1834 to the site of modern Port Simpson. Captain Simpson's remains were disinterred from their grave outside the palisades of the original fort and were removed to the new post. Dr. Tolmie gives a vivid picture in his diary of drunken Indians rushing in to pillage the abandoned fort.

There was no Indian village at the site of the second Fort Simpson, but Indians soon moved in to be close to the trading post. When the HBC closed down its establishment, there was a considerable Indian settlement, which exists to this day under the name of Port Simpson. (See also *Lach Goo Alams*.)

An interesting story gives Aemilius Simpson credit for the first apples to grow on our coast. At a dinner party shortly before he sailed from England, a lady put some appleseeds in his waistcoat pocket and told him, laughingly, that he must plant them when he reached his destination. Wearing the same waistcoat at Fort Vancouver, Simpson discovered the forgotten seeds, planted them, and saw apple trees soon grow.

PORT WASHINGTON, N. Pender I. (A-8). After Washington Grimmer (1851-1930), a native of London, England, who was the first postmaster on the island.

PORTAGE INLET, Victoria (A-8). For centuries there was an Indian trail across the strip of land between Thetis Cove and Portage Inlet. After the naval base was founded at Esquimalt, Royal Navy men wanting to visit Fort Victoria would sometimes follow the Indian example and portage their ships' boats across here when the sea was running high in the Strait of Juan de Fuca.

PORTAGE MOUNTAIN, Peace R. (H-8). So named in 1875 by A.R.C. Selwyn of the Geological Survey of Canada. Because of the rapids in the canyon of the Peace River before the building of the W.A.C. Bennett Dam, canoes were generally portaged past the mountain.

PORTEAU, Howe Sound (B-8). The name was suggested around 1908 by a Mr. Newberry of Deeks Sand and Gravel Company. It is an adaptation of the French *porte d'eau*, meaning a 'water gate.'

PORTE D'ENFER CANYON, N. Thompson R. (D-10). Named by Louis Battenotte, a Métis guide to Milton and Cheadle during their overland expedition of 1863. The French-Canadian voyageurs gave this name of 'Hell's Gate' to various stretches of dangerous rapids hemmed in by cliffs.

PORTER LANDING, N. end of Dease L. (K-4). After James Porter (1851-1926), who came to British Columbia as a child in 1853. He took part in the Cassiar gold rush and was appointed gold commissioner there.

PORTLAND CANAL, N. coast (H-4). Named in 1793 by Captain Vancouver in honour of William Henry Cavendish Bentinck (1738-1808), third Duke of Portland and Home Secretary from 1794 to 1801.

In 1903 the Canada-Alaskan Boundary Commission placed the boundary between Canada and Alaska along the middle of this canal.

PORTLAND ISLAND, SE of Saltspring I. (A-8). After HMS *Portland,* the flagship of Rear-Admiral Fairfax Moresby, Commander-in-Chief of the Pacific Station 1850-3. The whole of the island comes within Princess Margaret Marine Park. (When the Princess visited British Columbia in 1958, she was presented with Portland Island but returned it to the province for use as a park.)

POSTILL LAKE, NE of Kelowna (B-10). After the Postill family from Yorkshire. The father, Edward, died in 1873 before he could move onto the handsome large ranch that he had bought in the area, but his three sons took over its management.

POTATO RANGE, NW of Chilko L. (D-7). So called because of the abundance here of the 'Indian potato,' the pale round corm attached to roots of *Claytonia Lanceolata* or Spring Beauty, eaten by Indians (and grizzly bears).

POUCE COUPÉ, SE of Dawson Creek (H-9). This unusual name of 'Cut Thumb' comes from a Sekani trapper nicknamed 'Pouce Coupé' by the French-Canadian voyageurs because he had lost a thumb in an accident with his gun. Simon Fraser mentions the Indian Pouce Coupé as early as 1806.

CUT THUMB CREEK, which flows into the Parsnip River, is also named after him.

POUPORE, N. of Trail (B-11). After J.E. Poupore, a railway contractor and lumberman once active in the Kootenays.

POWELL RIVER, W. end of Malaspina Strait (B-7). After Dr. Israel Wood Powell (1836-1915), first McGill graduate in medicine to practise on the West Coast, first president of the Medical Council of British Columbia, and first Superintendent of Indian Affairs in British Columbia.

Born in Colborne, Upper Canada, Dr. Powell came to British Columbia, attracted by the Cariboo gold rush excitement, in 1862. The following year, already a busy medical man in Victoria, he entered politics as an advocate of responsible government and free education. His marriage to Jane Branks in 1865 was a singularly happy one. In 1867 he became superintendent of education for British Columbia. He was an ardent champion of the province's entry into the Canadian confederation.

As Superintendent of Indian Affairs for the province, he fought hard for better medical and educational services for the Indians, and in 1876 he took over the medical superintendency himself. In consequence of his active participation in the Victoria militia, we sometimes find him referred to as Lieutenant-Colonel Powell. In 1890 he was named first Chancellor of UBC (as yet nonexistent). In 1881 he made a tour of the BC coast aboard HMS *Rocket*, whose commander, Lieutenant-Commander V.B. Orlebar, named Powell River and Powell Lake in honour of his passenger.

POWERS CREEK, flows SE into Okanagan L. (B-10). This was originally Deep Creek, but in 1888 a good-natured young Englishman, William Powers, arrived via Montana, and the neighbours named it after him.

PPCLI RIDGE, W. of Tatlayoko L. (D-7). Named on 10 August 1989 to mark the seventy-fifth anniversary of the founding of the Princess Patricia's Canadian Light Infantry.

PREMIER RANGE, NE of Wells Gray Park (E-10). Mountains in this range are named after prime ministers of Canada: Sir John Abbott, R.B. Bennett, Sir Mackenzie Bowell, Mackenzie King, Sir Wilfrid Laurier, Arthur Meighen, Lester Pearson, and Louis St. Laurent. Some other prime ministers have mountains elsewhere in the province.

PRESIDENT RANGE, Yoho NP (D-11). After Lord Shaughnessy, president of the CPR 1898-1918.

PREVOST ISLAND, E. of Saltspring I. (A-8). After Captain James Charles Prevost, RN (1810-91). He first served on the Pacific coast in 1850 with the rank of commander on HMS *Portland*, flagship of his father-in-law, Rear-Admiral Moresby. He was here again in 1853 commanding HMS *Virago*. Promoted to captain, Prevost was on the Pacific Station again from 1857 to 1860, commanding HMS *Satellite*. He was a British commissioner for the settlement of the San Juan boundary dispute with the United States. In later years he rose to the rank of admiral. He was an earnest evangelical, deeply involved in the missionary activities of the Church of England on this coast.

PRICE ISLAND, Milbanke Sound, Central Coast (E-5). After Captain J.A.P. Price, on this coast in 1866-7 in command of HMS *Scout*. Later he was a senior officer at Hong Kong.

PRINCE GEORGE, central BC (F-8). Originally Fort George. This fort was founded by Simon Fraser of the NWC in 1807 and named after King George III, the reigning sovereign. The Carrier Indian name for this point at the junction of the Nechako and Fraser Rivers was 'Thle-et-leh,' meaning 'the confluence.' In 1862 a visitor to Fort George described it as a 'dreary Hudson Bay Company's trading post, infested with dogs.'

In 1910 there was a real estate boom in expectation of the coming of the GTPR (now part of CNR). The construction of the railway led to a prolonged and complicated struggle among three competing townsites. However, the GTP's own new townsite, Prince George, soon drew people away from rival Central Fort George and South Fort George. When the first civic elections were held in 1915, the name of Prince George was chosen over Fort George by a vote of 153 to 13.

PRINCE RUPERT, N. coast (G-4). In 1906, when the GTPR decided that this was the place for the western terminus of its transcontinental line, it offered a prize of $250 for the best name for its new city. It was stipulated that the proposed names must not exceed three syllables and ten letters. The winner out of some 5,000 entries was that of Eleanor MacDonald of Winnipeg, who suggested 'Prince Rupert,' after the first Governor of the HBC. Since her entry exceeded the set number of letters, the company awarded two other first prizes to the contestants who had suggested Port Rupert, which had the required number of letters.

Prince Rupert of the Rhine (1619-82), cousin of King Charles II, was a dashing leader of the Royalist cavalry during the Civil War. After the Restoration he was an important person at court and headed the syndicate to which, on 2 May 1670, the King granted a charter constituting them as 'The Company of Adventurers of England Trading into Hudson's Bay.'

PRINCESS LOUISA INLET, head of Jervis Inlet (C-8). Named after the mother of Queen Victoria. Its Sechelt Indian name, 'Sway-oo-lat,' means 'facing the rising sun's rays.'

PRINCESS MARGARET MARINE PARK, Portland I. (A-8). See *Portland Island.*

PRINCESS ROYAL ISLAND, Central Coast (E- and F-5). Named in 1788 by Captain Charles Duncan after his trading sloop *Princess Royal.*

PRINCETON, confluence of Tulameen R. and Similkameen R. (B-9). Close to here, on the banks of the Tulameen River, is an outcropping of red ochre, prized by the Indians for face paint. This deposit accounts for the early names of Vermilion Forks and Red Earth Forks given to the settlement that grew up near the junction of the Tulameen and Similkameen Rivers. It was also known as Similkameen and Allison's.

The name of Princeton dates from 1860, when Governor Douglas gave this name to the new townsite, which had been laid out below the forks. He chose this name in honour of the Prince of Wales (later Edward VII), who visited eastern Canada this year. 'Princetown' was a frequent early spelling of the name.

PRINCIPE CHANNEL, E. of Banks I. (F-4). Named Canal del Principe in 1792 by Jacinto Caamaño, commanding the Spanish frigate *Aranzazu.*

PRITCHARD, SW of Chase (C-10). When a post office was opened here in 1911, the original name of Pemberton Spur (after Arthur G. Pemberton, a local sawmill operator) was changed to Pritchard, after Walter P. Pritchard, who had farmed in the area since about 1904.

PROCTER, NE of Nelson (B-11). Originally Proctor, after T. Proctor, who had a ranch here. It became Procter later due to confusion with T.G. Procter, a Nelson real estate man.

PROPHET RIVER, flows N. into Muskwa R. near Fort Nelson (K-8). The Beaver Indians recognized certain people as 'dreamers' or 'prophets' who could foretell future events. This river may be named for a fairly recent prophet of the Beaver people, Notseta, the father of people still living on the Prophet River Reserve. Alternatively, it may be named for Deculla, a prophet of an earlier generation.

PROSPECT POINT, Vancouver (B-8). In the nineteenth century, this viewpoint in Stanley Park was known as 'Observation Point.'

PTOLEMY, MOUNT, Alberta-BC boundary, N. of Corbin (B-12). Originally Mummy Mountain since in profile it resembles the head and chest of a person lying flat and gazing upward. Ptolemy was the name of various Egyptian pharaohs.

PUKAIST CREEK, flows SW into Thompson R. (C-9). From the Thompson Indian word for 'white rock.'

PUKEASHUN MOUNTAIN, W. of Seymour Arm, Shuswap L. (D-10). Shuswap Indian word possibly meaning 'white rock.'

PULTENEY POINT, Malcolm I., N. coast VI (C-6). After Admiral Sir Pulteney Malcolm (1758-1838).

PUNCHAW LAKE, SW of Prince George (F-8). Derived from the Carrier Indian word meaning 'big lake.'

PUNTLEDGE RIVER, flows NE into Courtenay R. (B-7). This is from Pentlatch, the name of a band of Coast Salish Indians who used to live in the area but are now extinct. In field notes kept by Franz Boas, Pentlatch is said to be derived from words meaning 'abdomen' and, possibly, 'bury.'

PUNTZI LAKE, NE of Tatla L. (E-7). According to a missionary priest, Puntzi means 'blackberries' in the Chilcotin language, but Lieutenant H.S. Palmer, RE, who visited the lake in 1862, reported that the name meant 'small lake.'

PURCELL MOUNTAINS, E. of Selkirk Mtns. (B- and C-11). Named by Dr. James Hector after Goodwin Purcell (1817-76). Purcell, the last chieftain of the O'Leary line, was also Professor of Therapeutics and Medical

Jurisprudence at Queens University, Cork. He had been on the selection committee that had chosen the personnel for the Palliser Expedition, with which Hector was serving.

PURDEN LAKE, E. of Prince George (F-9). After H. Purden-Bell, CPR surveyor, who was this way in 1875

Q

QUADRA ISLAND, E. of Discovery Passage (C-7). After Don Juan Francisco de la Bodega y Quadra, Spanish naval officer, who took part in exploratory voyages off the BC coast in 1775 and 1779. In 1792, when Captain Vancouver met him at Nootka to effect the transfer to Britain of the Spanish base here, Quadra was in command of the naval establishment maintained at San Blas, Mexico. Although Vancouver was unable to reach agreement with Quadra about the expected transfer, the two formed a close personal friendship. As a symbol of their friendship, Vancouver, with the Spaniard's approbation, named Vancouver Island 'the Island of Quadra and Vancouver.'

In his account of his great expedition, Captain Vancouver has this to say of Quadra:

> The well known generosity of my other Spanish friends, will, I trust, pardon the warmth of expression with which I must ever advert to the conduct of Senor Quadra; who, regardless of the difference in opinion that had risen between us in our diplomatic capacities at Nootka, had uniformly maintained toward us a character infinitely beyond the reach of my powers of encomium to describe. His benevolence was not confined to the common rights of hospitality, but was extended to all occasions, and was exercised in every instance, where His Majesty's service, combined with my commission, was in the least concerned.

Quadra died in San Blas in September 1794, occasioning a renewed tribute from Vancouver.

Quadra Island received its name in 1903 after it had been established that it and Maurelle and Sonora Islands are actually three separate islands.

QUALICUM, NW of Nanaimo (B-7). Dr. Robert Brown, reporting on his explorations on Vancouver Island in 1864, mentions the beautiful tract of country extending past the Quall-e-hum River. In other early accounts, the name is also spelled Quallchum. The name is derived from a Nanaimo Indian term for 'place of the dog [chum] salmon.'

QUAMICHAN LAKE, N. of Duncan (A-8). From the Island Halkomelem name meaning 'humped back.' The shape of the hill behind the Quamichan Indian village reminded the Indians of a humpbacked person.

QUATAM RIVER, S. of Bute Inlet (C-7). From a Mainland Comox Indian word for 'stream.'

QUATHIASKI COVE, W. side of Quadra I. (C-7). From the Island Comox Indian word meaning 'island in the mouth,' referring to the island at the cove's entrance.

QUATSE RIVER, flows N. into Hardy Bay (C-6). *Quatse* is the Kwakwala Indian word meaning 'on the west side' (of Fort Rupert). It is also the Indian name for Port Hardy – which, like the river, is to the west of Fort Rupert.

QUATSINO SOUND, NW Vancouver I. (C-6). From the Kwakwala Indian word that has been translated variously as (1) 'people of the north country,' (2) 'the downstream people,' or (3) 'people who live on the other side' (of the island). It all depends on where one is.

QUAW, N of Prince George (G-8). This little station on the BCR is all that remains to commemorate the mighty Quaw (more often spelled Kwah), the most powerful Indian chief in central British Columbia when the first fur traders arrived early in the nineteenth century. At first hostile to the white men, Kwah later grew more friendly. It was his word that saved the life of the future Sir James Douglas when he was seized by the Indians at Fort St. James in 1828. Kwah's nephew and heir was holding Kwah's dagger at Douglas's breast and shouting, 'Shall I strike? Shall I strike?' when Kwah accepted the trade goods offered by Douglas's half-Indian wife and let him go.

QUEEN CHARLOTTE CHANNEL, Howe Sound (B-8). After HMS *Queen Charlotte,* Lord Howe's flagship when he won his great victory 'The Glorious First of June' in 1794.

QUEEN CHARLOTTE ISLANDS (E- and F-3 and 4). In 1785 a British syndicate headed by R.C. Etches formed the King George's Sound Company (see *Nootka Sound*). Armed with licences from the South Sea Company and the East India Company authorizing them to trade in furs between the North Pacific coast of America and China, they despatched two ships, the *King George* and the *Queen Charlotte,* to secure furs off this coast. In the summer of 1787, Captain George Dixon, after successfully trading for sea otter off both the west and east coasts of these islands, named them after his ship.

QUEEN CHARLOTTE SOUND, between VI and QCI (D-4 and 5). According to Captain Vancouver, this sound was named in 1786 after the consort of George III by Mr. S. Wedgborough, commanding the trading vessel *Experiment.*

Queen Charlotte (1744-1818) was the youngest daughter of a brother of the third Duke of Mecklenburg-Strelitz. The royal marriage having been arranged, she travelled to England in 1761 and married George III the day they met. She bore the King fifteen children, was thoroughly domestic in her interests, and never discussed matters of state with her husband. Horace Walpole described her thus at the time of her marriage: 'She is not tall nor a beauty. Pale and very thin; but looks sensible and genteel. Her hair is darkish and fine; her forehead low, her nose very well, except the nostrils spreading too wide. The mouth has the same fault, but her teeth are good. She talks a great deal, and French tolerably.' Queen Charlotte said that she knew no real sorrow in her marriage until the King went insane.

Also QUEEN CHARLOTTE STRAIT to the southeast.

QUEENS BAY [QUEEN'S BAY], W. side of Kootenay L. (B-11). Named in honour of Queen Victoria by the English who settled here around the end of the nineteenth century.

QUEENSBOROUGH (B-8). See *New Westminster.*

QUEEST CREEK, flows W. into Anstey Arm, Shuswap L. (D-10). This name was given to the creek by G.M. Dawson in 1877, and subsequently the nearby mountain was named Queest Mountain. The translation of this Shuswap Indian name is uncertain, but the often-quoted meaning of 'buffalo' is incorrect.

QUENNELL LAKE, N. of Ladysmith (B-8). Commemorates an RN deserter who became mayor of Nanaimo and somewhere along the way changed his name from Pannell to Quennell. He arrived in Nanaimo in 1864 and worked as a miner, steamboat hand, and farmer before founding a successful butcher business.

QUENTIN LAKE, Kwadacha Wilderness Park (J-7). After Quentin Roosevelt, US airman, killed at the front on 14 July 1918.

QUESNEL RIVER, flows NW and SW into Fraser R. (E-8). Also QUESNEL and QUESNEL LAKE. Returning from his journey to the mouth of the Fraser River, Simon Fraser noted in his journal for 1 August 1808, 'Debarked at Quesnel's River.' The Quesnel for whom he had named this river was Jules Maurice Quesnel, one of the two clerks of the NWC who accompanied him on his historic journey. Quesnel was at this time a young man of twenty-two. He left the fur trade three years later. In his final years, he played an active part in Quebec politics, being a member first of the Special Council of Lower Canada, and then of the Legislative Council of the united province of Canada. He died in 1842.

Quesnel River was at one time also known as Swift River. The ghost town of Quesnel Forks, or Forks City, was the earliest of the Cariboo gold camps. The town of Quesnel was once known as 'Quesnelle Mouth.'

QUILCHENA, Nicola L. (C-9). The origin of this Okanagan Indian word is not clear, but around 1920 the anthropologist James Teit wrote that it could be translated as 'red side or red bluff or side hills.'

QUINISCOE LAKE, Cathedral Park (B-9). Herb Clark, an early promoter of the Cathedral Lakes area, gave this name to the largest lake. It is the name of the Indian chief and hunter who is the subject of the poem 'Quin-is-coe' in Stratton Moir's book *In-Cow-Mas-Ket,* published in 1900. Randy Bouchard says that the Indian words that appear in this book are mostly Okanagan terms, so there is a chance that *quiniscoe* is also an Okanagan word.

QUINSAM LAKE, SW of Campbell River (B-7). From the Island Comox Indian name meaning 'resting place.' The area was important for the Island Comox people since a special mask, known as *xwayxway*, was believed to have fallen here from the sky.

QUISITIS POINT, W. point of Florencia Bay (A-7). From the Nootka Indian word meaning 'other end of beach.'

R

RACE ROCKS, SW of Victoria (A-8). This name originated about the time of the founding of Fort Victoria and was officially adopted in 1846 by Captain Kellett, surveying these waters in HMS *Herald*. He noted that 'This dangerous group is appropriately named, for the tide makes a perfect race around it.'

RADIUM HOT SPRINGS, S. of Golden (C-11). They were originally known as Sinclair Hot Springs (see *Sinclair Canyon*). In 1915 the name was changed to Radium Hot Springs on account of the alleged high radioactivity in the water here.

RAFT RIVER, flows SW into N. Thompson R. (D-10). Possibly the Overlanders of 1862 had raft trouble here, but the name may date from the 1870s, when the CPR had a base here for its survey parties.

RAINBOW ISLAND, E. of Bella Bella (E-5). Named after HMCS *Rainbow,* the first cruiser in the Royal Canadian Navy.

RAINBOW RANGE. There are two Rainbow Ranges in the province. (1) That in the southeastern part of Tweedsmuir Park (E-7) gets its name from the variety of colours in the volcanic rocks here. Since red predominates, the Indians called them 'the Bleeding Mountains.' (2) That in Mount Robson Park (F-10) is similarly named because of the prismatic range of colours offered by the rocks and vegetation.

RATCHFORD CREEK, flows SW into Seymour R. (D-10). After Joseph Ratchford, who apparently used this route during the Big Bend gold rush of 1865. Early in the 1870s, he had a mail contract with the CPR surveyors going through the country.

RATHTREVOR BEACH PARK, E. of Parksville (B-7). In 1885 W.H. Rath, formerly an officer in the Royal Navy, established Wildrose Farm here. A daughter later changed its name to Rathtrevor. In 1967 the provincial government, having secured the property, created the park. It added Crown foreshore to it in 1969.

RATTENBURY ISLAND, off Hecate I., Central Coast (D-5). After Francis Mawson Rattenbury (1867-1935), perhaps British Columbia's most distinguished architect. He designed the Parliament Buildings in Victoria (1893-7) and the Empress Hotel there, also the old courthouse in Vancouver (now the Vancouver Art Gallery). After his retirement, he lived in Bournemouth, England, where he was murdered by his chauffeur, who had become his wife's lover.

RAUSH RIVER, flows N. into upper Fraser R. (F-9). This was originally Rivière au Shuswap (the Shuswap Indians did extend this far north). This was shown on maps first as R au Shuswap, then as R au Sh, and finally as Raush River. The name Raush Valley was used when a post office was opened in the area in 1915.

RAY, MOUNT, E. of Clearwater L. (E-9). After John Bunyan (Johnny) Ray, who came into the Cariboo in 1908. Because of kindnesses shown to the Indians, they invited him to use their hunting grounds in this area. He also trapped, and established a farm.

RAYLEIGH, N. of Kamloops (C-9). When E.T. Webb and R.B. Homershaw settled here just before World War I, they were struck by the resemblance between a small hill here and the castle mound that is a feature of Rayleigh in Essex, their original home in England.

RAYMOND PASSAGE, W. of Campbell I. (E-5). Named for an American, Captain Raymond, who used this passage when bringing his sailing brig to Fort McLoughlin. HBC records mention this name as early as 1835.

READ ISLAND, E. of Quadra I. (C-7). After Captain W.V. Read, RN, of the Admiralty's Hydrographic Office. Named by Pender about 1864.

REBECCA SPIT, Quadra I. (C-7). After the coastal trading schooner *Rebecca,* thirty-five tons, launched in 1860.

RED PASS, E. of Tête Jaune Cache (E-10). So named because of the brilliant red rocks near its crest.

REDGRAVE, E. end of Kinbasket L. (D-11). After Stephen Redgrave, registrar and sheriff of the county court of East Kootenay, and mining recorder and stipendiary magistrate for East Kootenay late in the nineteenth century. Based at Donald, he won local fame as a Munchausen.

REDONDA ISLANDS, N. of Desolation Sound (C-7). *Redonda* is Spanish for 'round.' West Redonda Island and East Redonda Island (mistaken for a single island) were named Redonda Island by Galiano and Valdes in 1792.

REG CHRISTIE CREEK, flows W. into N. Thompson R. (D-10). After Harold Reginald Monro Christie, BCLS, killed in action in 1916.

REID ISLAND, off N. tip of Galiano I. (B-8). After Captain James M. Reid (1802-68) of the HBC's maritime service. In 1854 he had the misfortune to run his brigantine *Vancouver* onto Rose Spit, QCI. The Indians, intent on plundering the boat, so effectively prevented attempts to refloat her that Captain Dodd of the *Beaver* poured oil over her and set her on fire. This was the end of Captain Reid's sea-going career.

REIFEL ISLAND, Fraser delta (B-8). Originally Smoky Tom Island, it was purchased in 1929 by George C. Reifel, who took a great interest in the migrating waterfowl. In 1972 the Canadian Wildlife Service purchased much of the island. George H. Reifel, as a memorial to his father, donated to the cws the remaining 98.5 acres, which had been leased to the bc Waterfowl Society. The whole is now the George C. Reifel Waterfowl Refuge.

RELAY CREEK, N. of Carpenter L. (D-8). Possibly due to supplies having been 'relayed' to a mine in the area, with the party bringing in the supplies handing them over at this creek to other men who had come out from the mine to receive them.

RENNELL SOUND, W. coast of Graham I. (F-3). This name first appears on the map of the coast of North West America published with George Dixon's *Voyage round the World* (1789). Dixon almost certainly named this sound after James Rennell (1742-1830), the illustrious geographer and friend of Sir Joseph Banks.

RESOLUTION COVE, Bligh I., Nootka Sound (B-6). Captain Cook's larger vessel, hms *Resolution* (460 tons and twelve guns), underwent repairs here during the four weeks that Cook was at Nootka in 1778.

RETALLACK, NW of Kaslo (C-11). After John L. Retallack (1863-1924). In 1890, after serving as a sergeant in the nwmp, he came to Kootenay, where he was active in banking and mining, establishing the Whitewater mine here.

RETASKIT, N. side of Seton L. (C-8). Derived from a Lillooet Indian name, the most famous bearer of which lived around the end of the nineteenth century. He was probably James Nraiteskel ('Tyee Jim'), the chief who cosigned the Declaration of the Lillooet Tribe at Spences Bridge, 10 May 1911.

REVELSTOKE, Columbia R. (C-10). The locality was first known as 'The Eddy' because of a large swirl in the river that had eroded the right bank of the Columbia here. Marcus Smith referred to The Eddy in 1872. When a route was surveyed westward for the cpr, the place became known as 'Second Crossing' since the railway, cutting across the Big Bend of the Columbia, crossed the river first near Donald and a second time here.

In 1880 the original townsite was surveyed by A.S. Farwell, and for the next six years the settlement was known as Farwell. Having acquired the land where he expected the cpr to run its right of way, Farwell tried to make the railway pay an inflated price for it. The latter responded by laying out a new townsite on higher land to the east of Farwell's townsite and placing its station here.

The cpr, in no mood to immortalize Farwell, named its station 'Revelstoke' in honour of the first Lord Revelstoke, head of the British banking house of Baring Brothers. This firm had bought $15,000,000 of a cpr bond issue and so had averted the last financial crisis during the building of the transcontinental line. In June 1886 the post office of Farwell was renamed Revelstoke. The City of Revelstoke was incorporated in 1889.

RICHARD POINT, Nanoose Harbour (B-7). After Richard P. Wallis, a native of Cambridgeshire, who came to British Columbia in 1888 and purchased over 2,000 acres here in 1897.

RICHARDS, MOUNT, N. of Duncan (A-8). After Captain (later Admiral) George H. Richards, RN (1820-1900). From 1856 to 1863, commanding first HMS *Plumper* and then her replacement HMS *Hecate,* he conducted a most thorough survey of the coasts of the Crown Colonies of Vancouver Island and British Columbia. At the end of that period, he returned to England to become Hydrographer of the Navy, leaving his second-in-command, Commander Daniel Pender, to complete the survey using the historic *Beaver,* which the navy had leased from the HBC.

One of those remarkable Victorians compounded of energy and purpose, Richards was an English gentleman – and English gentlemen did not go around naming places after themselves. Accordingly, while he named places after Pender, he named nothing after himself, and the best that posterity has done for him is this one small 'mountain,' named in 1905 by Captain J.F. Parry of HMS *Egeria.*

RICHMOND, Lulu I. (B-8). In 1861 Hugh McRoberts, who had come to British Columbia from Australia, established Richmond Farm here, its name being chosen by one of his daughters, who took it from a favourite place in Australia. In September 1862 the *British Columbian* printed 'A Visit to Richmond,' a fact that negates the claim of Mary Boyd, wife of the first 'warden' or reeve of Richmond, that the settlement was named in honour of her birthplace in Yorkshire – the Boyds did not arrive in the area until 1863. Richmond became a city in 1990.

RICHTER PASS, W. of Osoyoos (B-10). After Francis Xavier Richter (1837-1910). A native of Bohemia, he came to British Columbia in 1860, helping to build the HBC post at Keremeos and taking charge of the company's pack-horses there. He began preempting land in the Cawston area about 1865 and ended up owning some 5,000 acres, which constituted the R Ranch (later the Cawston Ranch). In 1895 he bought a ranch near Keremeos, where he planted the first commercial apple orchard in the area.

RIDER, NW of McBride (F-9). This CNR station was named after Sir Henry Rider Haggard (1856-1925), a popular novelist who travelled along this line in 1916 when it was the old GTPR. Note also MOUNT RIDER and HAGGARD GLACIER in this area.

RIDLEY ISLAND, S. of Prince Rupert (G-4). After the Rt. Reverend William

Ridley, from 1879 to 1902 the first Anglican Bishop of Caledonia. Prince Rupert is now the cathedral city of this diocese.

RIONDEL, E. side of Kootenay L. (B-11). After Count Edouard Riondel, president of the French-owned Canadian Metal Company, which in 1905 acquired the famous Bluebell mine and renovated the smelter here.

RISKE CREEK, SW of Williams L. (D-8). Marcus Smith, travelling in this area in 1872, reported, 'right in our course lay a cultivated farm, to which we descended – 1,400 feet – by very steep slopes, and there met the owner L. W. Riskie, Esq., a Polish gentleman, by whom we were hospitably entertained.'

RIVERS INLET, Central Coast (D-6). Named by Captain Vancouver after George Pitt, first Baron Rivers (1721-1803), a writer and politician, chiefly famous for the beauty of his wife, Penelope. Horace Walpole described Lady Rivers as 'all loveliness within and without' and Lord Rivers as 'her brutal and half-mad husband.'

ROBB BLUFF, Comox Harbour (B-7). After Mrs. James Robb, who came to Vancouver Island in 1862 as matron aboard the 'brideship' *Tynemouth*. Under her supervision were 'sixty marriageable lassies' of varying degrees of respectability, intended to become mates for lonesome BC bachelors. Mrs. Robb was the first white woman in Comox.

ROBERTS CREEK, W. of Gibsons (B-8). After Thomas Roberts, who built a cabin for his family here on the Sechelt peninsula in 1889. This creek was the boundary between the Squamish Indians to the south and the Sechelt Indians to the north.

ROBERTS, POINT, BC-US boundary (A- and B-8). Also ROBERTS BANK. Point Roberts was named by Captain Vancouver after his 'esteemed friend' Lieutenant Henry Roberts, RN, who had been his comrade on Captain Cook's second and third voyages of discovery. Roberts was a cartographer and prepared the maps for the published account of Cook's three voyages. He was originally designated to command the expedition sent to the Pacific Northwest coast in 1791. However, by the time the ships were ready, another commission was in prospect for Roberts and the position was given to Vancouver, who was originally intended to be only the second-in-command.

ROBIE REID, MOUNT, N. of Alouette L. (B-8). After Robie L. Reid (1866-1945), prominent Vancouver lawyer and BC historian. Members of the undergraduate History Club at UBC used to be invited to his home to see his splendid collection of British Columbiana. After his death these books joined those of his friend Judge Howay in the Howay-Reid collection in the UBC library. (See also *Judge Howay, Mount.*)

ROBINSON CREEK, N. of Naramata (B-10). After John Moore Robinson (1855-1935), the founder of Peachland, Summerland, and Naramata. Born in Ontario of Irish descent, he went in 1879 to Manitoba, where he edited the *Portage La Prairie Review,* founded the *Brandon Times,* and became an MLA. He came to Rossland in 1890. Visiting a ranch near today's Peachland, he was so impressed by the peaches grown here that he founded his Okanagan towns and brought out hundreds of prairie families as settlers.

ROBSON, W. of Castlegar (B-11). The first settlement near the junction of the Columbia and Kootenay Rivers was named Sproat's Landing after Gilbert Malcolm Sproat, the gold commissioner in the district. In 1890 Sproat's Landing became East Robson, and the corresponding settlement on the other bank of the Columbia became West Robson, in recognition of John Robson, Premier of British Columbia from 1889 to 1892.

Robson made his reputation in British Columbia as the fiery editor of the *British Columbian,* which he founded in New Westminster in 1861. In his paper Robson launched violent attacks upon Governor Douglas, whom he saw as a 'czar' bent on subordinating the interests of the mainland to those of Vancouver Island. Robson later moved to Victoria and made a comfortable fortune in real estate. As premier he embarked on a program of protecting the province's natural resources from improper exploitation. Premier Robson died of blood poisoning, the consequence of jamming his finger in a carriage door while visiting London.

ROBSON BIGHT, S. side of Johnstone Strait (C-6). Renowned among naturalists because killer whales congregate here, this bight is named after Lieutenant-Commander C.R. Robson, commanding HM gunboat *Forward.* He died in 1861 after a fall from a horse.

ROBSON, MOUNT, NW of Yellowhead Pass (F-10). The highest mountain in the Canadian Rockies and a major problem in the study of BC place names. The name Robson's Peak was already in use when Milton and Cheadle saw it in 1863, a fact that makes very suspect the theory that the mountain was named after John Robson, who became Premier of British Columbia in 1889. Some questionable evidence indicates that the mountain may have been referred to as Mount Robinson as early as 1827. Actually both Robinson and Robertson were often given the slurred pronunciation of 'Robson.'

The most probable of some ten contending theories about Mount Robson's name is that it was named after Colin Robertson (1783-1842), an HBC officer who, after his retirement from the fur trade, became a member of the parliament of the united provinces of Upper and Lower Canada. In 1820 Robertson, in charge of the HBC post of St. Mary's, Peace River, sent a company of Iroquois fur hunters across the Rockies to the area around Tête Jaune Cache. This party, with Ignace Giasson in command and Pierre

Hatsinaton (perhaps the original 'Tête Jaune') as guide, must have passed close to Mount Robson and probably named it after Robertson.

ROCHER DÉBOULÉ RANGE, S. of Hazelton (H-6). This name, meaning 'fallen rock,' was originally applied to only one of the peaks here, that which had at its base a massive rock slide that had partly closed the Bulkley River to the passage of salmon. Simon McGillivray, who inspected the damage in 1833, noted especially 'two immense rocks' blocking the stream. He reported that the slide had occurred in 1824.

ROCK CANDY CREEK, N. of Grand Forks (B-10). Takes its name from a fluorspar mine whose ore has the crystalline appearance of rock candy.

ROCKY MOUNTAIN PORTAGE, W. of Hudson's Hope (I-8). First mentioned by Sir Alexander Mackenzie in the account of his journey to the Pacific in 1793. Here he speaks of coming 'to the carrying-place called Portage de la Montagne de Roche.' (See also *Portage Mountain*.)

ROCKY MOUNTAINS. The earliest reference to these is that of John Knight, Governor of York Factory, who states in his diary for 1716 that Indians had told him that very far to the west there were prodigious mountains. First mention of their present name is to be found in the journal of 1752 of Jacques Legardeur de Saint-Pierre, which refers to the 'Montagnes de Roche.' Rocky Mountains is a translation of the Cree Indian name for them, 'As-sin-wati.' Seen from the east across the prairie, they appear as a great rocky mass.

RODD HILL, Esquimalt (A-8). See *Fort Rodd*.

RODERICK ISLAND, W. of Finlayson Channel (E-5). Named about 1866 by Lieutenant Daniel Pender, RN, in honour of Chief Factor Roderick Finlayson (1818-92). Apparently Pender did not realize that adjacent Finlayson Channel had almost certainly been named after another HBC worthy, Chief Factor Duncan Finlayson.

RODNEY, MOUNT, head of Bute Inlet (C-7). After Admiral Lord Rodney (1719-92), the victor of the Battle of St. Vincent and of various actions against the French in the West Indies.

ROGER CURTIS, CAPE, Bowen I., Howe Sound (B-8). After Admiral Sir Roger Curtis, baronet (1746?-1816), flag captain of Lord Howe's flagship *Queen Charlotte* when that nobleman won his great victory called 'The Glorious First of June,' 1794.

ROGERS PASS, Selkirk Mtns. (D-11). Named after Major A.B. Rogers (1829-89), engineer in charge of the mountain division of the CPR from 1880 to 1885.

Although Walter Moberly denied Rogers's title to the discovery of Rogers Pass, the credit does seem to be properly his. The facts are as follows. Although Moberly, following his discovery of Eagle Pass between Shuswap Lake and Revelstoke, did in 1866 send his assistant, Perry, up the Illecillewaet River, there is no evidence that Perry continued his explorations beyond the valley of that river. Major Rogers, looking for a pass through the Selkirks in 1881, followed the lead given by Moberly and went up the Illecillewaet to its source in the Illecillewaet Glacier. He then scaled Mount Avalanche and from its heights saw how, by utilizing the valleys of Connaught Creek and Beaver River farther east, the CPR might have a pass. In August 1882, after further exploration travelling from the east, he certified the existence of the pass.

Major Rogers, an American with a degree in engineering from Yale, went by the ironic nickname of 'The Bishop.' He was described thus by J.H.E. Secretan, who knew him well:

> He was what we called a 'rough and ready' engineer, or rather pathfinder. A short, sharp, snappy little chap with long Dundreary whiskers. He was a master of picturesque profanity, who continually chewed tobacco and was an artist in expectoration. He wore overalls with pockets behind, and had a plug of tobacco in one pocket, and a seabiscuit in the other, which was his idea of a season's provisions for an engineer.
>
> His scientific equipment consisted of a compass and an aneroid slung around his neck. (*Canada's Great Highway*, pp. 186-7)

For his services the CPR gave Rogers a bonus cheque for $5,000, which he had framed. When Van Horne, the chairman of the CPR, asked Rogers why he hadn't cashed it, Rogers told him that he wasn't in the game for the money.

ROLLA, N. of Dawson Creek (H-9). Named in 1912 by L.H. Miller, after Rolla, Missouri.

ROLLEY LAKE, SW of Stave L. (B-8). After James Rolley, who homesteaded here in 1902.

RONAYNE, MOUNT, SW of Birkenhead L. (C-8). After John Ronayne, an Irish veteran of the Yukon gold rush who settled in the Pemberton area in 1906.

ROOSVILLE, E. Kootenay, BC-US border (B-12). When a post office was established here in 1908, it was named after Fred Roo, the postmaster, who had arrived in the country around 1899 and had built a store and hotel here.

ROSE POINT, NE of Graham I. (G-4). Named in 1788 by Captain William Douglas of the *Iphigenia*, after George Rose, MP. A writer on politics and economics, Rose took a special interest in trading voyages to the Northwest coast. A year earlier ROSE HARBOUR at the southern end of the Queen Charlotte Islands had been named after him.

To the Haidas, Rose Point was 'Nai-koon,' meaning, according to G.M. Dawson, something like 'long nose.'

ROSEBERY, Slocan L. (C-11). After Archibald Philip Primrose, fifth Earl of Rosebery (1847-1929). A leading Liberal, he was briefly Prime Minister of Great Britain (1894-5). He believed profoundly in the civilizing role of the British Empire.

ROSEDALE, E. of Chilliwack (B-9). When the post office was opened here in 1894, it was named after the wild roses that grew in abundance nearby.

ROSS BAY, Victoria (A-8). After Chief Trader Charles Ross of the HBC, in command of Fort Victoria when it was founded in 1843. He died the following year. The Straits Salish Indian name for Ross Bay, including Clover Point, was 'Wholaylch,' or 'pussy willows.'

ROSS PEAK, SW of Rogers Pass (D-11). After Sir James Ross, who was superintendent of construction for the CPR.

ROSSLAND, W. of Trail (B-11). After Ross Thompson, who in 1893 preempted a townsite here, which he called Thompson. Apparently because the post office department found that the name Thompson caused confusion with a Thompson Landing on the Arrow Lakes, it was changed to Rossland two years later.

ROY, E. side of Loughborough Inlet (C-7). When a post office was established here, D. McGregor, the first postmaster, named it after Rob Roy, the Scottish freebooter, whose name was originally Robert MacGregor.

ROYAL ROADS, S. of Esquimalt (A-8). Originally Royal Bay. The name probably has reference to Queen Victoria and her husband, Prince Albert, since this stretch of coast lies between the city of Victoria and Albert Head. 'Roads' is a variant of 'roadstead,' a place where ships may safely lie at anchor close to the shore.

ROYSTON, S. of Courtenay (B-7). William Roy and his family settled here in 1890. In 1910 Roy, collaborating with a real estate promoter named Frederick B. Warren, laid out a townsite here that they named Royston. There were two reasons for choosing this name: (1) it was based on Bill Roy's surname; (2) Warren came from Royston in Cambridgeshire.

RUBY CREEK, NE of Agassiz (B-9). Rubies and garnets of little value are to be found here and in nearby Garnet Creek.

RUPERT INLET, arm of Quatsino Sound (C-6). So named because this was the arm that led toward Fort Rupert.

RUSKIN, NW of Mission (B-8). In 1897 the Canadian Cooperative Society established a sawmill here. J.T. Wilband, the president, and Thomas Robinson, the secretary of the cooperative, named the new settlement after John Ruskin (1819-1900), the English author and critic who was one of the apostles of the cooperative movement. The mill went bankrupt about a year later.

RUSSELL, CAPE, Cape Scott Park (C-5). After Lord John Russell, Prime Minister of Great Britain (1846-52), created Earl Russell in 1861.

RUTH LAKE, NE of 100 Mile House (D-9). Arthur Williamson of North Vancouver named this lake after his daughter Ruth.

RUTLAND, suburb of Kelowna (B-10). After John Rutland, who came from Australia and established a wheat farm here in the 1890s.

RYKERTS [RYKERT'S], BC-US boundary (B-11). After J.C. (Charlie) Rykert, who came to British Columbia in 1881 as part of the NWMP escort for the Marquess of Lorne, then Governor-General. Rykert quit the force and remained in British Columbia. In 1883 he established a customs office here where Kootenay River reenters Canada, serving not only as customs officer but also as immigration inspector, gold commissioner, and registrar of shipping.

S

SAANICH PENINSULA, N. of Victoria (A-8). From the Saanich (Straits Salish) word meaning 'elevated' or 'upraised.' The word describes what Mount Newton looked like to the Indians approaching by sea from the east – someone lying down with his rump up. Another possible explanation is that Saanich refers to the appearance of the peninsula when seen through a mirage on a warm summer day. Under such conditions, it indeed looks 'elevated' or 'upraised.'

The Saanich Indians have four reserves on Saanich Peninsula: Paquechin (Coles Bay), meaning 'drop off,' so named because the shallow water in front suddenly deepens; Tsartlip (Brentwood Bay), meaning 'maple leaf'; Tsawout (East Saanich), meaning 'on top,' because houses are built on top of higher ground; and Tseycum (Patricia Bay), meaning 'place of clay.'

SABINE CHANNEL, NE of Lasqueti I. (B-7). Named in 1861 in honour of Major-General Edward Sabine, president of the Royal Society, an expert in magnetic research.

SAHALI, Kamloops (C-9). This suburb of Kamloops has been given a Chinook jargon name meaning 'above' or 'up.'

SAHTLAM, W. of Duncan (A-8). From the Island Halkomelem word meaning 'evergreen.'

ST. ELIAS MOUNTAINS, extreme NW corner of BC (L-1). From Mount St. Elias in Alaska, first sighted by Vitus Bering, a Dane in Russian service, on St. Elias's Day (16 July), 1741.

ST. EUGENE MISSION, N. of Cranbrook (B-12). When Father Leon Fouquet, OMI, founded this mission in 1874, he named it after a namesaint of the founder of his order, Charles Joseph Eugene de Mazenod (1782-1861), Bishop of Marseilles. (St. Eugene, Bishop of Carthage, died in exile in 505 AD.)

ST. JAMES, CAPE, S. tip of QCI (D-4). Named by Captain Dixon, who rounded this promontory on St. James's Day (25 July), 1787.

ST. LEON CREEK, E. side of Upper Arrow L. (C-11). An early hunter and trapper in this area had relatives living in St. Pol de Leon, Finistère, France. The hot springs here were discovered by Michael Grady, a young prospector from Ontario, who built a lodge on the site in 1906. He went bankrupt in World War I but never left the place until he had grown too old to look after himself.

ST. MARY RIVER, flows E. into Kootenay R. at Fort Steele (B-12). This name goes back at least to 1865, when it was mentioned in the *British Columbian,* the early newspaper published in New Westminster. The naming is often erroneously attributed to Father Fouquet, OMI, who did not visit Kootenay until 1874, when he founded St. Eugene's Mission on this river.

ST. VINCENT BAY, Jervis Inlet (B-7). After Rear-Admiral Sir John Jervis (1735-1823), who, having defeated the Spanish at the Battle of Cape St. Vincent, was created Earl St. Vincent in 1797.

SAKINAW LAKE, N. of Pender Harbour (B-7). From the Sechelt Indian word for 'neck' – perhaps referring to the narrow neck of land between the end of the lake and the sea.

SAKWI CREEK, flows SE into Weaver Cr. (B-9). From the Mainland Halkomelem word meaning 'steelhead place.'

SALMO, S. of Nelson (B-11). *Salmo* is the Latin word for 'salmon.' The Salmo River is referred to as the Salmon River as early as 1859. The settlement first began as Salmon Siding, during the building of the Nelson and Fort Sheppard (Shepherd) Railway.

SALMON ARM, Shuswap L. (C-10). Takes its name from Shuswap Lake's southwest arm, which in turn was named back in the days when the salmon swarmed up its creeks and river in such numbers that the settlers pitchforked them out of the water and used them for fertilizer on their fields.

SALTERY BAY, Jervis Inlet (B-7). After a fish-salting plant operated here many years ago.

SALTSPRING ISLAND, Gulf Islands (A-8). The name was applied early by HBC officers who were interested in the possibility of getting salt from the briny springs on this island. In 1859 Captain Richards named this Admiral Island and its highest peak Mount Baynes, in honour of Rear-Admiral Baynes, then in command of the Pacific Station. The name Saltspring Island continued to be used by the islanders, however, and in 1905 it was made the official name of the island.

SAMBO CREEK, flows N. into Dease R. (L-5). Sambo was 'a small jolly Indian,' a born clown and mimic, who worked for a trader at Lower Post.

SAMUEL ISLAND, between Mayne I. and Saturna I. (A-8). After Dr. Samuel Campbell, assistant surgeon, HMS *Plumper*. (See *Campbell Bay*.)

SAN CHRISTOVAL RANGE, Moresby I. (E-4). Named by Juan Perez in 1774 when he sighted these mountains from the sea.

SANDNER CREEK, flows S. into Christina L. (B-10). After Charles Sandner, who arrived from Chicago in 1896 to make his fortune prospecting. According to the Boundary Historical Society's fifth report, 'He found very little mineral riches, but the beauty and freedom he found caused him to stay ... until his death in 1934.'

SANDON, E. of Slocan L. (B-11). After John Sandon, a French-Canadian. Sandon and Eli Carpenter discovered ore in this area in 1891, and the townsite was staked the following year. Sandon drowned in Kootenay Lake in 1893.

SANDWICK, N. of Courtenay (B-7). Named after Sandwick parish in the Shetland Islands by Eric Duncan, first postmaster here and author of *From Shetland to Vancouver Island*.

SAN JOSE RIVER, flows NW into Williams L. (E-8). San Jose is the Spanish for 'St. Joseph,' and this stream takes its name from St. Joseph's Mission.

SAN JOSEF BAY, Cape Scott Park (C-5). Apparently unaware that in 1786 Captain Hanna had named this St. Patrick's Bay, Francisco de Eliza gave it its present name in 1791.

SAN JUAN RIVER, W. end of Strait of Juan de Fuca (A-7). Takes its name from Port San Juan (the Spaniards' Puerto de San Juan), into which it enters. Apparently the latter was so named because José Maria Narvaez had been there on the day of John the Baptist (24 June), 1789.

SANS PEUR PASSAGE, W. side of Hunter I. (D-5). Named after the Duke of Sutherland's yacht, in BC waters when World War II broke out. The *Sans Peur's* cruise around the world came to an abrupt end when she was transferred to the Royal Canadian Navy.

SANSUM NARROWS, W. of Saltspring I. (A-8). After Arthur Sansum, the great fat first lieutenant on HMS *Thetis* when she was in these waters in 1852-3. In the latter year, he died at Guaymas, Mexico, of heat apoplexy.

SAPPERTON, New Westminster (B-8). A number of the 'sappers' (Royal Engineers) despatched to British Columbia in 1858 and 1859 formed a building society, purchased lots, built themselves houses, and so founded the village that they called Sapperton. Sapperton, however, was first used by Colonel Moody as the name for the campsite that the Royal Engineers had cleared here, to the east of the original townsite of New Westminster.

SARAH ISLAND, E. of Princess Royal I. (E-5). After Sarah, second daughter of Chief Factor John Work of the HBC. In 1859 she married another HBC Chief Factor, Roderick Finlayson (note adjacent RODERICK ISLAND).

SARDIS, S. of Chilliwack (B-9). When Mrs. Vedder was asked to name the new community, she opened her Bible at random and, putting her finger down, lit on Sardis, seat of one of the seven churches of Asia (see *Revelations* 3.4).

SASAMAT LAKE, NW of Port Moody (B-8). The Spanish explorers noted that the Indian name for at least part, if not all, of what is now Burrard Inlet was Sasamat. Randy Bouchard and Dorothy Kennedy have pointed out to the authors that Sasamat may have been the Halkomelem name of an early Indian low-class village on the site of Ioco, near Port Moody. The name Sasamat can be translated as 'lazy people,' but it is not known why the inhabitants of this village were so named.

SASQUATCH CREEK, E. of S. end of Harrison L. (B-9). Stories about the Sasquatch, a giant hairy, wild, and apelike creature, are rife in the area. (See Belle Randall, *Healing Waters: History of Harrison Hot Springs and Port Douglas Area.*)

SATELLITE CHANNEL, S. of Saltspring I. (A-8). After HMS *Satellite,* on this coast 1857-60 under Captain J.C. Prevost, a commissioner for establishing the maritime boundary between the British territory and the United States. *Satellite* gave much useful service during the Fraser River gold rush, particularly as a guardship at the mouth of the Fraser River checking that miners going upstream had licences to prospect for gold.

SATURNA ISLAND, Gulf Islands (A-8). Named in 1791 after the *Saturnina,* a small Spanish naval ship of seven guns that, under the command of José Maria Narvaez, explored these waters. As early as 1862, Saturnina Island had become corrupted into Saturna Island.

SAUL LAKE, NW of Kamloops (C-9). Saul Thoma, an old Indian, liked to hunt and fish here.

SAVARY ISLAND, NW of Powell River (B-7). Captain Vancouver notes in his journal for June 1792 that he sailed past 'an island lying in an east and west direction, which I named Savary's Island.' He does not tell us who Savary was, and nobody has been able to identify him.

Because of its shape, the Indians called this island Ayhus, meaning 'double-headed serpent.' This serpent was trying to return to its cave on Hurtado Point when the Transformer came by and changed it into this island.

SAVONA, W. end of Kamloops L. (C-9). In January 1859 Donald McLean wrote from Kamloops to Governor Douglas:

> There is a native of Corsica here, named Savona. He is master of the French, Spanish and Italian languages, and would be willing to enter into the employ of the Government should his services be required as an Interpreter. He writes the above languages, as well as speaking them. Savona is married to a daughter of St. Paul [see *Lolo, Mount*], and can, I believe, be depended upon as a steady sober person – as he now makes use of no intoxicating liquors.

Later in 1859 Francis Savona (François Saveneux) established a ferry across the Thompson River where it flows out of Kamloops Lake.

At the north end of the ferry, close to the HBC wharf and warehouse, a settlement known as Savona's soon grew up. When Savona died in 1862, his wife took over the ferry, but in 1870 the government acquired it. When the CPR was built, the population moved to a new townsite, Port Van Horne, which the railway had built on the southern shore. The new name, however, soon gave way to Savona.

SAYWARD, N. coast of VI (C-7). After William Parsons Sayward. Born in Maine in 1818, he was a carpenter and lumber merchant in California before moving to Victoria in 1858. He became a very successful lumberman on Vancouver Island.

SCARLETT POINT, Balaklava I. (C-6). Named after Sir James Yorke Scarlett, leader of the Heavy Cavalry charge at the Battle of Balaklava, 1854.

SCARLETT O'HARA, MOUNT, part of Starbird Ridge (C-11). The red rocks on this mountain reminded somebody of the red hair of the heroine of *Gone with the Wind.*

SCHAFFER, MOUNT, Yoho NP (D-11). After Dr. Charles Schaffer and his wife, Mary, who from 1889 on made annual summer trips from Philadelphia to botanize in the Rockies. By the time of Dr. Schaffer's death in 1903, Mary had overcome her early distaste for 'roughing it' and become an enthusiast for the mountains. With her friend Mollie Adams, she was among the first women to make packhorse expeditions into the remote parts of the Rockies. In 1915 she settled in Banff, having married Bill Warren, a local guide. Mary illustrated *Alpine Flora of the Canadian Rockies* (1907) based on Dr. Schaffer's research and was herself the author of *Old Indian Trails of the Canadian Rockies* (1911).

SCHOEN LAKE PARK, NW of Strathcona Park (C-6). After Otto Schoen, 'a very very expert canoe man' who once trapped in the area.

SCOTCH CREEK, flows S. into Shuswap L. (C-10). After Scottish prospectors who worked this creek for gold in the 1860s. There is still a bit of placer mining done on the creek.

SCOTCH FIR POINT, entrance to Jervis Inlet (B-7). Named by Captain Vancouver as 'producing the first Scotch firs we had yet seen.'

SCOTT, CAPE, NW tip of VI (C-5). Apparently named in 1786 by James Strange during his trading voyage with the *Captain Cook* and *Experiment.* David Scott was the principal member of the Bombay syndicate that had backed Strange's expedition.

SCOTTIE CREEK, flows SW into Bonaparte R. (C-9). After William (Scotty) Donaldson, who was conducting a rowdy two-room roadhouse on the

Bonaparte River as early as October 1861. An Orkneyman who wore his Scottish bonnet in all seasons, Scotty was a hard drinker. Miners putting up overnight during the Cariboo gold rush were sometimes roused in the middle of the night by a hallucinating drunken Scotty, who, convinced that he was about to be robbed and murdered, would drive them out blanketless into the night. In 1872 Scotty sold out and moved to the Okanagan. (See next entry.)

SCOTTY CREEK, flows W. into Kelowna Cr. (B-10). After leaving the Cariboo, the drunken innkeeper William (Scotty) Donaldson (see preceding entry) preempted land here on 25 September 1872 but never obtained title. He died in the Okanagan in 1882 at the age of sixty.

SCUZZY CREEK, flows SE into Fraser R. S. of North Bend (B-9). From a Thompson Indian word meaning 'jump' or 'jump across' (with reference to falls on the creek). Andrew Onderdonk's famous little steamer *Skuzzy*, used in the construction of the CPR, was named after this creek.

SEABIRD ISLAND, Fraser R., NE of Agassiz (B-9). After the American paddlewheel steamer *Sea Bird*, which ran aground here on 24 June 1858. The river dropping, the steamer was left high and dry, and a set of ways had to be built to relaunch her. She was totally destroyed by fire off Victoria a couple of months later.

SEAFORTH CHANNEL, E. of Milbanke Sound (E-5). Apparently named by HBC officers after Lord Seaforth, Baron Mackenzie of Kintail (1754-1815), who if the title had not been attainted would have been ninth Earl of Seaforth. He rose to be a lieutenant-general in the British Army, one of his important services being the raising in 1793 of the Ross-shire Buffs, better known as the Seaforth Highlanders. This famous regiment having been disbanded in Britain, Vancouver's Seaforth Highlanders of Canada alone survive to maintain its traditions.

SEA OTTER COVE, Cape Scott Park (C-5). This is the Sea Otter Harbour that Captain James Hanna named in 1786 after his ship *Sea Otter*.

SECHELT, W. of Howe Sound (B-8). There are at least three versions of the meaning of this name: (1) it was the ancient name that the Sechelt Indians had for themselves; (2) it is derived from a word meaning to 'climb' or 'climb over,' possibly referring to a fallen tree over which people had to climb in order to get up from the beach; and (3) it is derived not from an Indian word but from the English 'sea shelter.'

The present village of Sechelt did not come into being until the arrival of Catholic priests in the 1860s. The Indian word for the site of the present village means 'outside on the ocean side,' while the name of the area fronting on Porpoise Bay means 'inside.'

SEDGWICK BAY, Lyell I., QCI (E-4). Named by G.M. Dawson after the Reverend Adam Sedgwick (1785-1873), the noted English geologist.

SEGHERS, MOUNT, N. of Hesquiat Harbour, VI (B-6). After Charles John Seghers, Roman Catholic Archbishop of Vancouver Island, with jurisdiction over Alaska. He was killed by a demented companion on the banks of the Yukon River in 1886 and interred in St. Andrew's Cathedral, Victoria, in 1888.

SELKIRK MOUNTAINS, between the Arrow and Kootenay Lakes, and from the Big Bend of the Columbia to the US border (B-, C-, and D-10 and 11). They were originally named Nelson's Mountains by David Thompson, after the hero of Trafalgar (we now have the subordinate NELSON RANGE). After the union of the NWC and the HBC in 1821, they were renamed in honour of Thomas Douglas, fifth Earl of Selkirk (1771-1820).

As a means of furthering his philanthropic schemes for resettling dispossessed Scottish crofters in British North America, Selkirk in 1811 secured control of the HBC and founded the Red River settlement. This project led to violent animosity on the part of the NWC that culminated in the Seven Oaks Massacre of 1816. In 1817 Selkirk personally reestablished the colony. Resulting lawsuits broke his health but not his spirit. After his death his close friend Sir Walter Scott said of Selkirk, 'I never knew in my life a man of more generous distinction.'

SELMA PARK, E. of Sechelt (B-8). After the All Red Line's excursion steamer *Selma*, originally the eccentric Marquess of Anglesey's yacht *Santa Cecilia*. Soon after the Union Steamship Company acquired both the ship and the park in 1917, it changed the *Selma*'s name to *Chasina* but kept the Selma Park name for the resort that it developed here.

SELWYN, MOUNT, S. of Peace Reach, Williston L. (H-8). Named after A.R.C. Selwyn (1824-1902), director of the Geological Survey of Canada from 1869 to 1895. In 1875 he took an expedition up the Peace River to see if a mountain there could be as incredibly precipitous a cone as an English illustrator of W.F. Butler's *The Wild North Land* had made it. He found that the mountain was indeed an impressive one but not at all like the artist had shown it. At the suggestion of Professor John Macoun, the expedition's botanist, the mountain was named for Selwyn.

Another Mount Selwyn (in the Selkirks) was named for him, as was SELWYN INLET in the Queen Charlotte Islands.

SEMIAHMOO BAY, S. of White Rock (B-8). This is the name of one of the subgroups of the Straits Salish. Their territory included the shore of Boundary, Mud, and Semiahmoo Bays and below the forty-ninth parallel.

SEMLIN, E. of Cache Creek (C-9). After Charles Augustus Semlin, who arrived in the area in 1864 and soon established himself as a successful rancher. He entered politics early, becoming a member of the provincial legislature in 1871. He was Premier of British Columbia from 1898 to 1900. He died in 1927, aged ninety-one.

SENANUS ISLAND, Saanich Inlet (A-8). A Saanich (Straits Salish) Indian word meaning 'chest' (of a person).

SETON LAKE, W. of Lillooet (C-8). A.C. Anderson said that he named this lake after a 'near relative and playmate of my early days, Colonel Alexander Seton of the 74th, whose heroic fate I also commemorate by naming the Birkenhead Strait after the ship in which he so nobly perished.' (See *Birkenhead River*.)

70 MILE HOUSE, N. of Clinton (D-9). Seventy miles from Lillooet on the old Cariboo Road.

SEYMOUR INLET, Central Coast (D-6). Named after Frederick Seymour, Governor of British Columbia from 1864 until his death, from acute alcoholism, aboard HMS *Sparrowhawk* at Bella Coola in 1869. MOUNT SEYMOUR and SEYMOUR CREEK north of Vancouver are also named after him. So was SEYMOUR CITY at the north end of SEYMOUR ARM of Shuswap Lake, an important centre during the Big Bend gold rush of 1865-6.

SEYMOUR NARROWS, N. of Campbell River (C-7). After Rear-Admiral Sir George Francis Seymour, commanding the Pacific Station 1844-8. Until the tremendous underwater blast that removed Ripple Rock in 1958, the navigation of these narrows was hazardous.

SHAKES ISLANDS, S. of Porcher I. (F-4). After a renowned chief of the Kitkatla Indians who died in 1901, Shakes being an anglicized version of his hereditary Indian name. Captain Walbran relates how he gave Queen Victoria a potlatch gift of $100, to which she responded with two valuable rugs and a steel engraving of herself. Years before his death, he erected a monument to himself. Its inscription read, 'In memory of William Ewart Gladstone Sheuksh, Great Chief of the Kitkatlahs.'

SHALALTH, N. side of Seton L. (C-8). From a Lillooet Indian word meaning 'lake.' Locally called 'Shalath.'

SHANNON CREEK, flows NW into Howe Sound (B-8). Also SHANNON FALLS PARK. After William Shannon (1843-1928), who manufactured bricks here before World War I.

SHATFORD CREEK, W. of Penticton (B-10). After Walter T. Shatford, local merchant and land developer in the 1890s and 1900s.

SHAUGHNESSY, MOUNT, Glacier NP (D-11). After Thomas George

Shaughnessy, first Baron Shaughnessy (1853-1923). He was an American railwayman recruited in 1882 over a glass of Milwaukee beer as general purchasing agent of the CPR. Shaughnessy was president of the company from 1898 to 1918. Knighted in 1901, he became a peer in 1916. SHAUGHNESSY HEIGHTS in Vancouver is also named after him.

SHAW SPRING, Thompson R. (C-9). Named in 1930 when W.H. Shaw applied for and received the water rights to this spring.

SHAWATUM MOUNTAIN, E. of Skagit R. (B-9). From the Thompson Indian word meaning 'we asked him.'

SHAWNIGAN LAKE, S. of Duncan (A-8). The name first appears in 1859. Linguists have not been able to identify it with any known BC Indian word – indeed, it may not be Indian at all. Mrs. Beryl Cryer, who came to the lake in 1892, maintained that it was a hybrid name based on those of two early settlers, Shaw and Finnegan.

SHEARWATER, Denny I. (E-5). After HMS *Shearwater*, which served on this coast as a ship of the Royal Navy from 1902 to 1915, when she was transferred to the Royal Canadian Navy. The Canadians used her as depot ship for the two submarines that British Columbia had bought in Seattle. In 1917 she conducted them to the east coast by way of the Panama Canal.

SHEBA, MOUNT, N. of Gun Cr. (D-8). This peak with its twin summits was originally Sheba's Breasts, but some prudish civil servant put a stop to that!

SHEEMAHANT RIVER, flows W. into Owikeno L. (D-6). From an Oowekyala Indian word meaning 'taken or taking down [from heaven or sky].' The name is appropriate since this river takes its rise from a large icefield.

SHEFFIELD, MOUNT, S. of Muskwa R., N. BC (J-7). After Bert Sheffield, a trapper who drowned about 1945. A very good woodsman, with only a .22 rifle he could live off the country for six months at a time.

SHEGUNIA RIVER, flows E. into Skeena R. (H-6). From the Nisgha Indian word meaning 'water from the place of spring salmon.'

SHEILA LAKE, S. of Wells Gray Park (D-10). Those who seek the austere beauty of the high country will find near timberline south of Trophy Mountain a metal plaque declaring that the lake below is named after Mrs. Sheila Leonard (1939-77). She so loved the country that after she died from cancer her ashes were scattered here.

SHELFORD HILLS, NW of Ootsa L. (F-6). After two brothers from England, Arthur and Jack Shelford, who settled in the area about 1909. In the next generation, Cyril Shelford became a provincial cabinet minister.

SHELLEY, NE of Prince George (G-8). After a railway contractor, not the poet.

SHERBROOKE CREEK, flows S. into Kicking Horse R. (D-11). After Sherbrooke, Quebec.

SHERIDAN LAKE, SE of 100 Mile House (D-9). After James Sheridan, first preemptor here.

SHERINGHAM POINT, S. coast VI (A-8). Named Punta de San Eusebio by the Spaniards in 1790 but given its present name in 1846 by Captain Kellett of HMS *Herald.* Commander (later Vice-Admiral) William L. Sheringham took part in various naval surveys, none of which apparently brought him to this coast.

SHINGLE CREEK, W. of Penticton (B-10). Earlier known both as Rivière aux Serpens and Beaver Creek. It gets its present name from the fact that early settlers cut their shingles from the cedars growing along this stream.

SHIRLEY, W. of Sooke (A-8). Named after the Hampshire village from which Edwin Clark, the first postmaster, came.

SHORTS CREEK, upper W. side of Okanagan L. (C-10). After Captain Thomas D. Shorts (1837-1921), who preempted here in 1883 and was captain of the first steamboat on the lake, the *Mary Victoria Greenhow,* 1886.

SHORTY STEVENSON, MOUNT, W. of Bear R. (I-5). After an old-time prospector from New Brunswick. A giant of a man, he had to have an outsize uniform ordered from Ottawa when he enlisted in World War I. Killed in action.

SHULAPS RANGE, between Carpenter L. and Yalakom R. (C-8). From the Lillooet Indian word for the ram of the bighorn sheep. (See *Yalakom River.*)

SHULUS, NW of Merritt (C-9). A Thompson Indian word meaning 'open area,' pronounced 'shoe loose.'

SHUMWAY LAKE, S. of Kamloops L. (C-9). After Ammi Warren Shumway, a Mormon who left Salt Lake City after the death of his wife in 1863. Arriving in British Columbia, he became a packer on the Cariboo Road, wintering pack animals in the upper Nicola Valley. He helped in the search for the McLean outlaws in 1879.

SHUSHARTIE BAY, S. side of Goletas Channel (C-6). Derived from the Kwakwala Indian word meaning 'where the cockles are.'

SHUSWAP LAKE, E. of Kamloops (D-10). Named after the Shuswap Indians, an Interior Salish band who ranged from west of the Fraser River east to the Columbia River, and from north of McBride to south of Kamloops. Shuswap may be derived from *sixwt,* meaning 'downriver.'

SHUTE PASSAGE, N. of Swartz Bay (A-8). After Captain James Shute, Royal Marines, on board HMS *Topaze,* on Pacific Station 1859-63.

SHUTTLEWORTH CREEK, flows W. into Okanagan L. (B-10). After Harry Shuttleworth, who preempted land near Keremeos in 1877. He was a stock-man, big-game guide, and hunter.

SHUTTY BENCH, N. of Kaslo (B-11). After John Shutty and his son Andrew, who arrived from Czechoslovakia about 1900. Having established a home-stead here, they were joined by the rest of the family, along with five other Slovak families.

SICAMOUS, Shuswap L. (C-10). From the Shuswap Indian term meaning 'narrow,' or 'squeezed in the middle.' This is a good description of this area where the growing delta of Eagle River has progressively narrowed what once was another arm of Shuswap Lake and formed Mara Lake and Mara Channel (into Shuswap Lake).

SICKER, MOUNT, S. of Chemainus (A-8). After John J. Sicker, a pioneer homesteader with a reputation for stubborn individualism.

SICK WIFE CREEK, flows E. into Vents R. (L-6). From the message found on a blaze on a tree near the trail here: 'no meat, wife sick, go home.'

SIDLEY, E. of Osoyoos (B-10). After Richard G. Sidley, the first settler. In 1905 he stated that he owned 12,000 acres. (For his radical politics, see *Anarchist Mountain.*)

SIDNEY, N. Saanich Pen. (A-8). The town of Sidney takes its name from nearby Sidney Island, originally known as Sallas Island but renamed Sidney Island in 1859 by Captain Richards of the survey ship *Plumper.* Apparently the Sidney he had in mind was Frederick W. Sidney, who entered the Royal Navy only a few months after Richards and like him joined the surveying branch.

The Saanich Indians knew the site of Sidney as 'Tseteenus,' meaning 'sticking out.'

SIFFLEUR LAKE, W. of Tahtsa L. (F-6). The *siffleur* is the marmot.

SIFTON PASS, Cassiar (J-6). This pass, linking the valleys of the Kechika and Fox Rivers, was named by Inspector Moodie of the NWMP in 1898 during his patrol from Edmonton to the Yukon. He named it after Clifford Sifton, the federal Minister of the Interior from 1896 to 1905. Sifton was knighted in 1915 and died in 1929. Also MOUNT SIFTON, north of Rogers Pass (D-11).

SIKANNI RANGE, headwaters of Omineca R. (I-6). According to Father Morice, the forebears of the Sikanni people originally lived on the eastern slopes of the Rocky Mountains but were gradually forced westward into the recesses of the Rockies by their enemies, the Beaver Indians, who had acquired firearms from the fur traders. Their name, Sikanni, means 'people on the rocks' – that is, the Rocky Mountains.

SIKANNI CHIEF RIVER, flows N. into Fort Nelson R. (K-9). Named for a Beaver Indian 'dreamer' or 'prophet,' Makenunatane, known to the white men as the 'Sikanni Chief.'

SILVERDALE, W. of Mission (B-8). Writing in response to an enquiry from Ottawa in 1906, J.A. Skinner, formerly the postmaster here, declared, 'I selected the name of Silverdale from a list of names of places in England, a small town in Staffordshire.' He noted that the name was appropriate for the Fraser Valley settlement since a Silver Creek is there and the settlement lies in a valley.

SILVER STAR MOUNTAIN, NE of Vernon (C-10). Formerly Aberdeen Mountain. Now named after the Silver Star Mining Co., which had a claim here.

SILVERTON, Slocan L. (B-11). After Silverton, Colorado.

SIMILKAMEEN RIVER, flows S. into USA (B-10). Alexander Ross, in his account of his visit to the Kamloops area in 1812, mentions returning to Fort Okanogan by way of the Samilkameigh River, and he lists the Samilkameigh as one of the twelve Indian groups making up the Okanagan nation. In his Columbia journal of 1825, Sir George Simpson mentions the Samilkameighs as living on the north side of the river of the same name.

Samilkameigh became corrupted into Similkameen by analogy with its tributary, the Tulameen River. A.C. Anderson, the pioneer HBC man, speaks of 'Similk-ameen – literally Salmon-river,' though he notes that there are no salmon in it. Actually the derivation of Samilkameigh is probably lost forever, for the word comes from the now extinct language known as Nicola-Similkameen, one of the Athapaskan language group.

SIMOON SOUND, Gilford I., N. of entrance to Knight Inlet (C-6). Named in 1863-4 when Rear-Admiral John Kingcome commanded the Pacific Station. Back in 1853 he had commanded the troopship *Simoon*.

SIMPSON PASS, Alberta-BC boundary (D-12). After Sir George Simpson (1792-1860), Governor of the HBC in Canada 1822-60. Simpson Pass through the Rockies was discovered by a party headed by Governor Simpson in 1841.

In his journeys across the continent, Simpson travelled with amazing rapidity. His arrivals at the company's forts were state occasions, intended to impress the Indians with the greatness of the honourable company. We are told of how he arrived at Fort St. James with a bagpiper playing and the British ensign displayed, while the fort fired a salute.

Father Morice is the authority for an amusing little story. To impress the

Indians, Simpson had a tiny music box attached to his dog's neck in such a way that when it was started the music seemed to come from the dog's throat. Morice found the Carrier Indians still speaking of Simpson as the 'Great Chief whose dog sings.'

SIN LAKE, Peace R. district (H-9). An evangelical Bible camp was once established on its shore, with converts being baptized by total immersion in its waters. Local residents, remarking that a great deal of sin must have been washed off into the lake, took to referring to it simply as 'Sin Lake.'

SINCLAIR CANYON, E. of Radium Hot Springs (C-11). Known to the Indians as 'the red rock gorge,' it takes its present name from James Sinclair (1805-56), who, engaged by the HBC, led a party of emigrants through this passage in 1841, conducting them from the Red River settlement to Oregon. During most of his career, Sinclair was a 'free trader,' operating outside the HBC's monopoly of the fur trade.

SINKUT LAKE, W. of Cluculz L. (F-8). From a Carrier Indian word possibly referring to a deposit of red ochre.

SIR ALEXANDER, MOUNT, headwaters of McGregor R. (F-9). After Sir Alexander Mackenzie, whose route in 1793 lay some distance west of this mountain. (See *Mackenzie*.)

SIR ALEXANDER MACKENZIE PARK, Dean Channel (E-6). Here Sir Alexander Mackenzie, having completed the first crossing of the continent north of Mexico, painted on a great rock, 'Alexander Mackenzie from Canada by land 22 July 1793.'

SIRDAR, N. of Creston (B-11). Named after Field Marshal Lord Kitchener (1850-1916), who in 1892 was made sirdar (Commander-in-Chief) of the Egyptian army. While sirdar he won the Battle of Omdurman (1898) and established British control over the Sudan as well as over Egypt.

SIR DONALD RANGE, Glacier NP (D-11). Also MOUNT SIR DONALD. After Sir Donald Smith, later Lord Strathcona and Mount Royal (1820-1914). For biographical detail see *Strathcona*.

SIR FRANCIS DRAKE, MOUNT, E. side of Bute Inlet (C-7). After the famous Elizabethan navigator, who in 1579 sailed up the Pacific coast, perhaps as far as Cape Blanco on the Oregon coast. (For the case for Drake having come as far north as the BC coast, see R.P. Bishop, 'Drake's Course in the North Pacific,' BCHQ 3 [July 1939]:151-82.) It was Bishop who was responsible for the naming of this mountain.

SIR SANDFORD RANGE, N. from Glacier NP (D-11). Also MOUNT SIR SAND-FORD. Named after Sir Sandford Fleming, KCMG (1827-1915). In 1871 Fleming was appointed engineer-in-chief of the CPR, and under his direction the route was surveyed from Fort William westward to the Rockies and through the two alternative passes, the Yellowhead and the Kicking Horse. He was the first to advocate twenty-four-hour-day standard time for railways.

SISKA CREEK, S. of Lytton (C-9). This word and its variant, *cisco,* are generally thought to be derived from the Thompson Indian word meaning 'uncle.' However, another translation has been proposed: 'lots of splits [small cracks] in the rocks.'

SITKUM CREEK, flows SE into West Arm, Kootenay L. (B-11). A Chinook jargon word meaning 'middle,' or 'half.'

SIWASH ROCK, Stanley Park, Vancouver (B-8). Originally Ninepin Rock. *Siwash* is a Chinook jargon word, derived from the French *sauvage,* meaning a 'native Indian.' The original Squamish Indian name for Siwash Rock means 'standing up.'

The Squamish Indians have a legend about the creation of Siwash Rock. The Transformers (three brothers) were paddling along in their canoe and saw a man bathing and scrubbing himself with hemlock boughs. When they asked him why he was bathing, the man answered that his wife had just given birth to his first son. The Transformers then changed the man into a special rock with a small tree on top of it – the hemlock he had been using to scrub himself with. This rock is known as Siwash Rock. The wife was also turned into stone, and this smaller rock can still be seen nearby.

SKAGIT RIVER, flows S. into USA (B-9). This is the name of a Coast Salish Indian band living along this river. The meaning of Skagit is uncertain, but it may come from a Straits Salish word meaning 'to hide or conceal.'

SKAHA LAKE, S. of Okanagan L. (B-10). Originally Lac du Chien, then Dog Lake, and finally in 1930 officially named Skaha Lake to recognize local usage. *Skaha* is the Shuswap Indian, not the Okanagan Indian, word for 'dog.' Angus McDonald, a pioneer HBC man, noted that the company's men killed and ate fat dogs here when they could not get venison or fish. Dogtown was an early name of Okanagan Falls.

SKAIST MOUNTAIN, headwaters of Skagit R. (B-9). This name has the same derivation as that of Skihist Mountain (q.v.).

SKEDANS, Louise I., QCI (E-4). Originally the village was called Koona ('Edge Village'), and Skedans was the name of a prominent chief here.

SKEENA RIVER, N. coast (G-5). First named Ayton's River by Captain Duncan of the *Princess Royal* in 1788 and later known either as Simpson's River or as Babine River. The name Skeena River first appears on the map in 1861.

This name is derived from two Tsimshian Indian words meaning 'water out of the clouds.'

SKIDEGATE, QCI (F-4). The Haidas had the practice of referring to a village sometimes by its own proper name and sometimes by the hereditary name of its principal chief. Skidegate comes from 'Skit-ei-get,' said to mean 'red paint stone,' and was the hereditary name of the principal chief of the village (the name of which can be translated as 'pool of boulders village'). Samuel Patterson records that in July 1805 his ship, the *Juno* of Bristol, 'anchored at Skittagates.' Mention of 'Skettegats' was made in 1826 by Chief Factor McLoughlin of the HBC.

SKIHIST MOUNTAIN, W. of Lytton (C-9). From a Thompson Indian name that can be translated as 'great crack between rocks' or 'split rock.' This fine mountain was an important one to the Indians, being in an area where the young people went to train for guardian spirit power. SKIHIST PROVINCIAL PARK on the Trans-Canada Highway east of Lytton is so named because of the fine view it affords of this distant mountain.

SKIMIKIN CREEK, NW of Salmon Arm (C-10). From a Shuswap Indian word meaning 'along the sidehill.'

SKINCUTTLE INLET, E. coast of Moresby I. (E-4). The Haida name was 'Suu kaalhi,' meaning 'lake inlet.' The white man's Skincuttle is from the Natives' name for Slug Island, a compound of words meaning 'seagull' and 'to scoop.'

SKINS LAKE [SKIN'S LAKE], N. of Ootsa L. (F-7). After Skin Tyee, the chief of the little Indian community that once lived here.

SKOOKUMCHUCK, N. of Cranbrook (B-12). In the Chinook jargon, *skookum-chuck* means 'turbulent water' or 'rapid torrent' (*skookum* = 'strong'; *chuck* = 'water'). Also SKOOKUMCHUCK NARROWS, Sechelt Inlet.

SKUMALPASPH ISLAND, E. of Nicomen I. (B-8). From the Halkomelem word meaning 'many big-leaf maples.'

SKUTZ FALLS, Cowichan R. (A-8). From the Cowichan Indian word meaning 'waterfall.'

SKWAAM BAY, W. side of Adams L. (D-10). From the Shuswap Indian word for 'bay.' This bay was a wintering ground for the Adams Lake Indians and a place where they hunted caribou.

SKWAHA LAKE, W. of Spences Bridge (C-9). The name of this lake and the nearby ecological reserve comes from the Thompson Indian language. Meaning not certain but may pertain to black Spanish moss.

SKWAWKA RIVER, flows SE into head of Jervis Inlet (C-8). From a Sechelt Indian word that may mean 'river splits.'

SKY PILOT MOUNTAIN, E. of Britannia Beach (B-8). Back around World War I, 'Sky Pilot' became a slang term for a chaplain or any other member of the clergy. Sky Pilot Rock at the east end of Desolation Sound was named after the United Church's mission boat *Sky Pilot.*

    The highest summit on the ridge west of Sky Pilot Mountain is known as The Copilot.

SLATECHUCK MOUNTAIN, NW of Skidegate Inlet, Graham I. (F-3). The name, a curious compound of the English word *slate* and the Chinook word *chuck* ('water'), comes from the stream running off the mountain where the Haidas find the argillite out of which they carve small totem poles and other artifacts. The Haida name for this mountain is Kaagan. (See *Kagan Bay*.)

SLESSE, MOUNT, SE of Chilliwack (B-9). This name (pronounced 'suh-lée-see') comes from the original language of the Chilliwack Indians and means 'fang.' It is very descriptive of the mountain's appearance.

SLOAN, MOUNT, W. of Bralorne (C-8). After David Sloan, wealthy mining engineer and manager of the Pioneer Mine. He died of injuries suffered in a plane crash at Alta Lake in 1935.

SLOCAN LAKE, W. of Kootenay L. (B- and C-11). Members of the Palliser Expedition in 1859 passed the mouth of the Schlocan or Sloghan River. This name is derived from the Okanagan Indian word meaning 'pierce, strike on the head.' This refers to the Indian practice of spearing or harpooning the salmon that used to be plentiful throughout the Slocan district.

SLOLLICUM PEAK, E. of Harrison L. (B-9). Takes its name from nearby Slollicum Lake, which name is derived from the Halkomelem word meaning 'supernatural creature lake.'

SMITH INLET, Central Coast (D-6). 'Smith's Inlet' was named in 1786 by Captain James Hanna of the *Sea Otter,* who unfortunately has left us no information as to Smith's identity.

SMITH ISLAND, mouth of Skeena R. (G-4). After Marcus Smith, CE (1815-1904), who surveyed possible routes for the CPR prior to construction of the railway.

SMITH RANGE, E. of Powell R. (B-7). After Private Ernest A. (Smoky) Smith of the New Westminster Regiment, who won the Victoria Cross, 1 October 1944, by holding a bridgehead across the Savio River in Italy.

SMITHERS, on Bulkley R. (G-6). After Sir Alfred Waldron Smithers (1850-1924), chairman of the GTPR when a divisional point was established here in 1913. The name has been celebrated in verse in *Punch*:

Then let me go to Smithers in the West,
    And on my gravestone let these words be read:
'Attracted by its name to this fair scene,
    He died a Smitherene.'

SNASS CREEK, flows S. into Skagit R. (B-9). *Snass* is the highly expressive Chinook jargon word for 'rain.'

SOB LAKE, W. of Cluculz L. (F-8). A survey party got so annoyed with a cantankerous homesteader here that they named S.O.B. Lake to record their opinion of him. Victoria felt that this would never do and removed the periods.

SOCKEYE, SE of Prince Rupert (G-4). This rail point is one of various places in British Columbia bearing this name. It comes from the Halkomelem language and is the name that these Indians applied to what they deemed the best species of salmon.

SODA CREEK, N. of Williams L. (E-8). Takes its name from the white alkali powder that dries on the rocks.

SOINTULA, Malcolm I., N. of VI (C-6). In December 1901 a Finnish cooperative colony was founded here. The name chosen for it is the Finnish word for 'harmony.' Unfortunately, like most utopias, the colony proved neither harmonious nor successful and was disbanded in 1905, with about half of the members leaving the island.

SOLSQUA, on CPR, NE of Sicamous (C-10). An anglicization of the Shuswap Indian word meaning 'water.'

SOMASS RIVER, Port Alberni (B-7). Derived from the Nootka Indian word meaning 'current in water.'

SOMBRIO RIVER, E. of Port Renfrew (A-7). Named Rio Sombrio ('shady river') by Quimper in 1790.

SOMENOS, N. of Duncan (A-8). The plural of a Cowichan Indian word meaning 'a resting place.' Here, on a quiet stretch of the Cowichan River, canoemen rested before resuming their paddling upstream.

SONGHEES POINT, Victoria Harbour (A-8). This is the name applied to the Straits Salish people who moved from a number of places to the Victoria area after the founding of Fort Victoria in 1843. 'Songhees' may have originally been the name of an Indian band formerly living at Parry Bay, or it may be derived from a Straits Salish word meaning 'people gathered from scattered places.'

SONORA ISLAND, N. of Quadra I. (C-7). Named in 1903 by the Geographic Board of Canada to commemorate the schooner in which Quadra sailed along the BC coast in 1775. Actually the *Sonora* was never on the east side of Vancouver Island.

SOOKE, W. of Victoria (A-8). Takes its name from the Sooke band of Straits Salish Indians. 'A most warlike and hardy race,' they were nearly annihilated in a combined attack of the Cowichans, Clallums, and Nitinats launched about 1848. The derivation of Sooke is unknown. It was early spelled 'Soke' and pronounced 'soak.'

SOPHIA RANGE, Nootka I. (B-6). In the 1850s the five pretty daughters of Captain Edward E. Langford were the belles of Victoria. Captain George H. Richards, RN, surveying the coast, named this mountain range after Sophia Elizabeth, the fourth of the Langford girls. (See also *Langford Lake*.)

SORRENTO, S. shore of Shuswap L. (C-10). Mistakenly thinking that the CPR was to be rerouted along the shore of Shuswap Lake, an eastern capitalist, J.R. Kinghorn, laid out a townsite here and named it after Sorrento, where he had spent his honeymoon. Sorrento, Italy, looks out toward the Isle of Capri, and Sorrento, British Columbia, looks out toward Copper Island, which reminded Kinghorn of Capri. The original name of Sorrento was Trapper's Landing.

SOUES CREEK, NW of Clinton (D-9). After Frederick Soues, at one time government agent at Clinton, who preempted land by the creek in 1869. Also MOUNT SOUES, north of Kelly Lake.

SOUTHEY POINT, N. end of Saltspring I. (A-8). Named in 1859 after James L. Southey, RN, secretary to Rear-Admiral Baynes. (See also *Saltspring Island*.)

SOUTHGATE RIVER, flows W. into Bute Inlet (C-7). After J.J. Southgate, a slightly mysterious Victoria capitalist and politician, who flitted between London and Victoria in the 1860s.

SOWAQUA CREEK, flows NW into Coquihalla R. (B-9). A Thompson Indian name meaning 'to drink water.'

SPAHATS CREEK, flows W. into Clearwater R. (D-9). *Spahats* is the Thompson Indian word for 'black bear.'

SPAHOMIN CREEK, S. end of Douglas L. (C-9). From the Thompson Indian word meaning 'shavings or cuttings,' as of wood.

SPAIST MOUNTAIN, E. of Spences Bridge (C-9). From the Thompson Indian word that means 'burnt.'

SPALLUMCHEEN, N. of Vernon (C-10). Derived from the Shuswap Indian word *spalmtsin*, meaning 'flat area along edge.' (The Okanagan Indian word *spelemtsin* has the same meaning, and from it has come Spillimacheen River [q.v.].) For many years the Shuswap River, which flows into Mara Lake, was known as the Spallumcheen River.

SPANISH BANK, Vancouver (B-8). Near here, on 22 June 1792, Captain Vancouver met two Spanish ships at anchor – the brig *Sutil* commanded by Galiano, and the schooner *Mexicana* commanded by Valdes. Vancouver has left us a description of the Spanish warships:

> They were each about forty-five tons burthen, mounted two brass guns, and were navigated by twenty-four men, bearing one lieutenant, without a single inferior officer. Their apartments just allowed room for sleeping places on each side, with a table in the intermediate space, at which four persons, with some difficulty, could sit, and were, in all other respects, the most ill calculated and unfit vessels that could possibly be imagined for such an expedition; notwithstanding this, it was pleasant to observe, in point of living, they possessed many more comforts than could reasonably have been expected.

Vancouver had been exploring northward in a yawl, while the *Discovery* and *Chatham* were left at anchor south of Point Roberts. He brought his ships up to join the Spaniards, and then the four ships travelled north in company.

SPARWOOD, W. of Crowsnest Pass (B-12). In September 1905 the postmaster at Sparwood wrote to the Chief Geographer at Ottawa: 'It is commonly reported here that, at the time of the construction of the Crows Nest Pass Railway, the engineers remarked that the size and quality of the trees here were such as were required for spars; hence the name "Sparwood."'

SPATSIZI PLATEAU, headwaters of Stikine R. (J-5). From the Tahltan Indian word meaning 'red goat.' High up on a mountain is a red sandstone formation, and the mountain goats, which roll in the dust nearby, become tinted red.

SPATSUM CREEK, flows SW into Thompson R. (C-9). Derived from the Thompson Indian word for Indian hemp *(Apocynum cannabinum)*. This plant was important to the Indians as a source of cordage fibres, especially for fishing lines and nets.

SPECTRUM RANGE, W. of Iskut R. (J-4). So named because of the wide variety of colours presented by the lava from extinct volcanoes.

SPENCES BRIDGE [SPENCE'S BRIDGE], Thompson R. (C-9). This settlement was originally known as Cook's Ferry, since a ferry was operated here by Mortimer Cook between 1862 and 1865. The ferry was replaced in the latter year by a toll bridge built under government contract by Thomas Spence.

SPEYUM CREEK, flows E. into Fraser R. N. of Yale (B-9). From a Thompson Indian word meaning 'large flat area.'

SPILLER CHANNEL, N. of Seaforth Channel (E-5). After Corporal Richard Spiller, Royal Marines, who served on HM hired survey ship *Beaver* from 1863 to 1870. He was Commander Pender's personal attendant.

SPILLIMACHEEN RIVER, SE of Golden (C-11). From the Okanagan Indian word meaning 'flat area along edge' or 'flat mouth.' (See also *Spullumcheen*.)

In 1901 the BC member of the Geographic Board of Canada wrote to Ottawa as follows: 'Spillimachene has been used in the official Reports for some time – previously it was Spillemcheen, but the former represents the pronunciation best. I would suggest that Spillimacheen be adopted as conforming with terminations of Similkameen, Tulameen, etc.'

SPIUS CREEK, flows N. into Nicola R. (C-9). This is a Thompson Indian name meaning 'twisted creek.'

SPOKSHUTE MOUNTAIN, W. of Ecstall R. (G-5). This is the old name of the site of Port Essington and comes from the Tsimshian Indian word meaning 'autumn encampment.' It was here that the Indians camped while on their way down the Nass River to winter on ice-free Metlakatla Channel.

SPRINGHOUSE, W. of Lac la Hache (D-8). In gold rush days, a roadhouse here was close to St. Peter's Spring.

SPROAT LAKE, W. of Port Alberni (B-7). This lake was known by its anglicized Nootka Indian name of Kleecoot until 1864, when Dr. Robert Brown renamed it in honour of Gilbert Malcolm Sproat (1834-1913). Sproat had arrived from England in 1860 with men and equipment to establish a sawmill at the head of Alberni Canal. An amateur anthropologist, he became keenly interested in the life of the Native population of Vancouver Island, publishing in 1868 his *Scenes and Studies of Savage Life*.

Sproat returned to England in 1865, and in 1873 he became British Columbia's first agent-general in London. Returning to British Columbia about 1876, he was appointed to the Indian Land Commission. From 1885 to 1890, 'Judge' Sproat was gold commissioner for the West Kootenay district. With his friend Farwell (see *Revelstoke*), he was active in the real estate business after his retirement from government service. He spent his final years in Victoria.

MOUNT SPROAT on Upper Arrow Lake is also named after him.

SPROULE CREEK, flows into Kootenay R., W. of Nelson (B-11). Named after Robert E. Sproule (or Sprowle), who in 1882 staked the very rich galena ledge, which he named the Bluebell claim. Unfortunately there was no mining recorder close enough to be reached during the seventy-two hours that the law then permitted a man to be absent from a claim for filing purposes. Later that year a young Cornish prospector, Thomas Hammil, working for the

American capitalist Ainsworth, staked the same area and induced the gold commissioner at Wild Horse to come to Kootenay Lake and record his claim. In the ensuing litigation, a decision was given at Nelson in favour of Sproule. Ainsworth's lawyers took the case to the Supreme Court in Victoria, where the verdict was reversed. Driven out of his mind with anger, Sproule went to the Bluebell claim, where Hammil had started work, and, shooting from ambush, fatally injured him. Sproule was captured while trying to escape to the United States, tried, and hanged.

SPUZZUM, Fraser Canyon (B-9). From the Thompson Indian word meaning 'little flat,' with reference to the site of the settlement. Not far south of Spuzzum, at Saw Mill Creek, was the boundary between the Coast Salish Indians and the Thompson Indians of the Interior.

SQUAKUM LAKE, SW of Harrison Hot Springs (B-8). (See *Errock, Lake*.)

SQUAMISH, Howe Sound (B-8). This town and SQUAMISH RIVER take their name from the local Indians. No meaning is known for Squamish.

Around 1912 some real estate promoters, thinking that a more 'civilized' name would attract customers, renamed Squamish 'Newport.' The name was never popular with the local people, and a few years later the PGE Railway held a contest, inviting school children throughout the province to select a new name and win $500. The prize-winning new name was the old name, Squamish.

SQUEAH MOUNTAIN, SE of Yale (B-9). From the Halkomelem word meaning 'waterfall.'

SQUILAX, Shuswap L. (C-10). The authors vividly remember visiting an old Indian, Michel Toma, who was young when the CPR was building its transcontinental line at Squilax. He mentioned that a black bear had wandered out from the woods and along the cleared strip where the tracks were to be laid. The men laying the tracks asked the Indian onlookers what they called a black bear in their language. 'Squilax,' replied the Indians, and so this station got its name. (Only in the Chase or eastern dialect of the Shuswap Indian language is this the word for black bear.)

STAMP RIVER, NW of Port Alberni (B-7). After Captain Edward Stamp, an Englishman who first visited Vancouver Island in 1857 and began placing orders for timber for the British market. He was responsible for building the sawmill on Alberni Inlet for a Scottish company in 1860 and managed the mill for its first couple of years. He remained in the lumber business and in 1865 built Stamp's mill, the first on the south shore of Burrard Inlet.

STANFORD RANGE, E. of Windermere L. (C-12). Named by Captain Palliser in honour of Edward Stanford, English geographer and map publisher.

STANLEY, W. of Barkerville (F-9). Founded during the Cariboo gold rush of the 1860s, this settlement was named after Edward Henry, Lord Stanley, who in 1869 became Earl of Derby. (See also *Van Winkle Creek.*)

STANLEY PARK, Vancouver (B-8). Originally known as Coal Peninsula because of the coal deposits within its boundaries. It was early set aside as a military reservation where fortifications could guard the entrance to Vancouver harbour. In 1886 the Vancouver city council, with admirable fore-sight, petitioned the federal government for a lease of the land. The lease was granted in perpetuity for a fee of $1 a year. In September 1888 Lord Stanley of Preston, Governor-General of Canada, opened the park named after him.

STARBIRD RIDGE, W. of Radium (C-11). After Thomas Starbird, a native of Massachusetts, who ran a guest ranch on Horsethief Creek before World War I. An expert prospector, he discovered the Lake of the Hanging Glacier.

STAVE RIVER, NW of Mission (B-8). Also STAVE LAKE. The HBC had a cooper-age at Fort Langley, where the first barrels in British Columbia were made. The staves for them were cut on the banks of this river. An earlier name was Work's River. The Indians knew the river as s'Hai-uks or Skeeacks.

STAWAMUS RIVER, flows into N. end of Howe Sound (B-8). Takes its name from a once populous village of the Squamish band that stands at the mouth of this river. According to Indian tradition, people from this village greeted Captain Vancouver off nearby Watts Point in 1792.

STEELHEAD, N. of Mission (B-8). In the 1920s a logging railway was used to get out the timber here, and 'Steelhead' indicated the end of the steel.

STEIN RIVER, flows E. into Fraser R. (C-9). This was the name of an old Thompson Indian village at the mouth of the river.

STELLAKO RIVER, joins François L. and Fraser L. (G-7). From two Carrier Indian words, the one meaning 'peninsula' and the other 'river.'

STEMWINDER PARK, NW of Hedley (B-9). Takes its name from nearby Stemwinder Mountain, which in turn is named after the Stemwinder claim at Fairview Camp. The stemwinder watch was regarded as a notable improve-ment on the old watches that had to be wound with a key, and 'stemwinder' came to be a synonym for 'first rate.'

STEPHEN, MOUNT, Yoho NP (D-11). After George Stephen (1829-1921). Starting his working life as a draper's apprentice in Aberdeen, Scotland, Stephen became a buyer for a kinsman operating a drapery business in Montreal. Taking over this firm in 1860, Stephen prospered because of his canny business sense. In 1876 he became president of the Bank of Montreal and in 1881 president of the CPR. Becoming a peer in 1891, he took his title from this mountain, which earlier had been named after him, and became Lord Mount Stephen.

STEPHENS ISLAND, SW of Prince Rupert (G-4). Named by Captain Vancouver in honour of Sir Philip Stephens, secretary to the Admiralty from 1763 to 1795.

STEVENS LAKE, S. of Murtle L. (E-10). After Whitney W. Stevens of the BC Forest Service, who made a forestry reconnaissance of the Clearwater Valley in 1921.

STEVESTON, Lulu I. (B-8). After William H. Steves. The *Daily Colonist* of 31 July 1889 reported that 'W.H. Steves of Lulu Island has laid out a portion of his ranch in town lots ... It is fully expected that the little village now clustered at Steves will soon grow into a thriving business centre.' Within a month or so, Steves had been changed to Steveston.

The father of W.H. Steves, old Manoah Steves, a New Brunswicker, had settled in 1877 on Lulu Island, where he was joined the following year by his wife and six children. W.H. Steves was his eldest son.

STEWART, head of Portland Canal (H-5). This community, adjacent to Hyder, Alaska, takes its name from Robert Stewart, first postmaster, who, with his brother John, founded the community in 1903-4.

STEWART CREEK, flows E. into Christina L. (B-10). After Angus Stewart, prospector, newspaperman, and – to use the ambiguous language of the Boundary Historical Society's fifth report – 'custodian' of a colony of remittance men on English Point.

STIKINE RIVER, flows across Alaska-BC boundary (I-4). From a Tlingit Indian word meaning 'the river,' in the sense of 'the great river.' In the face of Russian expansion in Alaska and the discovery of gold in the area, the Stickeen Territories were established under British sovereignty in 1862 and incorporated into British Columbia in 1863.

STONER, S. of Prince George (F-8). It has been variously claimed that this settlement was named after Walter Stoner, a pioneer settler who later opened a store in Hixon, and a Stoner who was a civil engineer employed on the original PGE survey for its rail route from Quesnel to Prince George.

STOUT, E. side of Fraser R. (B-9). The name of this CNR point is derived from the Thompson Indian word *pstéwt*, which means 'across the river.'

STOUTS GULCH [STOUT'S GULCH], SW of Barkerville (F-9). After Ned Stout (1825-1924). A native of Germany, he fled that country to escape military service and became one of the 49ers of the California gold rush. He found gold in this gulch in 1861.

STOYOMA MOUNTAIN, NE of Boston Bar (B-9). From a Thompson Indian word meaning 'very bad or rough mountain.'

STRACHAN, MOUNT, W. of Capilano R. (B-8). After Admiral Sir Richard J. Strachan (pronounced 'Strawn'), 1760-1828. Strachan was one of those men seemingly born lucky but finally felled by a piece of ill fortune. Inheriting a baronetcy from an uncle at seventeen, he was a naval lieutenant at nineteen and the captain of HMS *Naiad* at twenty-three. In 1805 a squadron under his command captured four French ships of the line after a spirited battle, a victory that won Strachan the thanks of Parliament and promotion to the rank of rear-admiral. Then Strachan was put in command of the naval forces supporting the army's disastrous invasion of the Island of Walcheren in the Low Countries. The consequent obloquy attached itself to Strachan, who was never given another command, though he received the customary promotions on the inactive list to vice-admiral in 1810 and admiral in 1821.

STRAITON, E. of Matsqui (B-8). After Thomas B. Straiton, pioneer settler, who with his brother established a general store here on Sumas Mountain. A post office was opened in 1904.

STRANGE ISLAND, Tahsis Inlet, Nootka (B-6). After James Strange, a Bombay merchant who became interested in the maritime fur trade. With his two ships and their captains, he visited Nootka in 1786.

STRATHCONA PARK, central VI (B-7). Established in 1911, this, the first of British Columbia's provincial parks, was named in honour of Donald Smith, Lord Strathcona and Mount Royal (1820-1914).

Born in Scotland (a nephew of John Stuart of Stuart Lake fame), Smith entered the service of the HBC in 1838. While still one of the company's traders in Labrador, he began to buy stock in both the HBC itself and in the Bank of Montreal. In 1869 Smith became head of the HBC's Montreal department and in 1871 a Member of Parliament. His financial interests widening, he became a railway magnate and one of the men chiefly responsible for the building of the CPR. He was knighted in 1886. Three years later he became Governor of the HBC, a post that he held until his death. In 1896 he was appointed Canada's high commissioner in England, and the next year he acquired his magnificent title of Lord Strathcona and Mount Royal.

An extremely wealthy man, Lord Strathcona was a generous contributor to various charities. At his own expense, he raised a regiment, Lord Strathcona's Horse, for service in the Boer War. Opinion today is still divided about Smith. Some see him as a hard, self-centred man always interested chiefly in the wealth and renown of Lord Strathcona. To others he is a great builder of empire and a noble philanthropist.

STRATHNAVER, N. of Quesnel (F-8). In the summer of 1911, the fourth Duke of Sutherland, who owned more acres than any other British nobleman, detrained at Ashcroft and, with three automobiles for his entourage and a fourth for their liquor, headed north to add to his holdings. Here he purchased 4,000 acres of his own choosing and named his new estate Strathnaver ('valley of the Naver'). Earlier 'Lord Strathnaver' had been a courtesy title used by the heirs apparent of the Earls (later Dukes) of Sutherland. The next year the first party of Scottish settlers was sent out to Strathnaver, but in 1913 the Duke died and the whole colonization project was abandoned.

STUART CHANNEL, W. of Thetis I. and Kuper I. (A-8). After Captain Charles E. Stuart (1817-63), an experienced HBC officer lent to HMS *Virago* as a guide by Governor Douglas in 1853. Later Stuart was in charge of the company's Nanaimo post. Finally he became a free trader.

STUART ISLAND, mouth of Bute Inlet (C-7). Named by Captain Vancouver, who also named Bute Inlet. Stuart was the family name of the Earls of Bute.

STUART LAKE, E. of Babine L. (G-7). Known to the Indians as Na-kas-le or Naukazeleh (sometimes shortened to Na'kal) and to Simon Fraser as Sturgeon Lake, it was early renamed in honour of John Stuart of the NWC, who wintered here in 1806-7. In 1808, aged twenty-nine, Stuart was one of the two clerks who accompanied Fraser on his famous journey down the Fraser River. In 1809 Stuart succeeded Fraser in the command of the New Caledonia district. He became a partner in the NWC in 1813 and, after the merger of 1821, a Chief Factor in the HBC. According to Father Morice, Stuart 'seems to have been one of those well-meaning men who, unconscious of their own idiosyncrasies, make life a burden to others.' Stuart retired to Scotland and died there in 1847.

STUIE, near mouth of Atnarko R. (E-6). In his book *Spatsizi*, Tommy Walker, a well-known professional guide who at one time operated Stuie Lodge, tells of a restless wolf that set out to seek new hunting territory. He picked up a small boxlike object in his mouth and padded away, descending deep canyons until he came to a place where the mountains were farther apart. Here he rested, dropping the box, which grew into a house, the beginnings of the village of Stooick (Stuie), which means 'a pleasing place of repose' or 'a quiet place.'

STUMAUN BAY, at head of Port Simpson (G-5). From the Tsimshian Indian word meaning 'humpback salmon' (pink salmon).

STUMP LAKE, S. of Kamloops (C-9).

> Stump Lake, or Lac des Chicots, ... derives its name from the fact that stumps and prostrate trunks of trees are found submerged along its edges, and even far out from the shore, showing that it cannot long have occupied this part of

the valley. The Indians, indeed, say that some among them still living can remember the time when no lake existed there. (G.M. Dawson, *Preliminary Report of the Physical and Geological Features of the Southern Portion of the Interior of British Columbia, 1877* [Montreal, 1879], 29B)

STUPENDOUS MOUNTAIN, Bella Coola R. (E-6). When it came to naming this mountain officially, somebody remembered Sir Alexander Mackenzie's description of it as he reached the end of his famous journey overland of 1793: 'Before us appeared a stupendous mountain, whose snow-clad summit was lost in the clouds.'

STURGEON BANK, mouth of Fraser R. (B-8). Named by Captain Vancouver 'in consequence of our having purchased of the natives some excellent fish of that kind, weighing from fourteen to two hundred pounds each.'

SUGAR LAKE, E. of Mabel L. (C-10). The story goes that a negro, Alexander Clark, and two prospectors were out in a canoe on this lake when a bag of sugar accidentally spilled overboard. 'This will sure be a sugary lake!' remarked Clark.

SUGARLOAF MOUNTAIN, S. of W. end of Nicola L. (C-9). So named not because of its shape but because on it grew some Douglas-fir trees that pro-duced the so-called 'wild sugar' or 'Douglas fir sugar' (melezitose, a very rare trisaccharide). The Thompson Indian name of the mountain can be trans-lated as 'tree milk.'

SUKUNKA RIVER, flows N. into Pine R. (H-9). The last two syllables of Sukunka, a Sekani Indian word, mean 'fire' and 'along [the river].'

SULLIVAN HILL, NW of Kimberley (B-11). Named for Pat Sullivan, from Bantry Bay in Ireland, who with three partners discovered the fantastically rich Sullivan mine here. Sullivan was killed soon after in a cave-in in Idaho while working to earn enough money to develop the new mine. His partners sold the Sullivan mine in 1896 for $24,000 and sent $6,000 of the proceeds to Sullivan's relatives in Ireland. The Consolidated Mining and Smelting Company (now Cominco) bought the mine in 1910.

SULLIVAN LAKE, N. of Heffley Cr. (C-9). After Michael ('Ten Percent') Sullivan (1838-1908), an Irish entrepreneur who, after participating in the Cariboo gold rush, bought land in this area. 'He did very well ranching, butter-making, intercepting the North River Indians on their way to the HBC post and buying their furs, selling horses to the railway surveyors, and owning a share in the first threshing machine. With his considerable earnings he became a money lender, and held mortgages on many properties' (Mary Balf, *Chase*, p. 26). See also *Vinsulla*.

SUMALLO RIVER, flows SE into Skagit R. (B-9). From a Thompson Indian word possibly meaning 'Dolly Varden fish numerous here in times past.'

SUMAS, E. of Abbotsford (B-8). From a Halkomelem word meaning 'big flat opening.' Sumas Lake, once extensive, shallow, and infested with mosquitoes, has been drained for many years, providing lush agricultural land.

The earliest mention of Sumas Mountain is to be found in Simon Fraser's account of the country he passed through on 30 June 1808: 'Continued our course with a strong current for nine miles [down the Fraser] where the river expands into a lake. Here we saw seals, a large river [the Chilliwack River?] coming in on the left, and a round Mountain a head [*sic*], which the natives call *shemotch*.' In the mid-nineteenth century, Sumas was often spelled Smess.

The Halkomelem name for Sumas Mountain means 'gap left when a large chunk broke away,' or 'divided head,' because of the mountain's seemingly separate parts.

SUMMERLAND, N. of Penticton (B-10). The townsite here was laid out and named in 1902 by John Moore Robinson, the founder of Peachland and Naramata. The name refers to the sunny Okanagan climate. West Summerland was originally known as Siwash Flats.

SUPERB MOUNTAIN, E. of Bute Inlet (C-7). Named about 1860 after HMS *Superb*.

SUQUASH CREEK, NW of Port McNeill (C-6). From the Kwakwala Indian word meaning 'where seals are butchered.' This place is historic, for here in 1835 coal was first discovered on Vancouver Island. For a short time, coal was mined until better seams were found up the coast at Fort Rupert.

SURREY, Lower Mainland (B-8). Probably so named because it lies south of New Westminster, just as the county of Surrey lies south of Westminster (part of London) in England. Surrey became a city in 1983.

SUSKWA RIVER, flows W. into Bulkley R. (H-6). From the Carrier Indian word for 'black bear river.'

SUTHERLAND CREEK, flows SW into Christina L. (B-10). After J. Sutherland, a cowboy-prospector who spent years prospecting the upper watershed of the creek and died in a cabin at its mouth.

SUTIL, CAPE, E. of Cape Scott (C-5). When Galiano and Valdes passed this way in 1792, they named this cape after Galiano's schooner.

SWAINE, CAPE, Athlone I., Milbanke Sound (E-5). Named by Captain Vancouver after his third lieutenant on HMS *Discovery*, Spelman Swaine.

SWANNELL RIVER, flows N. into Ingenika R. (I-7). Also SWANNELL RANGES. After Frank Cyril Swannell, DLS, BCLS (1880-1969), perhaps the greatest of the land surveyors who have explored and mapped British Columbia. His career began in the era of packhorses and canoes, that of Mackenzie, Fraser, and Thompson, and ended in that of aircraft and four-wheel drive vehicles. His meticulously kept journals, illustrated with the photographs that he took wherever he went, are now cherished items in the Provincial Archives of British Columbia.

SWANSON CHANNEL, SW of North Pender I. (A-8) After Captain John Swanson (1827-72) of the HBC's maritime service. He arrived on this coast in 1842, and among his commands were the *Beaver*, the *Labouchere*, and the *Otter*. Also SWANSON BAY on the central coast.

SWARTZ BAY, N. Saanich (A-8). This should be Swart's Bay, it being named after John Aaron Swart, who purchased land in the vicinity of the bay in 1876.

SWISS BOY ISLAND, Imperial Eagle Channel, Barkley Sound (A-7). In 1859 the brig *Swiss Boy*, bound from Port Orchard to San Francisco with a cargo of lumber, put into the sound to stop a leak. The local Indians attacked and looted the ship. Only the intervention of one of the chiefs saved the lives of the captain and crew.

SYLVESTER PEAK, S. of McDame (L-5). After R. Sylvester, who came into the country during the Cassiar gold rush of 1874 and established several trading posts that were subsequently taken over by the HBC. His prominent proboscis inspired the miners to name this landmark 'Sylvester's Nose.' G.M. Dawson considerately modified the name after finding Sylvester friendly and helpful.

T

TABOO CREEK, flows NW into Skaist R. (B-9). During gold rush days, a party of Chinese camped where the trail crossed this stream. One of their number dying here, he remained unburied for some time, the others refusing to touch his body.

TABOR LAKE, E. of Prince George (F-8). After Clement Tabor, the first landowner at the south end of the lake.

TACHIE RIVER, joins Stuart L. and Trembleur L. (G-7). From the Carrier Indian word meaning 'three outlets,' referring to the river.

TADANAC, N. of Trail (B-11). 'T' for Trail, plus 'Canada' spelled backward.

TAFT, NE of Sicamous (C-10). The CPR reports that it named its station here after 'Mr. Taft of Hood Lumber Co.'

TAGHUM, W. of Nelson (B-11). The Chinook jargon word for 'six.' The siding here was named Taghum since it is approximately six miles west of Nelson.

TAHINI RIVER, Alaska-BC boundary (L-1). From the Tlingit Indian name meaning 'king salmon.'

TAHLTAN RIVER, flows SE into Stikine R. (K-4). Named after the Tahltan Indians, whose name can be translated as 'something heavy in the water.' The Indian story behind this naming is as follows: during a salmon run, two women stood on opposite banks of the Tahltan River. One asked the other what she saw on the surface of the water. The other replied, 'Something heavy is going up the little water,' referring to the ascending salmon working their way up through the rapids of a smaller stream.

TAHSIS INLET, N. of Nootka Sound (B-6). From the Nootka Indian word meaning 'trail at beach' or, in the words of explorer John Buttle, 'where the water travel stops and they have to walk.' An old Indian route across Vancouver Island was up Tahsis Inlet and River, then down Nimpkish Lake and River. Jewitt, in the account of his captivity, mentions that early in September 1803 the Nootka Indians, following their usual practice, moved to their autumn and winter quarters at 'Tashees and Cooptee.'

TAHTSA LAKE, NW of Whitesail L. (F-6). According to Father Morice, this Carrier Indian name means 'water far off.'

TAHUMMING RIVER, flows S. into Toba Inlet (C-7). Named after an old village site on its banks that had a Mainland Comox Indian name meaning 'slide place.' Here it was possible to place a dugout canoe on the upper part of the beach and slide it down easily over the sand and mud into the water.

TAKAKKAW FALLS, Yoho NP (D-11). A Cree Indian word meaning 'it is magnificent.' The falls, dropping 1,248 feet (380 metres), are thought to be the third highest in Canada, after Della Falls and Hunlen Falls.

TAKLA LAKE, N. of Babine L. (H-6 and 7). This name is said to be a poor approximation of the Carrier Indian name for an early trading post near the northern end of the lake. This name means 'at the end of the lake,' similar to the French-Canadian *fond-du-lac.*

TAKU RIVER, flows SW across Alaska-BC boundary (K-3). May be from the Tlingit Indian word for 'goose,' one informant expanding the meaning to 'place where the geese sit down.'

TAKYSIE LAKE, S. of François L. (F-7). From the Carrier Indian word meaning 'three heads.'

TALTAPIN LAKE, S. of Babine L. (G-7). From the Carrier Indian word meaning 'a cross,' such as one sees in a church or on a grave.

TALUNKWAN ISLAND, E. coast of Moresby I. (E-4). The island's name was originally applied by the Haida Indians to only its most easterly point (Heming Head). The name means 'to slice something that is fat.'

TAMIHI CREEK, flows N. into Chilliwack R. (B-9). From the Halkomelem word for Mount McGuire. According to Dr. Brent Galloway, the name means 'deformed baby finishes.' Such infants were sometimes left exposed on the mountain to die.

TANGIER RIVER, flows S. into Illecillewaet R. (D-11). After the Waverley-Tangier group of mineral claims staked at its head about 1900.

TANGIL PENINSULA, E. coast of Moresby I. (E-4). Named by the Canadian Hydrographic Survey using a word taken from a Haida vocabulary compiled by G.M. Dawson. The word means 'tongue' and is descriptive of this long, narrow peninsula.

TANTALUS RANGE, N. of Howe Sound (B-8). According to the ancient Greek myth, to punish Tantalus the gods afflicted him with thirst and hunger while keeping him lashed to a tree whose fruit always eluded his grasp; similarly, water that reached to his neck always receded when he sought to drink.

Dr. N.M. Carter explained the naming of the range thus: 'the view of the exciting-appearing peaks in this unnamed and unclimbed range tantalized early climbers of Mt. Garibaldi and suggested the legend of King Tantalus.' (See 'Early Climbs in the Tantalus Range,' *Canadian Alpine Journal* [1964]:74-7.)

The first ascent of Mount Tantalus itself was finally made in 1911 by Basil Darling, J. Davies, and Alan Morkill.

TANU ISLAND, N. of Lyell I. (E-4). From the Haida word for eelgrass *(Zostera marina)*.

TAPPEN, N. of Salmon Arm (C-10). In 1884 George Tappan (note spelling), a subcontractor, had a camp here while laying track for the CPR.

TASEKO RIVER, flows NE into Chilko R. (E-8). From a Chilcotin Indian word that means 'mosquito river.'

TASHME, SE of Hope (B-9). When a relocation centre was built here for Japanese-Canadians evacuated from coastal British Columbia in World War II, the surnames of the three members of the BC Security Commission supplied the camp's name of Tashme: TA for Austin C. Taylor, SH for John Shirras, and ME for F.J. Mead. The whole had an Oriental sound.

TASU SOUND, W. side of Moresby I. (E-4). A much-shortened form of the Haida name for this inlet that can be translated as 'lake of food inlet.'

TA TA CREEK, NE of Kimberley (B-12). Many variants of the tale of Ta Ta Creek are told in East Kootenay, but basically it is as follows. A horse thief was being taken to Fort Steele by a posse when he persuaded the constable to let him have a fresh horse. A little later, where a stream ran through a heavily wooded steep gully, the thief suddenly put his spurs to his horse and got away. Before disappearing into the woods, he turned to his pursuers, lagging on their jaded mounts, and uttered the famous line: 'Ta ta, friends, I've business up the trail.'

TATCHU POINT, NW of Esperanza Inlet (B-6). Said to be derived from the Nootka Indian word *tatchtatcha*, meaning 'to chew.' This was a great fishing area, and much food was available for feasts.

TATLA LAKE, N. of Tatlayoko L. (D-7). This name comes from the Carrier Indian word meaning 'at the end of the lake.' (See *Takla Lake*.)

TATLATUI LAKE, head of Firesteel R. (I-6). From a Tahltan Indian word meaning 'headwater.'

TATLOW, S. of Smithers (G-6). After Robert Garnett Tatlow, BC Minister of Finance, killed in 1910 when his horse bolted, throwing him from his trap. Also MOUNT TATLOW (D-8).

TATOGGA LAKE, N. of Kinaskan L. (J-4). From the Tahltan Indian word meaning 'between two lakes.'

TATTON, NW of 100 Mile House (D-9). This BCR station is named after

Tatton Park, Cheshire, the home of Lord Egerton of Tatton, who bought the 105 Mile Ranch about the time in 1912 that the Marquess of Exeter was buying the Bridge Creek Ranch at 100 Mile House.

TAYLOR, confluence of Peace R. and Pine R. (I-9). According to the municipal clerk, this was named after D.H. (Herbie) Taylor, an HBC trader who homesteaded here in 1906.

TAYLOR RIVER, flows E. into Sproat L., VI (B-7). After Charles Taylor, who homesteaded at the head of Alberni Inlet in 1863.

TAYLOR RIVER, flows SE into Nass R. (I-5). After Kenneth C.C. Taylor, DSO, BCLS, killed in action in September 1916. He had surveyed in the Nass area in 1913-14.

TCHITIN RIVER, flows SE into Nass R. (H-5). From the Nisgha Indian word meaning 'fish weir river.'

TEDIDEECH LAKE, W. of Dease L. (K-4). From a Tahltan Indian word meaning '[where the] bear reaches down into the water with his paws' – presumably to catch fish.

TELEGRAPH COVE, E. of Port McNeill (C-6). A station was established here in 1911-12 when the federal government built a telegraph line connecting Campbell River with the northern end of Vancouver Island.

TELEGRAPH CREEK, flows S. into Stikine R. (J-4). Named in 1866-7 when the Collins Overland Telegraph to Asia was expected to cross the Stikine River at this point. The project was abandoned because of the success of the trans-Atlantic cable.

TELEGRAPH PASSAGE, S. approach to Skeena R. (G-4). Named in connection with the Collins Overland Telegraph, 1865.

TELKWA, junction of Bulkley R. and Telkwa R. (G-6). Apparently from the Carrier Indian word for 'frog.'

TENAS LAKE, expansion of Atnarko R. (E-7). The Chinook jargon word for 'small' or 'little.'

10 DOWNING STREET, NE of Pemberton (C-8). Many years ago a retired trapper was appointed census taker here. Apparently he took his position very seriously, for some wag chalked on his door '10 Downing Street,' thereby giving a name to the locality. Later an old-timer, Albert Gramson, bought the property – hence the present official name of Gramson's. The old name, however, was too good to let go, and the BCR's schedule still lists 'Gramson's (10 Downing Street).'

TERRACE, E. of Prince Rupert (G-5). One of the earliest preemptors here was George Little, who snowshoed in over the Kitimat Trail in 1905. When the surveys for the GTPR took the railway across Little's land, he did not fight it but offered to give free land for a station. The offer was accepted, and on his adjacent land Little laid out a townsite. The settlement was to have been named Littleton, but the post office department refused to accept the name, there already being a Littleton elsewhere in Canada. The name finally adopted, Terrace, comes from the series of terraces (benches) rising above the Skeena here.

Little was a tough-minded entrepreneur who prospered mightily. There was once a saying around Terrace, 'If you don't work for George Little, you don't work.' (See Nadine Asante's excellent *History of Terrace*, pp. 136-8.)

TERRY FOX, MOUNT, N. of Valemount (E-10). After the young man from Port Coquitlam who, with one leg amputated, ran halfway across Canada in 1980 in his Marathon of Hope to raise funds for cancer research. Renewed ravages of the disease forced him to abandon the Thunder Bay-Vancouver stretch of his run, and he died the following year, but Canadians contributed over $22 million to the cause he championed and learned anew what a hero is. This mountain, 8,656 feet (2,651 metres) in altitude, was the highest unnamed BC peak within sight of a public highway when the province made it a memorial to Terry Fox.

TESLIN LAKE, BC-Yukon border (L-3). From an Athapaskan Indian name said to mean 'long, narrow water.'

TÊTE JAUNE CACHE, W. of Mt. Robson Park (E-10). In June 1820 word reached the NWC's Fort St. James that an HBC party had crossed the Rockies for the first time and penetrated to the junction of the Fraser and Nechako Rivers. For their guide the newcomers had Pierre Bostonais, commonly known as 'Tête Jaune' ('Yellowhead'). (See David Smyth, 'Tête Jaune,' *Alberta History* 32, 1 [1984]:1-8.)

Bostonais, one of the Iroquois who had come west at this period, probably had some French blood. That he was chosen to guide the HBC party argues a prior knowledge of the country west of the Yellowhead Pass. It would seem, then, that in earlier years he had trapped and hunted in the Tête Jaune Cache area, caching his furs in a secure place there for transfer at the end of the season to an HBC post east of the Rockies.

Tête Jaune is pronounced 'Tee John' by British Columbians.

TEXADA ISLAND, S. of Powell River (B-7). The Spanish navigator Eliza shows Isla de Texada on his chart of 1791. The island had been named after Felix de Tejada (or Texada), a Spanish rear-admiral.

TEZZERON LAKE, N. of Fort St. James (G-7). From the Carrier Indian word meaning 'moulting lake,' where ducks and geese moult.

THETIS ISLAND, E. of Ladysmith (A and B-8). After HMS *Thetis*, a thirty-six-gun frigate on the Pacific Station 1851-3. Several features in Esquimalt harbour are named after the ship, as is, presumably, Thetis Lake near Victoria. (See also *Kuper Island*.)

THIBERT CREEK, flows into N. end of Dease L. (K-4). After Henry Thibert, a French-Canadian who arrived in the area in 1872 and the next year found gold on this creek.

THOMPSON RIVER, joins Fraser R. at Lytton (C-9). Named after David Thompson, the famous explorer, who in fact never saw it. The river was given its name by Simon Fraser, who, when he reached the junction of the Fraser with the Thompson on 20 June 1808, wrote in his journal: 'These forks the natives call *Camchin,* and [they] are formed by a large river which is the same spoken of so often by our friend the old chief. From an idea that our friends of the *Fort des Prairies* department are established upon the source of it, among the mountains, we gave it the name of Thompson's River.' Actually Thompson had built Kootenae House the previous year on the upper reaches of the Columbia River. By a pleasing symmetry, it was Thompson who later gave the Fraser River its name.

David Thompson (1770-1857) came to Canada in 1784 as an apprentice in the service of the HBC. Partly self-trained, he became an amazingly proficient surveyor and switched to the service of the NWC as an 'astronomer' in 1797. In 1807, en route to the upper Columbia, he became the first white man to cross the Rockies by Howse Pass. In 1811 he became the first white man to use Athabasca Pass. On this second journey, he took a somewhat roundabout route to get to the mouth of the Columbia, where he arrived at Fort Astoria only a few months after it had been founded by the American-owned Pacific Fur Company. Had Thompson followed his instructions more promptly, he would have arrived there first and deprived the Americans of a major part of that weak case by which they secured, from a not particularly concerned Great Britain, the country lying between the lower Columbia and the present international boundary.

A person who met Thompson in 1817 described him thus:

> A singular looking person of about fifty. He was plainly dressed, quiet and observant. His figure was short and compact, and his black hair was worn long all round, and cut square, as if by one stroke of the shears, just above the eyebrows. His complexion was of the gardener's ruddy brown, while the expression of deeply furrowed features was friendly and intelligent, but his cut-short nose gave him an odd look. His speech betrayed the Welshman.

The South Thompson was at one time known as the Shuswap River and the North Thompson simply as the North River.

THORMANBY ISLANDS, W. of Sechelt (B-8). After the racehorse Thormanby, winner of the Derby in 1860. One of a whole series of place names in the area, including Derby Point, Epsom Point, Oaks Point, and Tattenham Ledge, that reflect the interest in horse racing shared by the officers of HM survey ship *Plumper*. (See also *Buccaneer Bay* and *Welcome Passage*.)

THORNBROUGH CHANNEL, Howe Sound (B-8). After Admiral Sir Edward Thornbrough (1754-1834), who commanded HMS *Latona* at Howe's victory of 'The Glorious First of June,' 1794.

THRUMS, NE of Castlegar (B-11). After the Scottish village that is the scene of Sir James Barrie's *A Window in Thrums* (1889). Various quips were made about the window when the Sons of Freedom, a Doukhobor sect, went in for public disrobing around here.

THURLOW ISLANDS, off NE coast of VI (C-7). Named by Captain Vancouver after Edward, first Baron Thurlow (1731-1806). He became Lord Chancellor of England in 1778. Of him it was said, 'No man ever was so wise as Thurlow looks.'

THURSTON BAY, Sonora I. (C-7). Probably after Robert F. Thurston, who in 1917 became president of the Thurston-Flavelle Lumber Company.

THUTADE LAKE, head of Finlay R. (I-6). The noted surveyor Frank Swannell often jotted down meanings of Indian names that he put on the map. He wrote that the translation of this Sekani Indian name is 'long, narrow lake.'

THUYA CREEK, flows E. into N. Thompson R. (D-9). *Thuya* or *thuja* is the botanical name applied to certain kinds of cedar. *Thuja plicata* is Western red cedar.

THYNNE CREEK, N. of Otter L. (B-9). After J.C. (Jack) Thynne, a great-grandson of a Marquess of Bath. He came into the Nicola country in 1887. He was much in demand as a fiddler at dances.

TICKLETOETEASER TOWER, W. of Chilko L. (D-7). A climbing party from the Alpine Club of Canada playfully named peaks in the Capital Group of mountains after various persons and places named in the nonsense poem 'A Capital Ship for an Ocean Trip.' Two of the lines in it run: 'And pink and blue was the pleasing hue / Of the tickle-toe-teaser's claws.'

TIEDEMANN GLACIER, E. of Mt. Waddington (D-7). After Herman O. Tiedemann (1821-91), a German civil engineer who settled in Victoria in 1858, where he designed the first legislative buildings. He accompanied Alfred Waddington to Bute Inlet in 1862 to examine the proposed route to the Interior.

TILBURY ISLAND, Fraser delta (B-8). After Tilbury, England, where a fort once guarded the Thames approach below London. Later it became famous for its docks.

TIMOTHY LAKE, E. of Lac la Hache (D-9). Takes its name from adjacent Timothy Mountain, noted for the wild timothy hay on its slopes.

TINGLEY CREEK, flows NE into Fraser near Marguerite (E-8). After Stephen Tingley, handyman and driver for the BX (Barnard's Express). In 1868 he went to New Mexico and brought back 400 horses for the company's stock farm. He was the most famous of the coach drivers on the Cariboo Road.

TINNISWOOD, MOUNT, head of Jervis Inlet (C-8). After William Tinniswood Dalton (1854-1931), Vancouver architect and enthusiastic mountaineer, one of the party that made the first ascent of Mount Garibaldi.

TINTAGEL, Burns L. (G-7). Named after Tintagel, Cornwall. A monument, exhibiting a stone from Tintagel Castle, reminds passersby of the tie between King Arthur's Tintagel and British Columbia's.

TISDALL, SW of Pemberton (C-8). After Charles E. Tisdall, BC Minister of Public Works 1915-16.

TLELL, E. side of Graham I. (F-4). While one source says there is no translation for this very old Haida name, another says it means either 'land of berries' or 'place of big surf.'

TLUPANA INLET, NE of Nootka Sound (B-6). After the Nootka Indian chief, Clewpaneloo, whom Captain Vancouver mentioned in his journal.

TOAD MOUNTAIN, S. of Nelson (B-11). On 27 July 1887, a prospector, Charlie Townsend, sat down on a log here to fill out his location notice for a promising claim. When he reached the phrase 'situated on,' a great toad jumped into sight. He then wrote down 'Toad Mountain' since, up to then, this mountain had been nameless.

TOBA INLET, E. of Bute Inlet (C-7). Originally named Canal de la Tabla by Galiano and Valdes, who found a strangely decorated Indian wooden tablet or plank here while exploring in 1792. In 1795 the name was changed to Toba, apparently in honour of Antonio Toba Arredondo, the only officer with Malaspina in 1791 who had not had a place named after him.

TOBACCO PLAINS, E. of L. Koocanusa (B-12). The Indians used to grow a kind of native tobacco here. It was apparently a long-established name when Father De Smet used it in the mid-nineteenth century.

TOBY CREEK, W. of Invermere (C-11). After Dr. Levi Toby, from Colvile, Washington, who prospected here in 1864.

TOD INLET, E. side of Saanich Inlet (A-8). Named after John Tod (1794-1882). Born in Scotland, he joined the HBC in 1813 and began many years of service in isolated posts. To illustrate that isolation, Tod liked to tell how news of the Battle of Waterloo did not reach him until some three years after the famous victory – belatedly but loyally he then fired off a salute at his fort.

In 1823 Tod came to British Columbia, where he rose to the rank of Chief Trader. He was in charge at Fort Alexandria in 1841 when Samuel Black, the Chief Factor at Fort Kamloops, was slain. Tod organized the search for the murderer. (See *Tranquille*.)

The next year he built a new fort at Kamloops, where he was now stationed. In 1846, according to one of Tod's stories, a friendly Shuswap chief warned him that 300 braves intended to murder the whites when next they journeyed from the fort. Tod's response was to ride to the encampment of the would-be murderers, tell them that an epidemic of smallpox was coming their way, and add that out of kindness he had come to vaccinate them all. Perhaps he was a little rough with the lancet, but vaccinate them he did, and their gratitude brought an end to the 'Shuswap Conspiracy.' (Their swollen right arms must also have made them temporarily *hors de combat*.)

Tod retired in 1851 and settled in Victoria. He was too good a man to leave idle, and Governor Douglas appointed him first to the 'Council of Government' of Vancouver Island, and later to the Legislative Council. Douglas also made him a justice of the peace. Tod lived to a fine old age, shrewd and canny, with his mind clear to the last. Also MOUNT TOD, northeast of Kamloops.

TOD, MOUNT, NE of Kamloops (C-9). Site of a ski resort now called Sun Peaks. Mount Tod is the name of the height of land and Sun Peaks is the name of the community. (It is understandable why the developers wished a change of name – 'Tod' in German means 'death,' and they would be catering to German tourists.)

TODAGIN MOUNTAIN, SE of Eddontenajon L. (J-5). From a Tahltan Indian word meaning 'grass from base of mountain to top.'

TOFINO INLET, E. of Clayoquot Sound (B-7). Named in 1792 by Galiano and Valdes, after the Spanish Hydrographer Vicente Tofiño de San Miguel (1732-95).

TOKETIC, NE of Spences Bridge (C-9). This CPR point bears a Chinook jargon name meaning 'pretty.'

TOKUMM CREEK, flows SE into Vermilion R. (D-11). This seems to be derived from the Stoney Indian word meaning 'to misplace,' or 'something lost.' However, a somewhat similar Stoney Indian word means 'fox.'

TOLMIE, MOUNT, Victoria (A-8). Named as early as 1846, after Dr. William Fraser Tolmie (1812-88), a Scottish physician in the employ of the HBC, who first came to this coast in 1833. His diaries from 1830 to 1843 (published by

Mitchell Press in 1963) reveal a very serious-minded, not to say priggish, young Scot with keen scientific interests and a firm resolve never to take an Indian wife – 'it is only when I abandon the hope and wish of laying my bones in old Scotland that I will ever think of uniting myself in the most sacred of ties with a female of this country.' In 1850 he married Jane, the eldest daughter of Chief Factor John Work and his half-Indian wife, Josette.

In 1856 Dr. Tolmie reached the top of the ladder by becoming a Chief Factor himself and being appointed to the HBC's board of management in Victoria. He retired in 1860 but for the next five years served in the Legislative Assembly of Vancouver Island. His son, Simon Fraser Tolmie, was the Conservative Premier of British Columbia from 1928 to 1933.

The demolition in 1963 of 'Cloverdale,' the house built by Dr. Tolmie more than a century earlier, must be deplored by anybody with a feeling for British Columbia's history.

TOLMIE CHANNEL by Princess Royal Island is also named after Dr. Tolmie.

TOMLINSON, MOUNT, N. of mouth of Nass R. (H-5). After the Reverend Robert Tomlinson, an evangelical Anglican missionary, who began his labours on the north coast in 1867. (See *Cedarvale*.)

TONQUIN ISLAND, S. of Tofino (B-7). After the American vessel *Tonquin*, which brought around Cape Horn the men, material, and supplies for the founding in 1811 of Fort Astoria at the mouth of the Columbia River. Leaving there, she sailed to Clayoquot to trade for furs, where the insolence of Captain Thorn so outraged the Indians that they seized the ship and massacred all but three or four of the crew. These survivors, before escaping in the ship's boat, lit a fuse leading to the *Tonquin's* powder magazine. Almost 200 Indians perished when the ship blew up.

TOODOGGONE RIVER, flows E. into Finlay R. (J-6). From the Sekani Indian word meaning 'water's arms.'

TOPLEY, NE of Houston (G-6). Also TOPLEY LANDING on Babine Lake. When a post office was opened here in 1921, it was named after William J. Topley, an early settler. A one-room school followed, and old Mr. Topley, living in retirement in Alberta, took to writing to each pupil for his or her birthday. Moreover, he presented the little school with 500 books for a library.

TOPPING CREEK, flows E. into Columbia R., N. of Trail (B-11). After 'Colonel' Eugene S. Topping. In July 1890, when Deputy Mining Recorder in Nelson, he accepted the proposition of two prospectors that he personally pay the fees for recording their claims in return for being shown an adjacent strip that he could stake for himself. Thus, Topping acquired the fabulously rich Le Roi claim. (See also *Hanna Creek* and *Trail*.)

TOQUART BAY, NW side of Barkley Sound (B-7). From a Nootka Indian word meaning 'people of the narrow place in front,' or 'people of the narrow channel.'

TORPY RIVER, flows S. into Fraser R. (F-9). After Thomas Torpy, an American who worked as a tunnel construction foreman during the building of the GTPR, now part of the CNR.

TOW HILL, SW of Rose Point, Graham I. (G-4). Tow (which originally rhymed with 'cow') is derived from a Haida word that may mean 'any grease' or 'place of food.'

TOWDYSTAN, NE of Charlottle L. (E-7). From the Chilcotin Indian word meaning 'the water is dirty.'

TOWER OF LONDON RANGE, NW of Tuchodi Lakes (K-7). On 8 July 1960, as part of the British Army's 'Adventure Training Program,' two officers, a sergeant, and three corporals of the Royal Fusiliers (City of London Regiment) left the Tower of London on the first stage of their Canadian Rocky Mountains Expedition. Arriving at Dawson Creek, they were joined on 18 July by Lieutenant-Colonel S.W. Archibald, a veteran surveyor and a former commanding officer of the 3rd Battalion, Royal Canadian Regiment, and by his seventy-eight-year-old Cree assistant. Between then and the end of August, they explored around latitude 58°N. and longitude 125°W., scaling many peaks, erecting cairns, and naming (among other features) Lord Mayor Peak, Merchant Taylors Peak, and Constable Peak (for the constable of the Tower of London). Other peaks they named for towers in that fortress: for example, Beauchamp Peak, The White Tower, and Devereux Peak. However, as the author of the party's high-spirited *Final Report* did not fail to note, 'The Bloody Tower has not yet lent its name to a peak, although there are several that would undoubtedly qualify.'

TOWNER BAY, E. side of Saanich Inlet (A-8). After William Towner, a Kentishman, who came here from California in 1864, bringing with him some hops. With these he started the hop-growing industry in this province.

TRAFALGAR, CNR station N. of Hope (B-9). After Lord Nelson's great naval victory in 1805. Trafalgar Flat has an unusual Halkomelem name that means 'always rotten fish.' (The river current used to sweep spawned-out fish in here.)

TRAIL, Columbia R., NW of confluence with Pend d'Oreille R. (B-11). Originally known as Trail Creek Landing. Trail Creek got its name from the fact that the old Dewdney Trail followed it down to the Columbia River.

The history of Trail really began in 1891 when the first ore from the nearby Le Roi mine was taken to the landing here for shipment, via the Columbia, to the smelter at Butte, Montana. The same year 'Colonel' Eugene S. Topping had a townsite surveyed on the land that he and Frank Hanna had preempted at the mouth of Trail Creek. The settlement's future was assured when the American capitalist F. Augustus Heinze established a smelter here in 1895. Three years later Heinze's smelter was purchased by the CPR. It is now owned by Cominco, which the CPR sold to a consortium of companies.

The post office of Trail Creek, which opened in 1891, became Trail in 1897. Trail was incorporated as a city in 1901, Topping becoming the first mayor. (See also *Hanna Creek, Topping Creek.*)

TRANQUILLE, W. of Kamloops (C-9). Named after a local Indian chief whose quiet easy manner led the whites to refer to him as 'Tranquille.' His death in 1841 led to the murder at the HBC's post at Kamloops of Chief Factor Samuel Black (for the story see G.P.V. Akrigg and Helen B. Akrigg, *British Columbia Chronicle, 1778-1846: Adventures by Sea and Land*, pp. 325-7).

TRAPP LAKE, S. of Kamloops (C-9). After Thomas J. Trapp, an Englishman who started ranching in the area in 1874. After heavy losses in the bad winter of 1879-80, he moved to New Westminster and became a prosperous hardware merchant.

TREMBLEUR LAKE, NW of Stuart L. (G-7). So named because the lake is rarely still – it is located in a draw through which a wind generally blows. An earlier name was Traverse or Cross Lake, since it runs at right angles to all the other large lakes in the area.

TRÉPANIER CREEK, N. of Peachland (B-10). Behind this name probably lies a crude but successful operation of trepanning, performed upon the skull of a Shuswap chief named Short Legs, who had been very badly mauled by a bear. This trepanning was done by Alexander Ross of the NWC in 1817. The stream was formerly known as the Rivière de Jacques, Jacques having been one of Ross's men. The Okanagan Indian name for this creek means 'bald eagle nest.'

TREVOR CHANNEL, SE side of Barkley Sound (A-7). Frances Hornby Trevor, the seventeen-year-old bride of Captain Charles Barkley when he discovered Barkley Sound in 1787, was the first white woman to visit this coast.

TRIAL ISLANDS, Victoria (A-8). These islands got their name in the old days because they constituted a trial of one's skill in navigation. The trick was to round them in a small sailing ship and enter the Strait of Juan de Fuca, despite the tide rips off the islands and the prevailing westerly winds.

TRIBUNE BAY, Hornby I. (B-7). After HMS *Tribune* (Captain G.T.P. Hornby), a screw frigate mounting thirty-one guns, transferred to this coast from the China Station in 1859 to increase British strength during the San Juan crisis. The frigate left for England in 1860 but returned in 1864.

TRIMBLE LAKE, headwaters of Sikanni Chief R. (J-8). After Dr. James Trimble, for some years a surgeon in the Royal Navy before he took up practice in Victoria in 1858. He became a mayor of Victoria and was Speaker of the provincial legislature from 1872 to 1878. He was a courtly Irish gentleman with a fine presence and a warm heart.

Local people know this as Deadmen's Lake since two prospectors shot each other to death here around 1915.

TRINCOMALI CHANNEL, W. of Galiano I. (A-8). After HMS *Trincomalee,* on the Pacific Station 1852-6.

TRINITY VALLEY, NE of Vernon (C-10). Named for the three creeks and their valleys (Trinity, Vance, and Sowsap), which run north, south, and east from here.

TROITSA LAKE, S. of Tahtsa L. (F-6). After a village on the Dvina River, Latvia.

TROPHY MOUNTAIN, Wells Gray Park (D-10). Dan Case, a big-game guide who took parties in here, used to advertise that he went into the 'Trophy Area' – that is, where hunters found trophy-sized animals. Thus, he gave the mountain this name.

TROUP, NE of Nelson (B-11). After Captain James W. Troup, who managed the Columbia and Kootenay Steam Navigation Company's fleet from 1895 to 1898 and then transferred to Victoria to take charge of the CPR's coastal steamships.

TRUAX, MOUNT, S. of Carpenter L. (C-8). After Wesley Truax, a rancher and miner who preempted at the mouth of Truax Creek in 1912.

TRUTCH, Mile 200 on the Alaska Highway (J-8). Takes its name from Trutch Creek nearby, named after Sir Joseph W. Trutch (1826-1904). Trutch, an English civil engineer who had worked in California and Oregon, came to British Columbia in 1859. He was Chief Commissioner of Lands and Works and Surveyor-General of the Crown Colony of British Columbia from 1864 to 1871, when he resigned to become the highly successful first Lieutenant-Governor of the province of British Columbia.

TSACHA LAKE, part of West Road R. (F-7). According to G.M. Dawson, the lake takes its name from a Carrier Indian word meaning 'great stone, or mountain,' referring to a rocky mountain north of the lake.

TSAWWASSEN, S. of Fraser delta (B-8). This Halkomelem word means 'facing seaward.'

TSEAX RIVER, flows NW into Nass R. (H-5). From the Nisgha Indian word meaning 'new water,' possibly after a volcanic eruption had disturbed drainage patterns.

TSEHUM HARBOUR, E. side of Saanich Pen. (A-8). A Saanich Indian name meaning 'always clay place.' According to one account, the infusorial earth found here was used by the Indians to get the fat off animal skins. Formerly Shoal Harbour.

TSIMPSEAN PENINSULA, N. of Prince Rupert (G-4). Named for the Tsimshian Indians, who live in this region. Their name means 'people of the Skeena.'

TSITIKA RIVER, flows N. into Robson Bight (C-6). According to Franz Boas, the pioneer anthropologist, this name is derived from the Kwakwala Indian word meaning 'grey haired,' being a vivid metaphor to describe the forest left dead by a fire.

TSUIUS CREEK, flows W. into Mabel L. (C-10). Tsuius comes from the same Okanagan Indian word as Osoyoos (q.v.) and means simply 'the narrows,' in this instance referring to where Mabel Lake narrows at the creek's delta.

TSULQUATE RIVER, W. of Port Hardy (C-6). From the Kwakwala Indian word meaning 'place of warmth' – which has been explained either as 'where the sun first strikes' or as 'where there are hot springs.'

TSULTON RIVER, flows NE into Kokish R. (C-6). A Kwakwala Indian name that means 'black coloured.'

TSUNIAH LAKE, NE of Chilko L. (D-7). Based on a Chilcotin Indian name meaning 'spruce trees standing in a row.'

TSUSIAT RIVER, flows SW into Pacific (A-7). From the Nitinaht Indian word meaning 'water pouring down' (i.e., waterfall), a very suitable name as the river falls over a ledge and descends directly into the ocean.

TUAM, MOUNT, Saltspring I. (A-8). From the Straits Salish Indian word that means 'flanked by the sea' or 'facing the sea.'

TUCHODI RIVER, flows NE into Muskwa R. (K-8). From Slave Indian words meaning 'the place of big water,' presumably with reference to the Tuchodi Lakes, expansions of the river.

TUCK INLET, N. of Prince Rupert (G-4). Named after Samuel P. Tuck, land surveyor, who worked here in 1892. Tuck Inlet became important when the GTPR decided to place its western terminus, Prince Rupert, on its shore.

TUGULNUIT LAKE, S. of Vaseux L. (B-10). From the Okanagan Indian word meaning 'lake alongside river.'

TULAMEEN RIVER, flows into Similkameen R. (B-9). From a Thompson Indian word meaning 'red earth.' The source of the much-prized red ochre, for which Indians travelled from afar, was a steep bank of the river four miles north of Princeton. The settlement of Tulameen was earlier known as Campement des Femmes or Otter Flat.

TUMBLER RIDGE, N. of Quintette Mountain (H-9). The origin of this name is rather obscure but it may be derived from unstable strata on a nearby mountain.

TUMBO ISLAND, N. of E. end of Saturna I. (A-8). By brilliant research Peter Murray was able to track down the origin of this name, so long a mystery (see the *Islander*, 11 August 1991, M3). Murray found that the island was named in 1841 by Lieutenant Charles Wilkes, USN, after Tumbou on the island of Lakemba in Fiji. Wilkes had been in Fiji in 1840 and used several other Fijian names when charting the San Juan Islands.

TUMTUM LAKE, N. of Adams L. (D-10). A Chinook jargon term applied to all falls of water. There are falls in the Upper Adams River just below Tumtum Lake.

TUNSTALL BAY, Bowen I. (B-8). When a post office was opened here in 1909, George C. Tunstall Jr., the manager of Western Explosives Ltd., was appointed postmaster. In a letter that year to the Chief Geographer at Ottawa, he stated that Tunstall Bay was named after his father. Actually his father had little connection with Bowen Island. An Overlander of 1862, the senior Tunstall later became the government agent at Kamloops.

TUPPER, S. of Dawson Creek (H-9). Also TUPPER RIVER. After Frank Tupper, BCLS, who surveyed in the area.

TUPPER, MOUNT, Rogers Pass (D-11). Formerly Hermit Mountain, but renamed for Sir Charles Tupper (1821-1915), who was responsible for bringing Nova Scotia into Confederation despite the opposition of Joseph Howe.

Shortly after the change was made, the English traveller Douglas Sladen wrote:

> Sir Charles Tupper is one of Canada's greatest men, but his name is more suitable for a great man than a great mountain, especially since there is a very perfect effect of a hermit and his dog formed by boulders near the top of the mountain. The men in the railway camp have got over this difficulty with the doggerel:
>> That's Sir Charles Tupper
>> Going home to his supper.

TURNAGAIN RIVER, flows E. into Kechika R. (L-6). So named by Samuel Black of the HBC since this was the farthest point reached in his great journey of exploration in 1824.

TURNOUR ISLAND, off mouth of Knight Inlet (C-6). After Captain Nicholas E.B. Turnour, commanding HMS *Clio,* on the Pacific Station 1864-8.

TUTIZIKA RIVER, flows E. into Mesilinka R. (I-7). From the Sekani Indian word meaning 'water river along.'

TUTIZZI LAKE, headwaters of Tutizika R. (I-7). From the Sekani Indian word meaning 'long-water lake.'

TUTSHI LAKE, E. of Bennett L. (L-2). From the Tlingit Indian word meaning 'black lake.'

TUYA RIVER, flows S. into Stikine R. (K-4). Takes its name from Tuya Lake near its headwaters. Tuyas, flat-topped and steep-sided extinct volcanoes, are found west of Tuya Lake.

TWEEDSMUIR PARK, central BC (E- and F-6 and 7). This enormous wilderness reserve of more than 3,700 square miles was established in 1936 and named for Lord Tweedsmuir, then Governor-General of Canada. The first Baron Tweedsmuir (1875-1940) is better known as John Buchan, the author of such notable adventure stories as *The Thirty-Nine Steps* and *Greenmantle.* He was Britain's director of information in World War I, and in 1927 he became the MP for the Scottish universities.

TYAUGHTON CREEK, flows SE into Carpenter L. (D-8). Anglicized form of the Lillooet Indian word for the 'farthest place upstream that the salmon go' (i.e., their spawning ground).

TYEE LAKE, N. of Williams L. (E-8). In the Chinook jargon, *tyee* is the word for 'chief.' This was the favourite fishing lake of Chief William of the Soda Creek reserve, who was known locally simply as Tyee.

TYNEHEAD, N. of Cloverdale (B-8). Originally Tinehead, since the head of the Serpentine River is here, but converted to Tynehead by analogy with the River Tyne in northern England.

TZARTUS ISLAND, Barkley Sound (A-7). From the Nootka Indian word that means 'water flows inside at the beach.'

TZUHALEM, MOUNT, N. of Cowichan Bay (A-8). Named after one of the fiercest of the war chiefs of the Cowichans. In 1844 he led an attack on Fort Victoria after the HBC tried to collect damages for cattle that his band had slain. Because of his frequent murders, he was banished at last by his fellow Indians and took up residence in a cave on the side of Mount Tzuhalem. He had some fourteen wives with him, most of whom were widowed by him. Going to Kuper Island to acquire a new wife, Tzuhalem was slain by her husband before Tzuhalem could kill him.

U

UBYSSEY GLACIER, Garibaldi Park (B-8). A party drawn from the Varsity Outdoors Club of UBC, having made some first ascents in Garibaldi Park, in 1965 asked Ottawa to name two features VOC Mountain and UBC Glacier. The Canadian Permanent Committee on Geographical Names pointed out that it did not accept abbreviations as names, and accordingly the petitioners settled for Veeocee Mountain and Ubyssey Glacier. This glacier, then, is named for the university, not its student newspaper, the *Ubyssey*. (Not long afterwards initials were accepted as names.)

UCHUCKLESIT INLET, N. of entrance to Alberni Inlet (A-7). From the Nootka Indian word meaning 'there inside the bay.'

UCLUELET, NW side of Barkley Sound (A-7). From the Nootka Indian word meaning 'people of the sheltered bay,' referring to the well-protected landing place for canoes.

ULKATCHO, N. of Gatcho L. (E-7). The Carrier Indian name of this area means 'good feeding place, where animals get fat.'

UNCHA LAKE, S. of François L. (F-7). From the Carrier Indian word meaning 'it is vast.' Father Morice commented that this 'must have been the seat of a large population in olden times, if we are to judge from the trails made by the moccasined feet of Indians.'

UNION BAY, S. of Comox (B-7). When in 1894 the Dunsmuir interests, which had taken over the Union Coal Company a few years earlier, changed the name of the mining town from Union to Cumberland, they left unaltered the name of the company's coal port, Union Bay.

UPRIGHT MOUNTAIN, Mt. Robson Park (F-10). As the Geographic Board of Canada noted in its eighteenth report, 'The strata of the mountain have been upheaved to an almost vertical position.'

USK, NE of Terrace (G-5). Although local inhabitants would like to think that their settlement is named after the birthplace of David Lloyd George, it is more probable that this Usk is simply the Tsimshian Indian word for 'stink.'

USSHER LAKE, W. of Shumway L. (C-9). James T. Ussher, BC government agent at Kamloops, was murdered here in 1879 when he tried to arrest the McLean Gang on charges of horse stealing.

UZTLIUS CREEK, flows SW into Anderson R. (B-9). From the Thompson Indian word meaning 'water that boils,' referring to the rapids in the stream.

V

VALDES ISLAND, N. of Galiano I. (B-8). After Cayetano Valdes y Bazan, a Spanish naval officer. Valdes first visited this coast in 1791 as a lieutenant serving under Malaspina on the *Descubierta*. He returned in 1792 as captain of the *Mexicana* and, in association with his superior, Galiano, commanding the *Sutil*, explored a good part of the area lying between Vancouver Island and the mainland. Both Valdes and Galiano commanded warships that were captured by the British in the Battle of Trafalgar.

VALEMOUNT, SE of Tête Jaune Cache (E-10). In consequence of the building of the Canadian Northern (now Canadian National) Railway, a post office was opened here in 1913 with the name of Cranberry Lake. In 1918 it was renamed Swift Creek, and in 1928 Swift Creek became Valemount ('vale amid the mountains').

VALENCIENNES MOUNTAIN, Alberta-BC boundary (D-11). So named to commemorate the entry of Canadian troops into Valenciennes, France, 2 November 1918. Also VALENCIENNES RIVER.

VALHALLA RANGES, W. of Slocan L. (B-11). Named by G.M. Dawson after the hall of immortality where the souls of Norse heroes went after their death in battle.

VAN ANDA, Texada I. (B-7). For a great many years the post office here was mistakenly listed as Vananda. Only in 1992 was the error rectified. Edward Blewitt, a Seattle capitalist interested in copper deposits on Texada Island, named his mining company after his friend Carl Van Anda, a New York journalist in the 1870s.

VANCE CREEK, N. of Lumby (C-10). After Alexander Vance, manager of the BX (Barnard's Express) Ranch near Vernon, from about 1868 to 1885.

VANCOUVER, S. of Burrard Inlet (B-8). The City of Vancouver is named after the famous British navigator, Captain George Vancouver, RN, who in 1792 named and explored Burrard Inlet, on whose shore the city stands.

Born of Dutch stock in the little Norfolk town of King's Lynn on 22 June 1757, George Vancouver entered the navy in 1771. In the following years, he sailed with the great Captain Cook on the latter's second and third voyages of exploration. Promoted to lieutenant in 1780, Vancouver saw action against the French in the West Indies and attracted the attention of Commodore Alan Gardner, through whose influence he was appointed in 1789 second in command on a voyage of exploration to be made in the Pacific by Captain Henry Roberts.

Expectations of war with Spain over Nootka caused this expedition to be postponed, and new postings were given to those who had been assigned to it. When the project was renewed in December 1790, Vancouver was given command, with special instructions to receive from the Spaniards at Nootka the land that they had taken from the British there and were now to surrender.

In April 1792 Vancouver arrived off the shores of what was to become British Columbia. Sailing through the Strait of Juan de Fuca, he began that detailed survey of the coast which was to occupy him until his departure for England late in 1794. Much of this work was done in small open boats operating at considerable distances from Vancouver's ship, HMS *Discovery*, and her tender, the *Chatham*. Considering the difficulties under which Vancouver laboured, this was a tremendous achievement of careful, meticulous surveying. During these explorations he was a stern disciplinarian, as any effective naval commander had to be in those days.

Back in England in 1795, Captain Vancouver (who had been promoted to that rank during his absence) devoted himself to preparing for publication an account of his great expedition. He had just about completed reading the proofs when he died on 10 May 1798. On 18 May he was buried in the graveyard of Petersham church in Surrey. His *Voyage of Discovery to the North Pacific Ocean and round the World* was published later that year.

Captain Vancouver was not the first white man to visit the site of the future city. That distinction goes to a Spanish navigator, José Maria Narvaez, who was on the scene a year earlier. For Vancouver's meeting off Point Grey in 1792 with two other Spaniards, Galiano and Valdes, see *Spanish Bank*.

At the time of Captain Vancouver's visit, Indians lived in the Burrard Inlet area mainly on a seasonal basis, their permanent villages being elsewhere. The Musqueams had their village on the North Arm of the Fraser (as they do now), and the Squamish wintered in the Squamish and Cheakamus valleys. Once white settlement began around Burrard Inlet, more Indians moved here permanently.

The main Squamish Indian village sites in the Vancouver area were Sen7ákw (meaning possibly 'inside at the head'), extending roughly from the Vancouver Museum area east to the mouth of False Creek; Schílhus ('high bank'), at the end of Pipeline Road in Stanley Park; X̱wáy̓xway ('place of sx̱wáy̓xwi mask'), a very important village on the site of Lumberman's Arch; Temtemíxwtn ('place of lots of land') at Belcarra; Xwmelch'stn (referring to fish finning or rolling on the water surface), at the original mouth of the Capilano River east of Lions Gate Bridge; and Slha7án' ('against the edge of the bay'), at the mouth of Mosquito Creek.

White settlement in the Vancouver area began in 1862 when three Englishmen, John Morton, William Hailstone, and Sam Brighouse, having learned of the discovery of coal in Coal Harbour, went there and staked a claim to all the land lying between modern Burrard Street and Stanley Park,

English Bay, and Coal Harbour. After the trio had obtained title to the land (which became known as the 'Brickmakers' Claim,' since they later manufactured bricks here, the first on the mainland), they cleared some land in the northeast corner and built a shack and a barn. About this time farms were being started in the Marpole area by Hugh Magee, William Shannon, William Catchpole, Henry Mole, the McCleery brothers, and a man named Gariepy.

The year 1865 saw two new settlements on the south shore of Burrard Inlet, settlements that would grow, merge, and become parts of Vancouver. Early that year Douglas Road, a crude corduroy affair, was completed, linking New Westminster with Burrard Inlet. That summer the fine beaches and tranquil waters at 'The End of the Road' became a favourite place for picnickers from the little colonial capital on the Fraser. Summer cottages were soon built. The little resort was named 'New Brighton,' and that August Hocking and Houston built their Brighton Hotel, with elaborate grounds and a steamboat landing. Three years later a townsite was laid out at New Brighton and named 'Hastings' after Rear-Admiral the Hon. George F. Hastings, who had taken command of the Royal Navy's Pacific Station the previous year and was already proving himself a good friend to the colonists. Today New Brighton Park, lying between Exhibition Park and the Second Narrows, marks where this first resort area once stood.

Another event of 1865, more important than the founding of New Brighton, was the establishment of Captain Edward Stamp's sawmill. After an unsuccessful attempt to found a mill near the present site of Lumberman's Arch in Stanley Park, Stamp received a Crown grant of the land at the foot of Dunlevy Street, where the National Harbours Board now has its offices. Here on his 'Sawmill Claim' he built the Hastings Mill, which, for more than sixty years, was to export its lumber to all parts of the world.

In 1867, when the Hastings Mill finally came into operation, sometimes with as many as three tall-masted ships taking on lumber at its wharf, there was an obvious need for a place where thirsty millhands and sailors could slake their thirst. Thus, there arrived on the scene 'Gassy Jack' Deighton 'with his squaw, his yellow dog, and a barrel of whisky ... to lay the foundations of the future city of Vancouver' (*BCHQ* 11 [April 1947]:70). 'Captain' Deighton was an Englishman, born in Hull in 1830, who had served as a river pilot on the Fraser. He had picked up his nickname of 'Gassy' because of his loquacity and his tall stories. Now, in 1867, he built his Deighton House, complete with saloon, in a little clearing where today we have the intersection of Water and Carrall streets. Around Gassy Jack's establishment grew up a little settlement that, in honour of its founder, became known as 'Gastown,' a name that found its way onto the charts of the British admiralty before the residents, in a fit of respectability in 1870, renamed their hamlet 'Granville' in honour of the noble Earl who was Britain's Colonial Secretary.

The 1880s saw major developments. With the CPR ending its transcontinental line at Port Moody at the head of Burrard Inlet, property on the inlet was obviously going to be much more valuable. Perhaps with inside information that the railway would be extended west, Morton and Brighouse (Hailstone had gone back to England) in 1882 laid out a plan for 'the City of Liverpool,' covering the area now bounded by Burrard and Nicola Streets, Georgia Street, and Coal Harbour. Nothing came of this project.

In 1884 the CPR announced its decision to have its terminus at Coal Harbour rather than at Port Moody, and the following year L.A. Hamilton, making a survey for the railway, laid out a townsite that determined the street pattern for downtown Vancouver. In later years Hamilton recalled his conversations with the mighty Van Horne, the general manager and later the president of the CPR, about a name for the new city. Van Horne thought that Granville or Liverpool would not do. Neither would give the world any idea of the geographical location of the CPR's new terminus. Vancouver seemed to him the name – Vancouver Island had got 'Vancouver' firmly identified with Canada's west coast. Said Van Horne: 'Hamilton, this eventually is destined to be a great city in Canada. We must see that it has a name that will designate its place on the map of Canada. Vancouver it shall be, if I have the ultimate decision.' And Vancouver it became. Hastings and Granville survive only as street names. The City of Vancouver was incorporated on 6 April 1886, and on 1 May of that year the post office of Granville was renamed Vancouver.

And so Vancouver came into being – to the anger of those dwelling on Vancouver Island, to the mystification of logically minded foreigners who insist that Vancouver must be on Vancouver Island, and to the especial confusion of Americans, who are still apt to confuse Vancouver, British Columbia, with Vancouver, Washington – the latter having been founded in 1825, over half a century earlier.

Those who wish to know more about the history of Vancouver, British Columbia, are referred to Chuck Davis's *Greater Vancouver Book*, Eric Nicol's *Vancouver*, and Michael Kluckner's *Vancouver: The Way It Was*. (See also *Brockton Point; Burrard Inlet; Coal Harbour; Eburne; English Bay; Fairview; False Creek; Ferguson Point; Grey, Point; Jericho; Kerrisdale; Kitsilano; Lost Lagoon; Marpole; Spanish Bank; Stanley Park*.)

VANCOUVER ISLAND. Late in August 1792, Captain George Vancouver arrived at Nootka Sound with his ships *Discovery* and *Chatham* to take over, under the terms of the Nootka Convention, the land that the Spaniards had taken from the British several years earlier. Unfortunately Vancouver and the Spanish commander at Nootka, Don Juan Francisco de la Bodega y Quadra, were unable to agree as to just what area was to be handed over. They decided to leave the matter unsettled pending clarification from their home governments. Quadra was, in fact, doing his utmost at this late stage to keep a Spanish base at Nootka.

Despite this political disagreement, Vancouver and Quadra quickly established a close personal friendship. Together they made a trip up to Tahsis, of which Vancouver later recorded:

> In our conversation whilst on this little excursion, Senor Quadra had very earnestly requested that I would name some port or island after us both, to commemorate our meeting and the very friendly intercourse that had taken place and subsisted between us. Conceiving no spot so proper for this denomination as the place where we had first met, which was nearly in the centre of a tract of land that had first been circumnavigated by us, forming the southwestern sides of the gulf of Georgia, and the southern sides of Johnstone's straits and Queen Charlotte's sound, I named that country the island of QUADRA and VANCOUVER; with which compliment he seemed highly pleased.

Early admiralty maps show 'Quadra and Vancouver's Island,' which early HBC traders shortened to Vancouver's Island. By the mid-nineteenth century, this had become simply Vancouver Island.

VANDERHOOF, W. of Prince George (G-8). Named after Herbert Vanderhoof, a Chicago publicity agent. In 1908 the Canadian government, the CPR, the Canadian Northern Railway, and the GTPR engaged Vanderhoof to conduct a major campaign in the American press to draw settlers to western Canada. Making maximum use of farm journals of the day, and of his magazine *Canada West,* Vanderhoof was so successful that a second, similar contract was signed with him.

Vanderhoof had ambitious plans for the development of the settlement that was given his name in 1914. He is said to have planned, among other schemes, the building of a large hotel as a refuge for burned-out authors!

VAN HORNE RANGE, N. of Kicking Horse R. (D-11). This and the VAN HORNE GLACIER in Glacier National Park are named after Sir William Cornelius Van Horne (1843-1915), the American railroader who became general manager of the CPR in 1882, vice-president in 1884, president in 1888, and chairman of the board of directors in 1899. Like many other good Americans, he became an ardent Canadian citizen and garnered (among his rewards) a knighthood in 1894. Van Horne was one of those human bulldozers needed in any great construction project. His energy and determination were essential to the building of the CPR.

VAN WINKLE CREEK, flows N. into Lightning Cr. (F-9). The name, dating back to 1861, is said to have been taken from a rich Rip Van Winkle Bar on the Fraser, with the hope that the creek would prove rich also. The gold rush settlement of Van Winkle was superseded by nearby Stanley after the re-routing of the highway to Barkerville in 1885.

VARGAS ISLAND, Clayoquot Sound (B-7). Apparently named for the Vargas who regained New Mexico for Spain in 1693-4.

VASEUX LAKE, S. of Skaha L. (B-10). The French word *vaseux*, meaning 'muddy' or 'slimy,' was probably applied to the lake early in the last century by French-Canadian employees of the fur-trading companies. The name apparently refers to the amount of silt deposited in the lake.

VAUX, MOUNT, Yoho NP (D-11). Almost certainly named by James Hector after W.S.W. Vaux, FRS (1818-85), a numismatist on the staff of the British Museum. A close friend of Captain Palliser, Vaux helped to secure government funds to finance Palliser, Hector, and Sullivan while they prepared the report on the Palliser Expedition. Probably much of the work was done in Vaux's chambers at Lincoln's Inn Fields.

VAVENBY, N. Thompson R. (D-10). When a post office was established here in 1910, an early settler, Daubney Pridgeon, suggested that it be named after his birthplace, Navenby in Lincolnshire. Unfortunately the postal authorities misread his handwriting, and so it became 'Vavenby.'

VEDAN CREEK, flows N. into Big Cr., tributary of Chilcotin R. (D-8). After Louis Vedan, who died in the Old Men's Home, Kamloops, in 1954 at an age of more than 100, if we accept his own estimate. Earlier, on his remote ranch, he had so emancipated himself from an awareness of time that he would ask his rare visitor, Indian or white, not the day but the month. He usually got his hay in late and lost many cattle in consequence.

VEDDER RIVER, SW of Chilliwack (B-8). Named after Volkert Vedder, who arrived in the Hope district around 1860. In the 1870s he owned land by the Vedder River.

VENABLES VALLEY, N. of Spences Bridge (C-9). After Captain Cavendish Venables, formerly of the 74th Highlanders, who secured a military land grant in the area around 1861. An alcoholic, he habitually wore a greasy army blue coat with tarnished gold braid and a matching cap. Neighbours derisively called his estate 'Mount Freeze Out' since its altitude exposed it to frequent frosts. A visitor found him sitting down with his solitary Indian helper to a meal consisting of a soup plate full of radishes, a little salt, and a bowl of whisky.

VENN PASSAGE, W. of Prince Rupert (G-4). After the Reverend Henry Venn (1796-1873), honorary secretary (1841-73) of the Church Missionary Society in England. He was a man of outstanding character and notable administrative skills.

VENTEGO MOUNTAIN, N. of Glacier NP (D-11). *Ventego* is the Esperanto word for a storm, tempest, or squall.

VERENDRYE, MOUNT, W. of Vermilion R. (D-11). Named by G.M. Dawson after Pierre Gaultier de Varennes, Sieur de la Verendrye (1685-1749), the great French-Canadian explorer who sought with his sons to find the Western Sea, though it is doubtful that he ever got farther west than the Black Hills in the Dakotas.

VERMILION RIVER, Kootenay NP (D-11). From the ochre of ferruginous beds in the area. The colour coats the stones and marsh grass of the river flats.

VERNON, N. Okanagan (C-10). In 1862 Father Paul Durieu, OMI, built a cabin here, an out-station of Okanagan Mission, and thus he gave Vernon its first name of Priest's Valley. In 1887, after almost becoming either Centreville or Forge Valley, Priest's Valley became Vernon, in honour of Forbes George Vernon (1843-1911), Chief Commissioner of Lands and Works for British Columbia.

Vernon was born near Dublin. After a short period of service as an officer in the British Army, he came to British Columbia with his brother Charles in 1863, arriving in the Okanagan with Colonel C.F. Houghton, the first owner of the Coldstream Ranch. This magnificent ranch was bought by the Vernon brothers in 1869 and later passed into the sole ownership of Forbes George Vernon.

Vernon entered provincial politics and became a power in the land. In *The Valley of Youth*, C.W. Holliday preserves a lively recollection of Vernon's electioneering technique as practised one night in the barroom of the Ram's Horn at Lumby:

> Forbes George, a big genial Irishman with a merry twinkle in his eye, sized up his audience, and mounted a barroom chair – there was nothing else to mount, and there did not appear to be a chairman ... He mounted that chair, and, had a representative of the press been present he would have had little trouble reporting the speech, for, holding up his hand to silence the applause, 'Gentlemen,' he said, 'you boys all know me and know all about me, and I am quite sure none of you want to hear me make a speech, so all I will say at present is: Let us all go and have a drink.' (p. 308)

VESUVIUS BAY, Saltspring I. (A-8). After the Royal Navy's paddle sloop *Vesuvius*, which served in the Black Sea during the Crimean War.

VICTOR LAKE, W. of Revelstoke (C-10). Probably named after Victor, one of two Indians who accompanied Walter Moberly when he was in this area in 1865.

VICTORIA, S. end of VI (A-8). Named after Queen Victoria (1819-1901), who came to the throne in 1837.

The story of Victoria begins in 1837 when Captain McNeill of the HBC's maritime service, exploring the rocky and inhospitable coast of southern Vancouver Island, reported that he had found a good anchorage surrounded by several miles of plain with excellent soil. The harbour was known to the local Indians by a name variously transcribed as Camosun, Cammusan, or

Camosack. This name, according to Roderick Finlayson, who assumed command of Fort Victoria in 1844, meant 'the rush of water' and presumably referred to the race of the tide moving in and out of 'The Gorge.' Recently it has been suggested on linguistic grounds that Camosun was the name of the gorge itself and meant 'cut mouth.'

McNeill's discovery of a good harbour with adjacent arable land was particularly welcome to the HBC, which already feared that American encroachments would force the abandonment of Fort Vancouver on the Columbia and require the establishment of a new depot farther north, one that could be kept British and would have good land to replace the lush farms the company had developed at Fort Vancouver.

When Sir George Simpson, the Governor of the HBC, was out to the Pacific coast in 1841, he favoured founding a new depot at the southern tip of Vancouver Island. Accordingly in the summer of 1842 James Douglas, who enthusiastically approved the site, made a thorough survey of Camosun, drawing a map of the area between Esquimalt and Oak Bay. At the same time, the Council of the Northern Department, meeting at Norway House, accepted Simpson's recommendation of the previous year and ordered that a depot be established at the southern end of Vancouver Island with the 'least possible delay.'

On 14 March 1843, Douglas arrived off Clover Point. Going ashore the next day, he decided where to build the new fort, hired some of the local Indians to cut stakes for the palisades, and set six men to work digging a well. These preparatory steps taken, he sailed north to Fort McLoughlin near Bella Bella and to Fort Durham on Taku Inlet, Alaska, which the HBC had decided to abandon. On 3 June Douglas arrived back at Camosun aboard the *Beaver* with the men from the abandoned forts. These he left here under the command of Chief Trader Charles Ross to build the new post. Even while they were setting about their work, the Council of the Northern Department, meeting in Fort Garry, resolved on 10 June 1843 'That the new Establishment to be formed on the Straits de Fuca ... be named Fort Victoria.' Unfortunately Ross and his men, under the impression that the new fort was to be named for the sovereign's consort, not the sovereign, proceeded to name their new home Fort Albert. However, a message from London arrived saying that the name would not do, and with suitable ceremonies and firing of salutes the little post was renamed Fort Victoria. Meanwhile, early in 1844, Ross had died, and his second in command, Roderick Finlayson, aged twenty-six, had taken charge of the fort.

The following years saw a slow growth. Fields were cleared and yielded abundant crops of wheat. The first few settlers began to arrive from Britain on the Company's ships. In 1851-2 a townsite was laid out adjacent to the fort and given the name of Victoria. It is estimated that by 1857 there were a few hundred persons dwelling in the village. Then came 1858 and the gold rush on the mainland. Twenty thousand adventurers, merchants, and miscellaneous parasites arrived from California, and the village became an important city through which supplies were funnelled to the gold camps.

In 1866 the two colonies of Vancouver Island and British Columbia were united in a single new Crown Colony of British Columbia, and in 1868 the capital of British Columbia was transferred from New Westminster to Victoria. For the next few decades, first New Westminster and much later Vancouver agitated to have the capital brought back to the mainland. To settle the matter once and for all, in 1893 the government of Premier Theodore Davie began building palatial Parliament Buildings to replace the earlier 'birdcages.' After that monumental project was completed in 1897, there could be no more talk of moving the capital to the mainland. The investment was too great and immovable.

Victoria may not be the 'little bit of Olde Englande' of the tourist advertisements, but it does still have a way of life that does owe something to the palmy days of Victorian England. It has come a long way since New Year's 1854, when Robert Melrose wrote in his diary: 'New Year's Day, a day above all days for rioting in drunkenness, then what are we to expect of this young, but desperate Colony of ours; where dissipation is carried on to such extremities ... it would almost take a line of packet ships, running regular between here and San Francisco to supply this Island with grog, so great a thirst prevails amongst its inhabitants.'

(See also *Albert Head; Beacon Hill; Brotchie Ledge; Cadboro Bay; Clover Point; Cordova Bay; Craigflower; Douglas, Mount; Duntze Head; Elk Lake; Ellice, Point; Esquimalt; Fisgard Island; Fort Rodd; Foul Bay; Gonzales Bay; Gordon Head; James Bay; Jemmy Jones Island; Macaulay Point; Mary Tod Island; McNeill Bay; Oak Bay; Ogden Point; Race Rocks; Ross Bay; Royal Roads; Saanich; Tolmie, Mount; Trial Island; Work Point.*)

VICTORIA LAKE, SE of Port Alice (C-6). After Victoria Whalen, daughter of the president of Whalen Pulp and Paper Mills Ltd., the company that built the first mill at Port Alice (q.v.).

VIDETTE LAKE, headwaters of Deadman R. (D-9). The French word *vidette* signifies an outpost or mounted sentry. The French-Canadian engagés of the fur-trading companies used the word for a man guarding the horses during an encampment.

VINES LAKE, W. of McDame (L-5). After Lionel Vines, a bush pilot killed in a plane crash in 1941.

VINSULLA, N. Thompson R. (C-9). An anagram of Sullivan, after Michael Sullivan, a pioneer of the district. (See *Sullivan Lake.*)

VIRAGO SOUND, N. coast of Graham I. (G-3). After HMS *Virago*, a paddle sloop with six guns, which visited here in 1853 and named a number of features. (See G.P.V. and H.B. Akrigg, *HMS* Virago *in the Pacific 1851-1855*.)

VITAL RANGE, between Takla L. and Omineca R. (H-7). After Vital La Force, who prospected here with his partner, Michael Burns, in 1868-9, triggering the Omineca gold rush.

VOGHT CREEK, flows W. into Coldwater R. (B-9). After William H. Voght, gold-seeker and farmer, who settled in the Nicola Valley in 1873. He preempted land that would later be the townsite for Merritt.

VOLCANIC CREEK, N. of Grand Forks (B-10). After 'Volcanic' Brown, a colourful prospector who perished while seeking the legendary 'lost mine' on upper Pitt River.

VREELAND, MOUNT, headwaters of Parsnip R. (G-9). After Frederick K. Vreeland of New York, who travelled through this area tracing the route of Sir Alexander Mackenzie. When Vreeland learned that Ottawa proposed to name this mountain after himself, he wrote to the federal authorities declaring, 'I have no valid claim for the naming of this particular mountain,' and he asked that the name not be confirmed. No doubt feeling that Vreeland shone like a good deed in a naughty world, one in which many undeserving people strive to get mountains named after them, Ottawa retained Mount Vreeland.

W

WABRON, N. Thompson R. (D-10). This station is named after W.A. Brown, superintendent, Canadian Northern (now Canadian National) Railway.

W.A.C. BENNETT DAM, Peace R. above Hudson's Hope (I-8). After William Andrew Cecil Bennett, Social Credit Premier of British Columbia from 1952 to 1972. It has been remarked that, upon becoming premier, he treated British Columbia as an underdeveloped country and provided it with an entire new economic infrastructure. Part of this was his network of modern highways. Another part was a great scheme for developing new hydro power, which included the building of this dam.

WADDINGTON, MOUNT, NE of head of Knight Inlet (D-7). Height 13,104 feet or 3,994 metres. Don and Phyllis Munday, in 1927 the first to climb on this great mountain, called it Mystery Mountain (see W.A.D. Munday, *The Unknown Mountain*). The authorities decided, however, to name it after Alfred Waddington, an Englishman who arrived in Victoria in 1858 and was a notable champion of unsuccessful causes. He believed that Governor Douglas was entirely wrong in using the Fraser Canyon route for the wagon road to the Cariboo goldfields, and finally he launched a company of his own to establish the route, 175 miles shorter, by steamer to the head of Bute Inlet and then directly overland by way of the Homathko Valley and the Chilcotin. In 1864 the Indians massacred fourteen of his road builders. The first part of Waddington's route north of Bute Inlet proved extremely difficult, and only a few miles of road were ever built.

After this venture failed, Waddington became an ardent champion of a transcontinental railway that would use the Yellowhead Pass-Bute Inlet route. In 1867 he went to England seeking imperial support for this new project. He was in Ottawa seeking a Canadian charter for his envisioned railway when he died of smallpox in 1872.

Waddington was an incredible optimist, incapable of taking sufficient cognizance of realities. Sir John A. Macdonald called him 'a respectable old fool.' Perhaps it would have been best if the Geographic Board of Canada had stuck to an earlier plan to name this, the highest peak in the Coast Range, after G.M. Dawson (see *Dawson Creek*). WADDINGTON HARBOUR on Bute Inlet is also named after Waddington.

WAGLISLA, Campbell I. (E-5). This is a Heiltsuk (Bella Bella) Indian name meaning 'river delta' and is the name of a creek just south of Bella Bella village. At low tide the creek forms something of a little delta over the beach. Waglisla is also the Native name for the village site of New Bella Bella.

WAHKASH CREEK, flows W. into Knight Inlet (C-7). A Kwakwala Indian word meaning 'good water' or 'nice river.'

WAHLEACH ISLAND, lower Fraser R. (B-9). From the descriptive Halkomelem name for the island, indicating that it had 'willow on the back.'

WAIATT BAY, E. side of Quadra I. (C-7). From the Kwakwala Indian word meaning 'where there is herring spawn.'

WAITABIT CREEK, flows S. into Columbia R., N. of Golden (D-11). Early travellers descending the upper Columbia River would wait a bit, resting and trimming the loads in their canoes, before entering the rapids here.

WAKEMAN SOUND, N. of Kingcome Inlet (C-6). After Plowden Wakeman, clerk in the Esquimalt dockyard from 1866 until his death in 1872.

WALBRAN ISLAND, Rivers Inlet (D-6). After Captain John T. Walbran (1848-1913) of the Canadian Marine and Fisheries Service, for many years skipper of the dominion government steamship *Quadra*. His official duties with the *Quadra* took him to every part of the BC coast.

A well-educated man (he had attended Ripon Grammar School in Yorkshire), Walbran became fascinated by the names of the places his ship visited. Using extensive historical research, as well as personal correspondence and interviews with those who had firsthand information, he prepared his *British Columbia Coast Names,* published in 1909. That work must be mentioned with respect and affection by anyone interested in our BC place names. Walbran's notes were unfortunately destroyed after his death.

Readers wishing a more extended account of Walbran and his achievements should see the introduction, by G.P.V. Akrigg, prefixed to the 1971 reprint of Walbran's book.

WALES ISLAND, mouth of Portland Inlet (G-4). Takes its name from Wales Point, on its southern shore, named by Captain Vancouver after William Wales, astronomer and mathematician, 'to whose kind instruction, in the early part of my life, I am indebted for that information which has enabled me to traverse and delineate these lonely regions.' Wales had accompanied Captain Cook on his second great voyage, on which the youthful Vancouver had also served.

WALHACHIN, E. of Ashcroft (C-9). The name is a Thompson Indian word apparently meaning 'close to the edge' and not 'round stones.' Before World War I, an American real estate man, C.E. Barnes, established a settlement above the Thompson River here. Subsequently control passed to the Marquess of Anglesey. The upper-class British were not the right kind of settlers, the soil was unsuitable for fruit ranching, the irrigation flumes were poorly built, and the drain on its manpower during World War I seriously weakened the colony. It died in 1922 after a colourful life of about twelve years. The days of the Walhachin Hunt, afternoon tea at the luxury hotel, and dancing to Paderewski's piano were over.

Before the enterprising Barnes introduced the name Walhachin, the locality was known simply as Penny's (Penney's), after Charles Penney, who had homesteaded here.

WALKER, MOUNT, E. of Bella Coola R. (E-6). After T.A. (Tommy) Walker, big-game guide and author of *Spatsizi*.

WALL CREEK, Cathedral Park (B-9). From the perpendicular walls of its canyon.

WALLACE ISLAND, Trincomali Channel (A-8). After Captain Wallace Houston, HMS *Trincomalee,* on this coast in 1853.

WALLIS POINT, Nanoose Harbour (B-7). After Richard P. Wallis of Thorney Abbey, Cambridgeshire, who settled here in 1888, acquiring an estate of over 2,000 acres.

WANETA, junction of Pend d'Oreille R. and Columbia R. (B-11). Waneta, the name of the district near the international boundary, is derived from an Okanagan Indian word possibly meaning 'burned area.'

WANNOCK RIVER, drains Owikeno L. into Rivers Inlet (D-6). Derived from an Oowekyala Indian word sometimes mistranslated as 'poison river,' though it literally means 'river spirit.'

WAPTA LAKE, Kicking Horse Pass (D-11). This is the Stoney Indian word for 'river.'

WAPUTIK MOUNTAINS, Alberta-BC boundary (D-11). From a combination of two Cree Indian words that taken together mean 'white goat' – the mountain goat.

WARDNER, SE of Cranbrook (B-12). After James F. (Jim) Wardner, a prominent mining man who founded the town in 1896. He established a number of townsites in both Canada and the United States – Wardner, Idaho, among them.

WARE, confluence of Finlay R. and Kwadacha R. (J-7). Founded by the HBC in 1927 as Whitewater Post. In 1938, when a post office was established here, this became Fort Ware, being named after William Ware, manager of the HBC's BC district from 1927 until his retirement in 1932.

WARFIELD, W. of Trail (B-11). Named after an associate of F.A. Heinze, the builder of the original smelter at Trail.

WARNEFORD RIVER, flows S. into Kwadacha R. (J-7). After a young Canadian pilot, Lieutenant Reginald J. Warneford, VC, who destroyed a German zeppelin on 7 June 1915 with a single bomb that he dropped on it. He was killed ten days later.

WARRIOR POINT, Patricia Bay (A-8). The Saanich Indians report that in the old days they always kept a sentry posted here to watch for the approach of hostile bands.

WASA, N. of Cranbrook (B-12). Named by Nils Hanson after Vasa, a coastal town in his native Finland. Nearby Wasa Lake was formerly Hanson Lake.

WASHINGTON, MOUNT, NW of Courtenay (B-7). After Rear-Admiral John Washington (1800-63), FRS, Hydrographer of the Navy 1855-63 and secretary of the Royal Geographical Society.

WASHWASH RIVER, flows W. into Owikeno L. (D-6). From the Oowekyala Indian word meaning 'coho all over.'

WATERLOO MOUNTAIN, SW of Duncan (A-8). Named by W.A. Robertson when he climbed the mountain in 1865 on the fiftieth anniversary of the Battle of Waterloo. A descendant of his scaled it in 1915 on the centenary of the battle, and another descendant made the climb in 1965, establishing a family tradition.

WATSON ISLAND, between Kaien I. and Port Edward (G-4). After Alexander Watson, colonial treasurer of Vancouver Island in 1866 and subsequently general inspector of the original Bank of British Columbia.

WAWA LAKE, between Charlotte L. and Anahim L. (E-7). This is the Chinook jargon word meaning 'talk,' 'conversation,' 'tale,' or anything to do with speech whether written or articulated. A very apt name for this lake where Indians used to meet once a year to exchange news and items of interest.

WEAVER CREEK, NW of Harrison Hot Springs (B-9). After Thomas Weaver, who settled here in 1897. He died in 1947.

WEBSTERS CORNERS [WEBSTER'S CORNERS], E. of Haney (B-8). After James M. Webster, who arrived from Ontario in 1883 and became the first postmaster here. He named the place after himself.

WEDEENE RIVER, N. of Kitimat (G-5). From the Tsimshian Indian word for 'big valley.'

WEEDON CREEK, flows NE into Crooked R. (G-8). After Flight Lieutenant Weedon of the RCAF, engaged in the PGE's reconnaissance survey of 1929.

WEEPING WILLIE CREEK, S. of Skidegate Channel (F-3). 'Weeping Willie' was the nickname of Bill Martin, who hauled logs around here. He died in 1983.

WELCH PEAK, NW of Chilliwack L. (B-9). The firm of Foley, Welch, and Stewart held mining property in this area, and three of the peaks were named after the partners.

WELCOME PASSAGE, between S. Thormanby I. and the mainland (B-8). Named by Captain Richards and his officers on HMS *Plumper* in 1860 after they had received the welcome news that Thormanby had won the Derby.

WELLINGTON, N. of Nanaimo (B-7). Named after the famous British soldier and statesman, Arthur Wellesley, Duke of Wellington (1769-1852), victor of Waterloo and later Prime Minister of Britain. MOUNT WELLINGTON, Jervis Inlet, is also named for him.

WELLS, NW of Barkerville (F-9). After Fred M. Wells. In 1932, after ten years of prospecting in the area, Wells discovered the Cariboo Gold Quartz Mine. His mine introduced a new era in the history of gold mining in the Cariboo since there had hitherto been little lode mining. The town of Wells, which did not exist before 1932, had a population of 3,000 by 1940.

WELLS GRAY PARK, central BC (D- and E-9 and 10). Named after Arthur Wellesley (Wells) Gray, born in New Westminster in 1876. He became mayor of that city in 1913. Entering provincial politics he became Minister of Lands in 1933 and held that position until his death in 1944. Gray was a fine person, not just another hack politician, and proved to be an excellent Minister of Lands.

WENKCHEMNA PEAK, Alberta-BC boundary (D-11). Stoney Indian word for 'ten.' The mountain is suitably named, for it is the tenth peak in the Valley of the Ten Peaks.

WERNER BAY, E. coast of Moresby I. (E-4). G.M. Dawson named this bay after Abraham Gottlob Werner (1749-1817) of the Mining Academy, Freiburg, Germany, the first geologist to work out a systematic classification of minerals.

WESTBANK, W. of Kelowna (B-10). When a post office was opened here in 1902, it was given this name, indicative of its location on Okanagan Lake. The title was suggested by John Davidson, who had arrived in the district in 1892.

WESTBRIDGE, confluence of Kettle R. and West Kettle R. (B-10). When the post office here was established in 1900, a bridge was being built across the West Kettle River.

WESTHAM ISLAND, Fraser delta (B-8). Named by Harry Trim, who came from Westham, Sussex.

WESTHOLME, N. of Duncan (A-8). After the completion of the E & N Railway in 1886, Captain C.E. Barkley, RN (retired), took to walking over to the little station at Hall's Crossing and handing his letters to the mail clerk on the train when it stopped there. Neighbours began asking the captain to take their letters also, and soon the postal authorities took to supplying him with a mailbag. When a post office was opened, Captain Barkley was appointed postmaster, and the post office took the name of his house, Westholme, in which it was located.

According to a grandson, Captain Barkley named his house Westholme because it was his 'home in the west,' but it may have been the name of a family property in England. In 1909 Captain Barkley perished when a smaller house to which he had recently moved was consumed by fire. This Captain Barkley was, incidentally, a grandson of Captain C.W. Barkley, who in 1787 discovered and named Barkley Sound (q.v.).

WEST ROAD RIVER, flows E. into Fraser R. N. of Quesnel (F-8). Named by Alexander Mackenzie in 1793. When he turned westward from the Fraser River to strike overland to the Pacific, he travelled along stretches of this river. Mackenzie's name was in abeyance for many years, during which, because of its colour, this was known as the Blackwater River.

WEST VANCOUVER, Burrard Inlet (B-8). Although West Vancouver did not officially exist until the West Capilano District seceded from the Municipality of North Vancouver in 1912, the term had been used unofficially earlier. Thus, in 1910 John Lawson, the 'father of West Vancouver,' and his brother-in-law, W.C. Thompson, incorporated the West Vancouver Transportation Company to provide a ferry service with Vancouver.

WESTWOLD, SE of Kamloops (C-10). As early as 1826, this was named Grande Prairie. In 1900 the post office here was named Adelphi, but the new name never really supplanted Grande Prairie. In 1926, when Adelphi was finally dropped, the name of Grande Prairie was no longer available, there now being a Grande Prairie post office in Alberta. Westwold was then suggested by L.R. Pearse, an Englishman long resident in the area. 'Wold' is an old English word for a high open plain. The wold here was west of the railway station.

WHALETOWN, Cortes I. (C-7). There was a whaling station here in the 1870s. Whaletown Bay was known to the Klahuse Indians as Teck-tum, or 'sharpening place,' for, before contact with whites, Indians butchered whales here.

WHALLEY, SW of Port Mann (B-8). After Arthur Whalley, who settled here with his family in 1925. When the Pacific Stages bus line established a stop at the intersection of Ferguson and Bergstrom Roads, it was called Whalley's Corner. The name of Whalley was officially adopted for the community in 1948. Whalley post office was renamed North Surrey in 1966 and Surrey in 1969.

WHATSHAN LAKE, W. of Lower Arrow L. (C-10). This name was in use as early as 1866. Its derivation is uncertain.

WHIFFIN SPIT, Sooke Inlet (A-8). This unusual name immortalizes John George Whiffin, a clerk on HMS *Herald*, here in 1846.

WHIPSAW CREEK, SW of Princeton (B 9). Early gold hunters, needing lumber with which to build sluice-boxes and flumes, whipsawed their lumber in sawpits close to this stream. In this primitive method of sawing, a log was rolled onto two skids over a pit. One man stood on top of the log and another down in the pit, and they pulled the saw up and down. Progress was slow – 100 feet board measure was a good day's work.

WHISKERS POINT PARK, McLeod L. (G-8). After a whiskered squatter who settled here around 1914, expecting the PGE (now BCR) line to be built through here.

WHISTLER MOUNTAIN, N. of Squamish (C-8). Formerly London Mountain. Its present name comes from the numerous whistlers (marmots) on its slopes.

WHITE LAKE, N. of Salmon Arm (C-10). There is a pronounced showing of lime on the bottom near the shore.

WHITE PASS, Alaska-BC boundary (L-2). Named after Thomas White, Minister of the Interior 1885-8.

WHITE MAN PASS, SE of Mt. Assiniboine (C-12). The first white man to use this pass was James Sinclair, who led a party of emigrants through it in 1841 on their way to Oregon. He was followed in 1845 by Father De Smet, who set up a wooden cross at the summit.

WHITE ROCK, Semiahmoo Bay (B-8). This is named after a large white rock on the beach. According to an Indian legend, it was hurled across Georgia Strait by a young chief who had agreed with the girl whom he loved to make their home where the white rock landed.

WHITESAIL LAKE, N. of Tweedsmuir Park (F-6). This is a translation of the Indian word for the whitecaps blown up on the lake by the strong westerly winds.

WHONNOCK, W. of confluence of Fraser R. and Stave R. (B-8). From the Halkomelem word meaning 'place where there are [always] humpback salmon.'

WHYAC, S. of Nitinaht L. (A-7). From an old Nitinaht Indian word meaning 'open mouth,' used for the openings of fish traps. The Indians applied this name to the village because they thought the nearby canyon resembled a trap opening.

In 1864 Dr. Brown of the Vancouver Island Exploring Expedition described the village: 'Why-ack is a large fortified village protected by pickets from the sea dashing in breaks on the beach, or rushing through the narrow entrance of the inlet; so difficult is it to land that the Nittenahts [Nitinats] carry it with a high hand over the neighbouring tribes.'

WHYMPER, MOUNT, headwaters of Chemainus R. (A-7). Named in 1864 by Dr. Brown after Frederick Whymper, artist and explorer who was a member of Brown's expedition to explore Vancouver Island.

WHYMPER, MOUNT, Kootenay NP (D-11). Edward Whymper, the famous alpinist who in 1865 led the first party to ascend the Matterhorn, was also, in 1901, the first man to climb this mountain.

WHY NOT MOUNTAIN, W. of Chilko L. (D-7). Named in 1976. 'Why Not?' was the motto of the women's liberation movement.

WHYTECLIFF, Howe Sound (B-8). The admiralty survey of the area, taking note of the white rock of the promontory here, named it White Cliff Point. In 1909 Sir Charles Tupper and Colonel Albert Whyte of the West Shore and Northern Land Co. Ltd. opened their White Cliff City townsite. In 1914 Colonel Whyte used his influence with the PGE to have the company name its station here not White Cliff but Whytecliff.

WICKANINNISH BAY, Pacific Rim NP (B-7). It appears that the name Wickinanish's Sound was given in 1787 by Captain Barkley, a maritime fur trader. Wickaninnish was the hereditary name of the great chief of the Clayoquot Indians. A few years later, the same chief plotted to capture Captain Gray's ship *Columbia* by bribing a Hawaiian member of the crew to wet the priming of the firearms (which were constantly kept loaded) so that defence would be unavailing. Fortunately the plot was discovered and the would-be murderer frustrated.

Apparently the local Indians were known by the same name as their chief, for John Jewitt, in his account of his captivity in 1803-5, speaks of 'the Wickanninish, a large and powerful tribe, who made their canoes with greater skill than the other Indians and used sails as well as paddles.'

WIGWAM RIVER, flows W. into Elk R. (B-12). So named because Indians travelling to and from the North Kootenay Pass used to camp beside the stream.

WILD HORSE RIVER, flows SW into Kootenay R. (B-12). David Thompson, who called the stream Skirmish Brook, mentioned the herds of wild horses to be found in this country as early as the beginning of the nineteenth century. The stream seems to have been given its present name by Jack Fisher, leader of the party of miners who found gold here in 1863. His eye was caught by a white horse on a hillside, and consequently he named the stream either Stud Horse Creek or Wild Horse Creek. The latter name soon became universally used. The Geographic Board of Canada later promoted the creek to a river.

The gold camp of Fisherville grew up near the stream and prospered until the miners realized that it was situated on top of rich placer deposits. The houses were then hurriedly dismantled or, in some cases, burned down and a new settlement, Wild Horse, built on another site.

WILLIAM HEAD, S. tip of VI (A-8). Named after Rear-Admiral Sir William E. Parry (1790-1855), the Arctic explorer. Note the proximity of Parry Bay.

WILLIAMS CREEK [WILLIAM'S CREEK], Barkerville (F-9). Known as Humbug Creek until William (Dutch Bill) Dietz, a Prussian prospector, struck gold here in 1861. Although he was the first person to find gold on the creek, Dietz did not go very deep, and neither did those who immediately followed him. When shafts were sunk to deeper levels, they reached fantastically rich deposits. Dietz, a victim of rheumatism, died in poverty in Victoria in 1877.

WILLIAMS LAKE [WILLIAM'S LAKE], S. of Quesnel (E-8). The name of Williams Lake dates back at least to 28 April 1860, when John Telfer was granted a preemption 'situated at Williams Lake.' This lake is named for Chief William, at one time the leader of the Sugar Cane Reserve Indians. During the 'Chilcotin War' of 1864, Chief William blocked the endeavours of the Chilcotins to have the other Indians in the Cariboo join them in a general massacre of the whites.

WILLIBERT, MOUNT, Alaska-BC border (I-4). After Willibert Simpson, on the staff of the Alaskan Boundary Commission 1893-5. He became secretary of the International Boundary Committee and of the Geodetic Survey of Canada.

WILLOW RIVER, flows into Fraser R. NE of Prince George (G-8). So named because of the many willow swamps in its valley.

WILMER, N. of Invermere (C-11). Originally Peterborough, the name of this settlement was changed to Wilmer in 1902 in honour of Wilmer Wells, the provincial Minister of Public Works, who had been a sawmill operator in this part of the province.

WILSON CREEK, SE of Sechelt (B-8). Although this is sometimes said to have been named after Canon G.H. Wilson of St. Michael's Anglican Church, Vancouver, who built a cottage on a lot that he bought here in 1914, the original Wilson of Wilson Creek appears to have been James Wilson, who in the 1890s was a blacksmith in the Burns and Jackson logging camp here.

WILSON LANDING, W. side of Okanagan L. (B-10). After Harold Fitz-Harding Wilson, who settled here in 1900.

WINDERMERE LAKE, upper Columbia R. (C-11 and 12). This was David Thompson's 'Kootenae Lake,' near which he founded Kootenae House in 1807. It was subsequently known as Lower Columbia Lake. It was given its present name in 1883 by G.M. Sproat because of its resemblance to Lake Windermere in the English lake district.

WINDY JOE MOUNTAIN, Manning Park (B-9). C.P. (Chess) Lyons recalled the naming of this mountain:

> It was the first lookout that Bob Boyd, the early park ranger, could get up easily, and the wind blew up there like mad across the top of this. One of the trappers that had been in when we came in, Joe Hilton, started to work for us and became the lookout man, so the combination of the wind blowing through there and Joe being up on it got to be the reference of Windy Joe Mountain.

WINFIELD, N. of Kelowna (C-10). Formerly Alvaston, but renamed after Winfield Lodge, the home of Thomas Wood, justice of the peace and rancher, who settled here in 1871.

WINGDAM, E. of Quesnel (F-9). The Wingdam mine was in operation here from about 1898 to 1938. The early miners used wingdams (log cribbing filled with rock) to divert streams so that they could work the riverbeds below the dams.

WINLAW, S. of Slocan (B-11). Named after John B. Winlaw, who built a sawmill here around 1900 and in 1903 became the first postmaster.

WINSLOW CREEK, flows S. into end of Stave L. (B-8). After Rainsford H. Winslow, BCLS, killed in action in September 1918.

WINTER HARBOUR, Quatsino Sound (C-5). Winter Harbour and Forward Inlet provide the first haven for ships putting into Quatsino Sound for shelter from storms raging in the open Pacific. Such storms are, of course, most frequent during the winter.

WIRE CACHE, N. Thompson R. S. of Avola (D-10). In 1874 the federal government signed a contract with F.J. Barnard to build a telegraph line linking Cache Creek with Edmonton. When the contract was cancelled in 1878, a great amount of wire was left abandoned here, and Barnard, unpaid, sued the government.

WISTARIA, N. of Ootsa L. (F-6). When a name was needed for a post office here, a Mrs. William Harrison suggested that of the flowering vine.

WISUKITSAK RANGE, E. of Fording R. (B-12). After the mythical Old Man of the Cree Indians.

WITCH TOWER, THE, Glacier NP (D-11). This shoulder of Mount Fox is remarkable for its fantastic rock shapes. To quote the vivid description of A.O. Wheeler, 'The configuration suggests a number of hideous old giant bel-dames leaning from the parapet of a rock-tower and scattering vituperation broadcast over the earth.'

WITTY'S LAGOON, SW of Esquimalt (A-8). After John Witty, who purchased the adjacent property in 1867.

WIWAXY PEAKS, Yoho NP (D-11). This name is derived from the Stoney Indian language. Its exact meaning is not clear, but it has been said to mean 'windy.'

WOKAS LAKE, expansion of Quinsam R. (B-7). From the Kwakwala Indian word meaning 'good water.'

WOLFENDEN, MOUNT, S. of Quatsino Sound (C-6). After Lieutenant-Colonel Richard Wolfenden (1836-1911), who arrived in British Columbia in 1859 as a corporal in Colonel Moody's contingent of Royal Engineers. He remained in the colony after his unit was disbanded and became the King's Printer in Victoria. Active in the militia, he became in 1889 Lieutenant-Colonel of the BC Provincial Regiment of Garrison Artillery.

WONOWON, NW of Fort St. John (I-9). At Mile 101 on the Alaska Highway.

WOOD LAKE, N. of Kelowna (C-10). This should be Wood's Lake. It is named after Thomas Wood, who in 1871 established a large ranch on the eastern shore of this lake.

WOOD RIVER, flows S. into Kinbasket L. (E-10). Known originally as the Portage River or the Great Portage River. The early fur brigades, leaving their boats on the Columbia at Boat Encampment, carried their baggage along this river on their way via Athabasca Pass to the eastern slope of the Rockies and the North Saskatchewan River.

Father De Smet reported in 1846 an interesting custom observed by the voyageurs:

> We saw 'May-poles' all along the old encampments of the Portage. Each trav-eller who passes there for the first time selects his own. A young Canadian, with much kindness, dedicated one to me which was at least 120 feet in height, and which reared its lofty head above all the neighboring trees. Did I deserve it? He stripped it of all its branches, only leaving at the top a little crown: at the bottom my name and the date of transit were written.

It was the thick growth of these trees that gave the stream its later name of Wood River.

WOODBURY CREEK, flows SE into Kootenay L. (B-11). After Charles J. Woodbury, who explored this creek in 1883 with George Ainsworth.

WOODFIBRE, Howe Sound (B-8). Named by Sir George Bury, president of Whalen Pulp and Paper Company, when the company's mill was built here in 1920.

WOODPECKER, S. of Prince George (F-8). Takes its name from Woodpecker Island in the Fraser River here. The island is said to have been given its name in the old days when the sound of Chinese workers chopping wood for the river steamers that put in here for fuel was likened to the sound of woodpeckers.

WOODWARDS LANDING [WOODWARD'S LANDING], S. side of Lulu I. (B-8). After Nathaniel Woodward, who with his son Daniel settled here in 1874.

WOOLSEY CREEK, flows S. into Illecillewaet R. (D-11). After David Woolsey, who from 1889 on worked various mining claims in the area.

WORK POINT, Victoria Harbour (A-8). After Chief Factor John Work (1792-1861). Born John Wark in County Donegal, Ireland, he subsequently changed his name to Work. He entered the service of the HBC in 1814 and was posted to York Factory. Transferred to the Columbia Department in 1823, he spent the rest of his life there.

Work was on the expedition in 1824 to select the site for Fort Langley. He became a Chief Trader in 1830. For some reason he incurred the hostility of Dr. McLoughlin and for many years was left in the isolation of Fort Simpson. After McLoughlin's resignation in 1846, Work resumed his advancement in the company. That year he was promoted to Chief Factor, and in 1849 he became a member of the Board of Management for the Columbia Department. In the same year, the Church of England solemnized his marriage to Josette Legace, the half-Indian woman who had become his devoted 'fur trade wife' more than twenty years earlier.

In 1853 Work took up residence at Fort Victoria as a member of the Board of Management of the Western Department. That year Governor Douglas appointed to the Legislative Council of Vancouver Island 'John Work Esqre, a gentleman of probity and respectable character, and the largest land holder on Vancouver Island.' Chief among Work's holdings was his farm at Hillside, where he had his family home. He died in Victoria, still in the service of the HBC, in 1861.

H.D. Dee, in his article 'An Irishman in the Fur Trade' (*BCHQ* 7 [October 1943]:268), pays tribute to Work's 'physical courage, his great endurance, and endless loyalty.' Various of Work's journals have been published. WORK CHANNEL, north of Prince Rupert, is also named for him.

WOSS, SE of Nimpkish L. (C-6). From the Kwakwala Indian word meaning 'a river flowing through flat ground.'

WREDE CREEK, flows NE into Ingenika R. (I-7). Wrede was a prospector-trapper who vanished after heading up the Ingenika River in 1896 after leaving word that he would be back at Fort Graham in about five months. In 1898 his remains were found in his home camp, with a pair of homemade crutches nearby. He had cut himself badly with an axe.

WROTTESLEY, MOUNT, W. side of Howe Sound (B-8). After John, Lord Wrottesley (1798-1867), the astronomer who was president of the Royal Society from 1854 to 1857. Captain G.H. Richards, RN, named the mountain about 1860.

WYCLIFFE, NW of Cranbrook (B-12). After John Wycliffe (1330?-84), English church reformer and translator of the Bible. He has been described as 'the morning star of the Reformation.'

WYNNDEL, N. of Creston (B-11). Named for one of the early fruit growers in the district.

X

XENIA LAKE, NW of Christina L. (B-10). Xenia is one of only two place names in all of British Columbia to begin with an *x*. Somebody who knew classical Greek could have applied this name in the sense of 'pertaining to hospitality.' Xenia is also a botanical term.

Y

YACULTA, SW corner of Quadra I. (C-7). This is the name of the Cape Mudge Indian village and is an anglicization of the Comox people's pronunciation of Lekwiltok. (See *Yuculta Rapids*.)

YAHK RIVER, E. of Creston (B-12). Although the Kootenay Indian word for 'arrow' is somewhat similar to Yahk, the true meaning of Yahk appears to be 'female caribou.'

YAKOUN RIVER, flows N. into Masset Inlet (F-3). Various meanings have been claimed for this Haida name. 'Straight point' seems to be favoured, though one also hears 'in the middle' or 'on the east.'

YALAKOM RIVER, flows SE into Bridge R. (C-8). Derived from the Lillooet Indian word for the ewe of the bighorn sheep. (See *Shulaps Range*.)

YALE, Fraser Canyon (B-9). Known as The Falls during the early nineteenth century because of the rapids upstream in the Fraser. The present name dates back to 1847, when Ovid Allard was sent with a party from Fort Langley to establish a new HBC post here. This post, though it had neither a stockade nor bastions, was named Fort Yale, after James Murray Yale, 'a courageous, peppery little man,' the officer in charge at Fort Langley.

J.M. Yale (1796-1871) entered the service of the HBC in 1815. He was at Fort Langley from 1828 to 1859, being given the command of the fort in 1834, and was promoted to Chief Trader in 1844. Very aware of his short stature, Yale tried to avoid standing close to the commanding height of Governor Douglas, but Douglas, aware of this foible, took a sly delight in moving close to Yale.

During the days of the Fraser gold rush, the town of Yale had a lurid reputation. D.W. Higgins, who lived there in 1858, has left the following description of it in that year:

> A city of tents and shacks, stores, barrooms and gambling houses. The one street crowded from morning till night with a surging mass of jostling humanity of all sorts and conditions. Miners, prospectors, traders, gamblers and painted ladies mingled in the throng.
>
> In every saloon a faro-bank or a three-card-monte table was in full swing, and the hells were crowded to suffocation. A worse set of cutthroats and all-round scoundrels than those who flocked to Yale from all parts of the world never assembled anywhere. Decent people feared to go out after dark. Night assaults and robberies, varied by an occasional cold-blooded murder or a daylight theft, were common occurrences. Crime in every form stalked boldly through the town unchecked and unpunished. The good element was numerically large; but it was dominated and terrorized by those whose trade it was to bully, beat, rob and slay. (*The Mystic Spring*, p. 31)

At the head of navigation on the Fraser, Yale remained important as long as goods were transshipped there from the river steamers to the wagon trains going up the Fraser Canyon to the Cariboo goldfields. With the completion of the CPR, such transshipments ceased, and Yale almost became a ghost town. Mrs. St. Maur, passing through on the train in 1887, observed: 'the railway has changed everything, and consequently many nice-looking little wooden houses, with their patches of garden, are closed and deserted. A feeling of sadness came over me at the sight of the pretty little desolate homes' (*Impressions of a Tenderfoot*, p. 73).

The Halkomelem Indian name for the site of Yale means 'short-leaved willow trees' (Sitka willows).

YARROW, SW of Chilliwack (B-8). Writing on 7 September 1910, Michael Urwin, an official of the British Columbia Electric Railway Company, included Yarrow in a list of names of 'our oldest important shareholders' and suggested that stations on the new Fraser Valley line be named after them.

YELLOWHEAD PASS, Alberta-BC border, W. of Jasper NP (E-10). Also YELLOWHEAD LAKE. Both are named after Pierre Bostonais, nicknamed Tête Jaune ('Yellowhead'), who in 1820 guided one of the first HBC parties to penetrate west of the Rockies. (See *Tête Jaune Cache*.)

In the years that followed, the pass was frequently referred to as the Leather Pass since the HBC posts in New Caledonia used this route to bring in most of their leather, as much as forty packs of dressed moose skins annually. Leather was needed in large quantities since the rough terrain of the country required as many as eight pairs of moccasins per man for a single expedition through the mountains. The Yellowhead or Leather Pass was also known as Caledonian Valley.

Yellowhead Lake was originally known as Buffalo Dung Lake.

YEO ISLAND, N. of Seaforth Channel (E-5). After Dr. Gerald Yeo, surgeon on HMS *Ganges,* on the Pacific Station 1857-60.

YMIR, S. of Nelson (B-11). In 1897, when D.C. Corbin, president of the Nelson and Fort Shepherd Railway, put a station here, he named it after the nearby Ymir Range. The famous geologist G.M. Dawson had earlier named these mountains after Ymir, the father of the giants according to Norse legend.

YOHO NATIONAL PARK, E. of Golden (D-11). *Yoho* is a Cree Indian exclamation of astonishment (here used in the sense of 'How magnificent!').

YORKE ISLAND, W. of Hardwicke I. (C-7). The family name of the Earls of Hardwicke.

YOUBOU, Cowichan L. (A-7). Not an Indian word but a compound of the names of two officers of the Empire Lumber Company in 1914 – Yount, the general manager, and Bouten, the president of the company.

YUCULTA RAPIDS, E. of Sonora I. (C-7). Named after the local Indians. Captain Barrett-Lennard, who published in 1862 an account of his voyage around Vancouver Island in a yacht, observed of the 'Ucultas' (sometimes spelled Euclataws or Lekwiltok) that they were 'reputed the worst Indians anywhere to be found about here, plundering and killing those of the northern tribes, whenever they met with them.'

Yuculta is said to be derived from a Kwakwala Indian word for a large sea worm that cannot be killed by cutting it in pieces, for the separate pieces go wriggling off through the water, being unkillable.

YUEN LAKE, E. of Bear L. (I-6). Named in 1914 by Frank Swannell, the noted surveyor, after Nep Yuen, the Chinese cook with his party. Yuen had originally come to Canada as one of the hundreds of Chinese coolies imported to build the CPR. Close to Yuen Lake is Nep Peak.

YUKNESS MOUNTAIN, Yoho NP (D-11). Perhaps from the Stoney Indian word for 'standing.' Another possibility is that Yukness means 'sharpened, as with a knife,' a word descriptive of the peak.

YUQUOT, Nootka I. (B-6). From the Nootka Indian word meaning 'village exposed to the winds,' or 'windy place.'

Z

ZEBALLOS, NE of Esperanza Inlet (B-6). Takes its name from Zeballos Inlet, named in 1791 by Malaspina after one of his officers, Lieutenant Ciriaco Cevallos.

ZUCIARTE CHANNEL, Nootka Sound (B-6). Not a Spanish name but a corrupted form of the name of a band of the Muchalat Indians.

ZYMOETZ RIVER, flows into Skeena River E. of Terrace (G-5). Formerly this was Copper River. The meaning of Zymoetz is uncertain, but it may come from a Tsimshian word meaning 'palm of the hand.'

Other books by G.P.V. Akrigg and Helen B. Akrigg:

1001 British Columbia Place Names
British Columbia Chronicle 1778-1846: Adventurers by Sea and Land
British Columbia Chronicle 1847-1871: Gold and Colonists
Published by Discovery Press

HMS Virago *in the Pacific 1851-1855*
Published by Sono Nis Press

Other books by G.P.V. Akrigg:

Jacobean Pageant, or The Court of King James I
Shakespeare and the Earl of Southampton
Published by Harvard University Press and Hamish Hamilton

Letters of King James VI & I (edition)
Published by the University of California Press

Set in Scala by George Vaitkunas
Printed and bound in Canada by Friesens
Proofreader: Camilla Jenkins
Cartographer: Eric Leinberger
Designer: George Vaitkunas